A Conceptual History of Space and Symmetry

Pietro Giuseppe Fré

A Conceptual History of Space and Symmetry

From Plato to the Superworld

Springer

Pietro Giuseppe Fré
University of Torino
Turin, Italy

ISBN 978-3-319-98022-5 ISBN 978-3-319-98023-2 (eBook)
https://doi.org/10.1007/978-3-319-98023-2

Library of Congress Control Number: 2018950942

This Springer imprint is published by the registered company Springer Nature Switzerland AG
The registered company address is: Gewerbestrasse 11, 6330 Cham, Switzerland

This book is dedicated to my family, namely to my beloved daughter Laura, to my darling wife Olga, to my young son Vladimir and to my former wife Tiziana, Laura's mother, with whom, notwithstanding our divorce, the ties due to a common daughter and to more than twenty five years spent together remain strong. This work is also dedicated to my dear Franca, to whom, since my mother's death, I am tied by a profound filial affection and by the sharing of deep intellectual interests. My feelings of love are hereby transmitted to all of them, with all the hopes and the worries of an old father who stubbornly believes that culture and science are just one thing and encode, for mankind, the unique escape route from global disasters.

Preface

This book forms a twin pair with another book [90] by the same author, which is of a different, thoroughly mathematical character, while the character of the present volume is historically and philosophically oriented. These two pieces of work constitute a twin pair since, notwithstanding their different profiles and contents, they arise from the same vision and pursue complementary goals.

The vision, extensively discussed in the first chapter of the present book and throughout several other chapters, consists of the following main conceptual assessments:

1. Our current understanding of the Fundamental Laws of Nature is based on a coherent, yet provisional, set of five meta-theoretical principles, listed by me as A)-E) and dubbed the current *episteme*. This episteme is of genuine geometrical nature and can be viewed as the current evolutionary state of Einstein's ideas concerning the geometrization of physics.
2. *Geometry* and *Symmetry* are inextricably entangled, and their current conception is the result of a long process of abstraction, traced back in the present work, which was historically determined and makes sense only within the *Analytic System of Thought* of Western Civilization, started by the ancient Greeks.
3. The evolution of *Geometry and Symmetry Theory* in the last 40 years has been deeply and very much constructively influenced by *Supersymmetry/Supergravity* and the allied constructions of *Strings and Branes*. My reader is not supposed to know what supersymmetric field theories and supergravity are, since this book has the ambition to be written for a much more general public than that formed by theoretical physicists specialized in the *super-world*. It suffices to be alerted that, since the seventies of the twentieth century, an entire new theoretical world has opened up by the introduction of a new symmetry principle, indeed dubbed *supersymmetry*, that can be enforced on field theories, in particular on Einstein Gravity, at the price of adding to every physical field a new copy of the same, named its *superpartner* with a spin shifted by $\frac{1}{2}$. This entrains that the superpartners of *bosons* are *fermions* and vice versa. In a

nutshell, supersymmetry is the invariance of such new type of field theories under the exchange of bosons with fermions.

4. Further advances in Theoretical Physics cannot be based simply on the Galilean Method of Interrogating first Nature and then formulating a testable theory that explains the observed phenomena. As stated later on in the present work, one ought to Interrogate also *Human Thought*, by this meaning frontier-line mathematics concerned with geometry and symmetry in order to find there the threads of so far unobserved correspondences, reinterpretations and renewed conceptions.

The complementary pursued goals are as follows:

(a) In the case of the present book

- the historical and conceptual analysis of the process mentioned in point (2). of the above list which led to the current episteme.
- the philosophical argumentation, on historical basis, of the assessment made in point (4). of the above list.

(b) In the case of book [90], the mathematical full-fledged illustration of the main developments in geometry and symmetry theory that occurred under the fertilizing influence of *Supersymmetry/Supergravity* and that would be inconceivable without the latter.

Repeating in a slightly different from the arguments advocated particularly in Chapter One of the present work, I think that what is currently practiced in the whole world as Fundamental Physics or Mathematics is based on the Greek view of the *episteme* and it is meaningful only inside the Analytic System of Thought founded by the Ancient Greeks. To recuperate a full-conscience of this fact is mandatory in order to continue on the difficult but exciting path we are confronted with.

The twin pair of which this book is a member is viewed by the author as his limited, humble contribution to the promotion of a new season of more scholarly teaching of *physical-mathematics*.

To my reader

Ideally, this book has been written for an audience wider than that composed by mathematicians and theoretical physicists. My ambitious aim was that of attracting the attention of all finely educated persons with an interest in the history of *Ideas*, mathematical conceptions being an integral part of the general *episteme*. Yet, mathematics is also technical and it is very difficult to single out the main ideas and tell the story of their evolution without mentioning some fundamental definitions and some formulae. Coping with this essential difficulty, well known to everyone who tries to explain the theories of the world to wider audiences, I have tried my best, where formulae and definitions were unavoidable, to convey the underlying concepts by means of wordly explanations and with the help of numerous *ad hoc* created images. I have tried to intertwine storytelling and philosophic argumentations with some essential technical material introduced according to the abovementioned method, hoping to keep the attention of all my dear readers alive throughout the chapters and up to my epilogue that expresses the results of long self-debating meditations.

Spes, ultima dea.

Turin, Italy
June 2018

Pietro Giuseppe Fré

Acknowledgements

With great pleasure, I would like to thank my collaborators and colleagues Pietro Antonio Grassi, and Mario Trigiante for the many suggestions and discussions we had during the writing of the present book. Similarly, I express my gratitude to the Editors of Springer-Verlag, in particular to Maria Bellantone, for their continuous assistance, constructive criticism and suggestions.

My thoughts, while finishing the writing of this long essay, that mostly occurred in Moscow in my last 2 years (2016–2017) of service as Scientific Counsellor of the Italian Embassy in Russia, were frequently directed to my late parents, whom I miss very much and I will never forget. To them I also express my gratitude for all what they taught me in their life, in particular to my father who, with his own example, introduced me, since my childhood, to the great satisfaction and deep suffering of writing books.

Furthermore, it is my pleasure to thank my very close friend and collaborator Aleksander Sorin for his continuous encouragement and for many precious consultations.

Contents

Chapter 1
The Episteme

Ce sont tous ces phénomènes de rapport entre les sciences ou entre les différents discours dans les divers secteurs scientifiques qui constituent ce que j'appelle épistémè d'une epoque.
Michel Foucault

The present book is a conceptual history, namely it aims at tracing back the development of some fundamental ideas. Those addressed here are the ideas of *Space* and of *Symmetry* that lay the foundations of our present understanding of the Physical World and of its conceptualization in Geometry. Such a tale belongs to the history of Mathematics and Physics but also to the General History of Thought, namely of Philosophy.

The motivations to write down this conceptual outline originate from a set of considerations on the present state of Culture, Science and Education on which I have long pondered, coming to much worried conclusions.

Let me expose these thoughts.

I begin by noting a matter of fact. In the teaching of *literature*, *philosophy* and other *disciplines* pertaining to the sphere of the so named *humanities*, the historical perspective is, worldwide, an essential part of the game. While introducing the reading of literary masterpieces, no one ever omits to locate historically and geographically the author, to illustrate the cultural environment in which he operated and to connect his work to the long–time political and social trends that shaped the world around him. The same is true for the teaching of the *Arts*. In philosophy, economics and political sciences it is absolutely standard to trace the historical development of the main ideas and conceptions. On the contrary the contemporary teaching of mathematics and physics is distinctively characterized by an almost complete absence of any historical perspective. Students learn theorems named after someone without ever learning who that someone was. Worse than that, the architecture of the mathematical or physical theories is typically presented in a self-contained way with no tracing of the development of ideas and conceptions leading to them.

The reasons for this state of affairs are many fold. On one side there is the obvious and legitimate need to compress the scientific achievements of the past into modern, logical and economical notations, in order to transmit them in a reasonable time and, most important of all, in order to integrate them into the current structure of Science.

P. G. Fré, *A Conceptual History of Space and Symmetry*,
https://doi.org/10.1007/978-3-319-98023-2_1

Secondly, behind this worldwide practice of teaching, there is the tacitly accepted positivistic attitude, shared by most scientists, by public opinion makers, by most politicians and by the man in the street, that Science is continuously progressing, that it mainly consists of the discovery of truths which, once uncovered, will last for ever, being great if they are beneficial to humanity. Such views are particularly strong with regard to mathematics where the common message conveyed to the new generations is that a theorem is just a theorem, namely an imperishable truth. No question about that, yet a theorem has hypotheses that are related to definite problems, historically determined within a conceptual framework, also pertaining to the History of Thought. Furthermore the very question to which a theorem gives an answer has no absolute meaning; it is meaningful only inside the conceptual framework mentioned above, where the abstract entities dealt with in the considered propositions are defined.

Last but not least, the causes for the current trend in the tuition of fundamental sciences lies in the generalized impoverishment of classical education of the entire society in general and of the scientific community in particular. Studying the biographies of the great mathematicians of the past, also of the recent past, one discovers that almost all of them had received a solid classical education, had learned efficiently Latin and Greek and sometimes, like Cayley, Hamilton and André Weil, were endowed with special talents for modern and ancient languages. Now we are facing a near future, in at most a couple of generations, when most educated people and, in particular, scientists will completely ignore classical languages, will have no familiarity with the history of philosophy and even more seriously a very superficial knowledge of the history of science. This development, which projects ominous shadows on the future of Modern European Civilization has received a fatally strong acceleration through the so named *Bologna agreements*, their consequences being the bureaucratization of European University Studies and the reduction of structured knowledge to pills of information to be quickly swallowed and readily forgotten. For a European Federalist like myself, who always dreamt of the United States of Europe as of the *Land of Utopia*, it is a bitter disappointment to see that the downgrading of European Culture has been precisely promoted and it is being successfully realized just within the framework of European Union Plans aimed at the cultural integration of the European Society.

For many years I pondered on these evolutions in Education and Culture and a great influence has been exerted on my ideas by the brilliant analysis of Hellenistic Science contained in the remarkable essay *La Rivoluzione Dimenticata* (The Forgotten Revolution) by Lucio Russo.[1] The author, professor of Mathematical Physics in the Second University of Rome, but also a refined classical philologist, after giving a vivid and exhaustive fresco of the advances of natural sciences and technology in the third and second century before Christ, mainly centered in Alexandria and in the other hellenistic capitals, concludes that Modern Science was born at least twice, which means that it died at least once. The causes of the first death of Scientific Thought are carefully analyzed by Russo who tends to blame it on the Romans. Whether our Latin ancestors should carry the entire responsibility for this epochal

[1] Lucio Russo, *La Rivoluzione Dimenticata*, Feltrinelli 1996.

disaster, which prepared the stage for the long intermission of the Dark Ages, is disputable, yet Russo's analysis of the mechanisms leading to the death of Science is magistral and, in my opinion, undisputable. At the very root of the process we find the emphasis on *applied science and technology* based on existing *theories*, at the expenses of the development of *new ideas*. The second stage of the process is the almost inevitable consequence of the first, namely a shift of gear in the Educational System: specialized programmes aimed at the creation of *specialists* substitute *formative schools* aimed at the education of *free thinkers*. Tightly untangled with the second is the third stage of the process: the writing of textbooks tuned up to the task of quick specialization which substitute the original works of the scientists and any ponderous conceptual synthesis containing discussions and elaborations. The fourth and final stage of this destructing path occurs in a couple of generations when the overwhelming majority of educated people are the product of the system established in the first three steps. Then almost no one exists who is still able to read and properly understand the original sources of living science and all what remains are the pills of all–purpose information written by people who are no longer scientists, just only *specialists*. At this stage we have the *Death Certification of Scientific Thought*.

Russo illustrates this point quite extensively and provides a definitely convincing evidence by showing that, with few important exceptions, all what we know about Ancient Science is through indirect reports contained in compilations that were written by non-scientist erudites, hardly knowing the topic they talked about.

To this I can add a personal remark. The absence of historical perspectives in the post-hellenistic scientific teaching, which was clearly an allied aspect of the negative path eventually leading to the final death of Science, is under our eyes in the clearest possible way. While we know with good precision the birth and death dates, together with extensive details of their lives, for *Plato, Aristotle, Aeschylus, Sophocles, Euripides, Virgil* and many other Greek and Latin writers, information about the life of *Euclid* and of many other ancient scientists is very scan. In some cases, as in that of *Diophantus*, the father of algebra, our ignorance reaches its peak: up to the end of the XIXth century, his life-time was determined only within a spread of four hundred years! A fortuitous discovery of a Byzantine document in Spain allowed to reduce the uncertainty. After that discovery we have some ground to think that *Diophantus* lived in the third century after Christ.

Another source of reflections provided by the reading of Russo's book is encoded in the quite circumstantial remarks made by the author on the relation between *the practice of democracy*, diffuse among the Ancient Greeks, and the birth of *Mathematics* as a discipline made out of propositions and proofs.

The present trends are essentially based on the already stigmatized neo-positivistic attitude towards Science that implicitly assumes scientific theories to be philosophically neutral statements, independent from any System of Thought, which are either false or provisionally true. A quite orthogonal view point to which I entirely subscribe is expressed, for instance, in the interesting article quoted below.[2]

[2] *Culture and Systems of Thought: Holistic Versus Analytic Cognition* by R. E. Nisbett et al. published in Psychological Review 2001, Vol. 108, No. 2. 291–310.

In this paper the System of Thought of the Ancient Greeks is compared with that of the contemporary Ancient Chinese, the first dubbed *Analytic*, the second *Holistic*. The authors discuss how Western Mathematics and Science are natural consequences of the Greek Analytic System while the Chinese Holistic System was apt to produce many advances in Technology, yet not the type of theoretical inductive and deductive science of the West. The authors also show that the deep differences in the way of thinking of contemporary individuals whose cultural origin is either in the West or in the Far East are still marked by this ancient dichotomy. Furthermore they trace back the difference between the two systems in social organization and locate the origin of the Greek Analytic System in the practice of Democracy of the Ancient Greeks and in the Individualism characterizing their attitude toward life.

Having sufficiently explained the ideas and the vision by which my story telling is inspired let me turn to describe its focus and my pursued goals.

As it follows from the above argumentation, it is my deep belief that mathematical and physical knowledge is also historical knowledge as much as literary and philosophical knowledge is. The mathematical ideas develop through a historical process which is strongly rooted in the general development of Culture and Society, not too differently from what happens with humanities. It is the goal of this book to enlighten the historical process which led to the contemporary visions about *Space* and *Symmetry* that are utilized by modern theoretical physics and in particular by such abstract and advanced descriptions of the Physical World as those provided by Supergravity and its High Energy precursor, i.e. Superstring Theory.

Let me once again resort to Ancient Greece and advocate the concept of *episteme* (see Fig. 1.1) whose status with respect to the Physical World I would like to summarize.

I start by noticing that, as a theoretical physicist, I consider myself very fortunate to have witnessed, in my own life-time, the following series of experimental discoveries:

1. The detection of the $W^{\pm}$ and Z particles, definitely confirming that fundamental non gravitational interactions can be described by gauge theories.
2. The detection of the Brout Englert Higgs boson, definitely confirming that gauge theories can be spontaneously broken by scalar fields falling into non symmetric extrema of some potential.
3. The direct detection of gravitational waves emitted in the coalescence of two black-holes which, not only confirms the general structure of General Relativity, but directly tests the dynamics encoded in Einstein Equations, namely in a set of purely geometrical differential equations.

Trying to summarize the implications for the *episteme* of the last thirty–three years of experimental physics culminating in the above three discoveries we can say the following.

Leaving apart the issue of *quantization*, that we can generically identify with the *functional path integral over classical configurations*, we have, within our Western Analytic System of Thought, a rather simple and universal scheme of interpretation of the Fundamental Interactions and of the Fundamental Constituents of Matter based on the following few principles:

Fig. 1.1 The personification of the *Episteme* in Celsus library in Ephesus

(A) The categorical reference frame is provided by Field Theory defined by some action $\mathscr{A} = \int_{\mathscr{M}} \mathscr{L}(\Phi, \partial\Phi)$ where $\mathscr{L}(\Phi, \partial\Phi)$ denotes some Lagrangian depending on a set of fields $\Phi(x)$.

(B) All fundamental interactions are described by *connections* **A** on principle fibre-bundles $P(\mathrm{G}, \mathscr{M})$ where G is a Lie group and the base manifold $\mathscr{M}$ is some *space-time* in $d = 4$ or in higher dimensions.

(C) All the fields Φ describing fundamental constituents are *sections* of *vector bundles* $B(\mathrm{G}, V, \mathscr{M})$, associated with the principal one $P(\mathrm{G}, \mathscr{M})$ and determined by the choice of suitable *linear representations* $D(\mathrm{G}) : V \to V$ of the structural group G.

(D) The spin zero particles, described by scalar fields ϕ^I have the additional feature of admitting non linear interactions encoded in a scalar potential $\mathscr{V}(\phi)$ for whose choice general principles, supported by experimental confirmation, have not yet been determined.

(E) Gravitational interactions are special among the others and universal since they deal with the tangent bundle $T\mathscr{M} \to \mathscr{M}$ to space-time. The relevant connection is in this case the Levi–Civita connection (or some of its generalization with torsion) which is determined by a metric g on $\mathscr{M}$.

In enumerating the principles (A)–(E) I have purposely used the mathematical language of fibre-bundles and emphasized the concepts of *Lie Group* and of *connection*, in order to advocate my main argument, which is the following. The conceptual basis of Modern Fundamental Physics, including Einstein General Relativity, is deeply rooted in the evolution of the mathematical conception of *Symmetry* and in its relation with the mathematical conception of *Geometry*. The frequently asserted statements that modern physics has *geometrized* our understanding of the world have no real meaning outside the mathematical framework alluded to in the enumeration of principles (A)–(E).

As I am going to outline later on, the mathematical theory of fibre-bundles, of connections and of characteristic classes was essentially ready and already brought almost to perfection in the middle of the 1950s, yet it took at least another forty years before that language and those conceptions were completely assimilated into the fabrics of theoretical physics and were recognized to be not just auxiliary instruments rather a relevant part of the very essence of the physical *episteme*.

A quick look at the list of principles (A)–(E) immediately reveals that, notwithstanding their simplicity and unifying power, they can be only provisional. There are still too many *ad hoc* choices which strongly demand some deeper unifying principle able to predict them from above. Most prominent among these choices are those of the structural group G, of the representations $D(\mathrm{G})$ and of the potential $\mathcal{V}(\phi)$, the latter choice including also, in some extended sense, the determination of quark and lepton masses. In the physical literature of the last forty years, what I have described in the above way is referred to as the problem of *grand-unification* or of *super-unification*.

In the same forty years an enormously extended set of developments have taken place in the quest for unification, starting from the new idea of *Supersymmetry* which, as the word reveals, is an extension of the notion of *Symmetry*, meaning by that Lie Algebras. The reason why Supersymmetry, which leads to the fields of *Supergravity*, *Superstrings* and *Brane-Physics* entrains so many structural and ramified implications is because it tackles with one of the most fundamental and, in my opinion, not yet fully penetrated, principles of physics, namely the distinction among fermion and bosons, intertwined, by means of the spin–statistic theorem, with Lie algebra theory, the distinction between two groups of representations, the vector and the spinor ones, being a distinctive property of the $\mathfrak{so}(n)$ Lie algebras, unexisting for the others.

The largest part of the developments mentioned above, related with Supergravity/Superstrings, have a distinctive geometric/algebraic basis. Entire chapters of algebraic geometry and of algebraic topology have been integrated by these developments into the fabrics of theoretical physics, while some new geometries have been introduced into the fabrics of mathematics. Furthermore the very way to analyze and interpret mathematical structures is sometimes redirected by the influence of Supergravity/Superstrings. Two or three examples suffice to illustrate what I mean. Exceptional Lie algebras that, up to the mid 1960s were considered by the majority of physicists like mathematical curiosities, have been promoted to the role of primary actors on the stage of the *super-world*. Special Kähler Geometries, never defined by pure mathematicians have by now entered, with full-rights, the mathematical club, revealing their relation with other geometries, already introduced by mathematicians, like HyperKähler geometry and quaternionic Kähler geometry. The notions of momentum-map, Kähler and HyperKähler quotients find a deep interpretation in the context of supersymmetric field theories and connect with some of the most brilliant mathematical achievements of the last few decades like the Kronheimer construction of ALE manifolds.

Relying on the above arguments and explanations I can now more appropriately restate the topic whose conceptual development constitutes the target of my story telling. This is *the entire scope of Group Theory and of the Differential Geometry of Coset Manifolds* from the basic initial definitions to the most advanced items utilized in the current research in Supergravity, that I have briefly sketched in the above lines.

In view of the provisional stationary point reached by the *episteme* with points (A)–(E), I chose to tell my story from a mathematical/geometric viewpoint, seeking through history the threads that led to the present views on geometry as the *science of space* which eventually is also the *science of phenomena substantiating space*, namely of physics.

When I started the organization of my material according to these principles, all the stages of the conceptual development I wanted to report were not yet completely clear to myself, yet, as I went on with my work, I had the pleasure to realize that a grandiose historic–philosophical fresco was unfolding in front of me: the main turning points in the History of Thought concerning *Symmetry* and *Space* progressively acquired an alive dramatic character, contributed by their actualization in history and in the life of great thinkers. Furthermore I captured glimpses of the remote roots of many conceptions that were previously all squeezed onto a sort of *intellectual flatlandia*, deprived of historical depth. The understanding of a field of mathematics which I know well and practice everyday in my research activity acquired a completely new quality by means of this historic-philosophical revisitation.

I think, that it is time to reshape and update our teaching of fundamental physics in view of the impressive, although provisional, advances made by the *episteme* in the last thirty years, which I tried to summarize in points (A)–(E). This involves reshaping our teaching of mathematics and in particular of mathematics for physics majors, which is sometimes extremely obsolete and inadequate. Such a reshaping cannot avoid involving the historic–philosophical dimension and should be a step on a reversed path, hopefully reinstating classical education to its proper place in the Western Culture.

Chapter 2
Symmetry and Mathematics

Mathematics, rightly viewed, possesses not only truth, but supreme beauty, a beauty cold and austere, like that of sculpture.
Bertrand Russell, Philosophical Essays (1910)

2.1 Setting the Stage: Symmetry and Beauty

The word symmetry comes from the Greek $\sigma\upsilon\mu\mu\varepsilon\tau\rho\acute{\iota}\alpha$ which is composed of two words $\sigma\acute{\upsilon}\nu$ (with) and $\mu\acute{\varepsilon}\tau\rho o\nu$ (measure). Literally $\sigma\upsilon\mu\mu\varepsilon\tau\rho\acute{\iota}\alpha$ indicates the adequate proportion of the different parts of something, material or immaterial, and it is well represented in classical Greek literature. For instance from Plato we have $\eta\ \nu\upsilon\kappa\tau\acute{o}\varsigma\ \pi\rho o\varsigma\ \eta\mu\acute{\varepsilon}\rho\alpha\nu\ \sigma\upsilon\mu\mu\varepsilon\tau\rho\acute{\iota}\alpha$, the right proportion of night to day time, or $\tau\grave{o}\ \sigma\acute{\upsilon}\mu\mu\varepsilon\tau\rho o\nu\ \kappa\alpha\grave{\iota}\ \kappa\alpha\lambda\acute{o}\nu$, namely what is proportionate is also beautiful.

The last sentence reflects not only Plato's specific views but a widespread conception which inspired the whole of Greek culture in the Vth century BC and crystallized into the *Canon* of *Classicism* for all the Arts. Indeed the *Canon*, from the word $\kappa\alpha\nu\acute{\omega}\nu$ (the rule), was the title of a treatise written by the famous Vth century sculptor Polykleitos, who exemplified his theory in a bronze statue, the *Doryphoros* (the Spear Bearer). Both the treatise and the statue are unfortunately lost, but a roman marble copy dating about 120 BC has reached us from Pompeii and it is preserved in Naples National Archaeological Museum (see Fig. 2.1).

A quotation from the treatise has survived in the book *De Placitis Hippocratis et Platonis* by Galen, the famous medical writer of the IInd century A.D. which reads as follows: *Chrysippos holds beauty to consist not in the commensurability or "symmetria" of the constituent elements of the body, but in the commensurability of the parts, such as that of finger to finger, and of all the fingers to the palm and wrist, and of those to the forearm, and of the forearm to the upper arm, and in fact, of everything to everything else, just as it is written in the Canon of Polyclitus. For having taught us in that work all the proportions of the body, Polyclitus supported*

P. G. Fré, *A Conceptual History of Space and Symmetry*,
https://doi.org/10.1007/978-3-319-98023-2_2

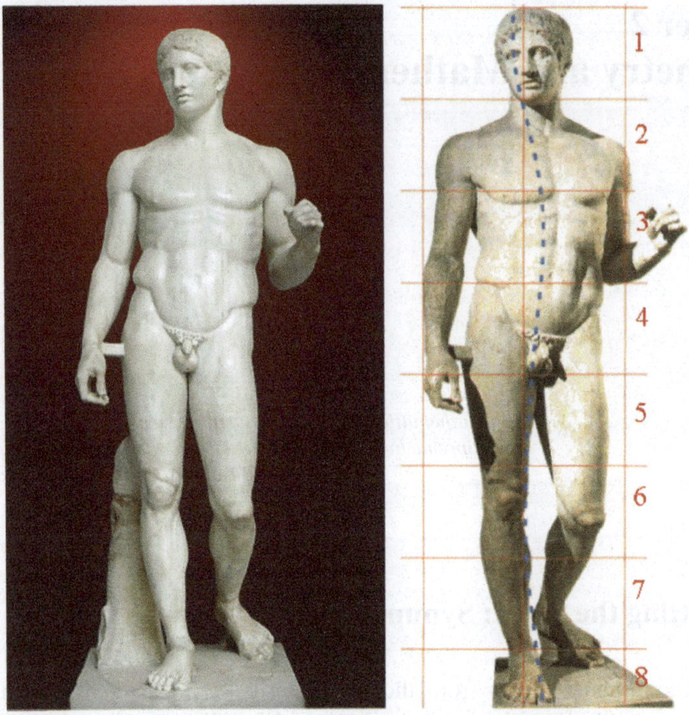

Fig. 2.1 The Canon of Polykleitos was exemplified by the statue the Doryphoros

his treatise with a work: he made a statue according to the tenets of his treatise, and called the statue, like the work, the 'Canon'.

Looking at Fig. 2.1 we easily realize what symmetry essentially meant for the classical Greeks: it meant measure–ratios provided by rational numbers $\frac{p}{q}$ where both the numerator p and the denominator q are small integer numbers from the set $\{1, 2, 3, 4, 5, 6, 7, 8\}$. The extension of the head should be $\frac{1}{8}$ of the full extension of the body, the extension of the torso $\frac{3}{8}$ and so on. That such rational numbers correspond to *beauty* in figurative arts is probably a conceptual extension of what Pythagoras (see Fig. 2.2) had discovered already in the VIth century BC about music, namely that the sounds that are nice to our ears are those produced by strings whose lengths are in simple ratios like $2 : 1$, $3 : 2$ and $4 : 3$. Indeed there are reasons to think that Polykleitos, as many other intellectuals of his own age and as Plato himself after him, was influenced by Pythagorean philosophy in which numbers were identified as the true substance of the existing things.

Before analyzing such Pythagorean ideas let us remark that there is another obvious influence to which Polykleitos was responding in conceiving his Canon, namely just the observation of the human body, whose actual structure he was trying to rationalize. Not every human has a head that is exactly $\frac{1}{8}$ of its full stature but, fortunately, there are no humans whose head is $\frac{1}{2}$ or $\frac{7}{3}$ of the same. Hence, in fixing a Canon

Fig. 2.2 A head-sculpture of Pythagoras: Roman copy from Musei Capitolini di Roma of a lost Greek original

of symmetry, the sculptor was actually inventing a code able to describe reality and this code was indeed *mathematical*. In some sense he was prefiguring the Platonic philosophy of Ideas: the fixed simple ratios correspond to the *Idea of a Human*. An individual, namely an actualization of the *Idea of a Human* is the more beautiful as closer his/her proportions are to those encoded in the Idea.

Behind this way of thinking we easily see the standard procedure utilized in modern science to analyze complex phenomena: you make a mathematical model, which is supposed to capture the essential dynamical degrees of freedom inherent to the phenomenon under investigation, then you compare actual existing phenomena with this mathematical model. Those phenomena that behave closer to the model are the *Ideal Ones*. Examples are ready to hand like *Ideal Gases* or *Ideal Fluids*. The difference between Modern Science and Classical Greek Philosophy is that we no longer attach any moral or aesthetical value to the adherence of Reality to the Idealistic Model: quite the opposite. If Reality is too much away from the Model it is the latter that comes under stress and has to be discarded. Yet we preserve the option to give an aesthetical evaluation of the Models: if they have no symmetry in the Greek sense of an adequate proportions between their parts, that are the assumptions on one hand and their consequences on the other hand, if there is no due economy in their founding principles, then the Models are not beautiful and, we tend to think, they are hardly true.

As we see, the identification of the *True* with the *Beautiful* and the guiding principle that *Symmetry* is the keyword for both, is valid in contemporary science just as it was in Classical Antiquity, although what symmetry means has undergone a very significant evolution in the last 2500 years. Illustrating this evolution and introducing the reader to the *Mathematics* involved in the contemporary conception of *Symmetry* is the mission of the present essay.

Although the alluded above evolution has been quite extensive and ramified, yet it is honest to say that some fundamental constants are there and it is proper to emphasize them at once. What has remained substantially unchanged from Antiquity to the present time is the association of the process of *understanding* with the conception of *numbers* and the *numerical encoding* of what we call *symmetries*.

Reviewing the origin of words is illuminating. The word *Mathematics* comes from the Greek $\tau\grave{o}\ \mu\acute{\alpha}\theta\eta\mu\alpha$ which means the object of study, what has to be understood, derived from the verb $\mu\alpha\nu\theta\acute{\alpha}\nu\omega$ (I investigate). The mathematician $o\ \mu\alpha\theta\eta\mu\alpha\tau\iota\kappa\acute{o}\varsigma$ is just the scientist.

Coming back to the Pythagoreans, in his *Metaphysics* Aristotle says:

...the so-called Pythagoreans applied themselves to the study of mathematics, and were the first to advance that science; insomuch that, having been brought up in it, they thought that its principles must be the principles of all existing things.

In his brilliant history of Greek Mathematics [112], Sir Thomas Heath says:

May we not infer from these scattered remarks of Aristotle about the Pythagorean doctrine that "the number in the heaven" is the number of visible stars, made up of units which are material points? And may this not be the origin of the theory that all things are numbers...?

There is in this sentence the echo of a deep historical truth that was pointed out more than one hundred years ago by the great German scholar of Antiquity, Theodor Mommsen (see Fig. 2.3). On the basis of Indo-European linguistics he conceived his proof that Astronomy and Arithmetic have a very ancient origin that precedes the agricultural revolution. He remarked that the roots associated with the names

Fig. 2.3 Christian Matthias Theodor Mommsen (1817–1903) and the Neolithic monument of Stonehenge, whose astronomical significance has been deeply studied

of the numbers, of the stars and of the planets are the same in all Indo-european languages, so as are the roots associated with the animals and the items in pastoral life, while there are several variants for the names of the plants, of the tools and of the products of farming. It is quite probable that the birth of the abstract notion of *unity*, which is the very beginning of Number Theory and of Mathematics is rooted in the observation of the sky that was practiced by all ancient populations of the world.

We quote again Sir Thomas Heath:

Aristotle observes that the One is reasonably regarded as not being itself a number, because a measure is not the things measured, but the measure or the One is the beginning (or principle) of number. This doctrine may be of Pythagorean origin; Nichomachus has it; Euclid implies it when he says that a unit is that by virtue of which each of the existing things is called one, while a number is 'the multitude made up of units'; and the statement was generally accepted.

It is often emphasized that Greek Mathematics lacked the notion of Zero which is of Indian origin and was transmitted to the West by the Middle Age Arabs. We see here that even the number One was not immediately a number and it took a good deal of abstraction to define the natural numbers which appear to be a fundamental achievement of Human Thought and the basis of our rational understanding of the external World.

Hence we see in *symmetrical* or *harmonic* a synonym of *easily countable* and hence of understandable. It is the human satisfaction, streaming from comprehension, what makes the object of observation *beautiful*.

We perceive in this sequence of arguments that symmetry principles are the basic weapons utilized by the Human Mind to grasp the structure of Reality.

2.2 Symmetry Principles

What are at the end of the day symmetry principles? Loosely speaking they are a powerful predictive instrument. Let us consider the pictures of some leaving beings that are displayed in Fig. 2.4. In the first two photographs, that of the leave and that of the dog, we have instances of a reflection symmetry with respect to a vertical plane that cuts the object in two halves. Thanks to this symmetry, if we observe the left half of the image, we no longer have to observe the right half: we are essentially able to predict it from the knowledge we have already acquired of the left one and this is quite economical for data storing in our mind. We perceive this reduction of efforts and it is in this that probably resides our appreciation of beauty. Arguing along these lines the picture of the starfish is even more economical. We can just observe one of the five arms and we can predict the other four. There is a larger degree of symmetry. Even larger is the predictive power inherent to the symmetries that characterize the tiled walls of the Alhambra Palace in Grenada, two examples of which are shown in Fig. 2.5. It is evident that from a finite portion of the figure we can predict it in the entire plane yet the prediction rules now become more and more complicated and most important of all we start wondering *how many of these patterns are possible?*

Fig. 2.4 Symmetries in the living beings

Fig. 2.5 Tessellations at the Alhambra

The moment we formulate such a question we have made a gigantic step forward on the road of abstract thinking.

Indeed it took a very long time, essentially up to the beginning of the XIXth century, before the proper mathematical framework was found where such questions as the above one might be asked and answered.

In order to understand the developments that lead to that fundamental advance we have to consider the history of Geometry and Algebra.

2.3 Geometry and Algebra up to the Birth of Group Theory

The origins of Geometry are quite ancient, although not as ancient as the origins of Number Theory. It appears that both the Egyptians and the Babylonians developed empirical formulae to calculate the area of surfaces of various shapes. The motivation was related with the collection of tributes. In ancient Egypt, landowners were supposed to pay taxes to Pharaoh's officers according to the extension of their property, usually rectangular shaped. Every year the periodic floods of the Nile changed the utilizable area for cultivation and the proprietor asked for a tax-reduction accordingly: Pharaoh's prospectors had to verify the adequateness of such claims and they needed such empirical ready to use formulae to accomplish their duties.

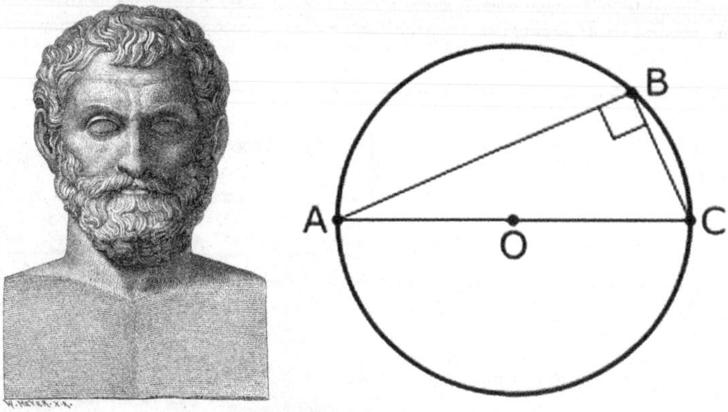

Fig. 2.6 Thales Milesius (624 BC–546 BC): the first Greek philosopher, according to Aristotle, and the first Greek geometer. Thales' theorem states that the angle in B of the triangle inscribed in a circle as shown in the figure is of 90 degrees

According to tradition *Thales Milesius*, the first Greek philosopher and mathematician, learnt about geometrical formulae in Egypt and in Babylon but he was the first who conceived the deductive method and provided what we call a *mathematical proof* of a *theorem*. Indeed the proposition about the inscription of a rectangular triangle in a circle shown in Fig. 2.6 is referred to as Thales' theorem.

Sir Thomas Heath quotes Proclus' summary about Thales:

...first went to Egypt and thence introduced this study (geometry) into Greece. He discovered many propositions himself, and instructed his successors in the principles underlying many others, his method of attack being in some cases more general, in others more empirical.

Next Heath quotes Plutarch who included Thales among the Seven Wise Men:

he was apparently the only one of these whose wisdom stepped, in speculation, beyond the limits of practical utility: the rest acquired the reputation of wisdom in politics.

In his impressive Essay on Hellenic and Hellenistic Science, Lucio Russo [153] puts into evidence the role in the development of the conception of a mathematical proof that was played by the practice of democracy in the Greek *πόλις* and by the frequent need of the Greek citizen, the *πολίτης*, to advocate publicly his own case in front of juries.

The Greek word for proof, *απόδειξις*, comes from the verb *αποδείκνυμι* which means *I present, I submit*. Russo illustrates the close relation between the geometrical *απόδειξις* and Rhetoric, quoting Aristotle's *Ars Rhetorica* where the *enthymemes* of rhetors are shown to be just *syllogisms*: then he recalls the impressive sentence by Quintilian: ...*nullo modo sine geometria esse possit orator*, there can be no orator without geometry [149].

Ἐπὶ τῆς δοθείσης εὐθείας πεπερασμένης τρίγωνον ἰσόπλευρον συστήσασθαι.
Ἔστω ἡ δοθεῖσα εὐθεῖα πεπερασμένη ἡ ΑΒ.
Δεῖ δὴ ἐπὶ τῆς ΑΒ εὐθείας τρίγωνον ἰσόπλευρον συστήσασθαι.

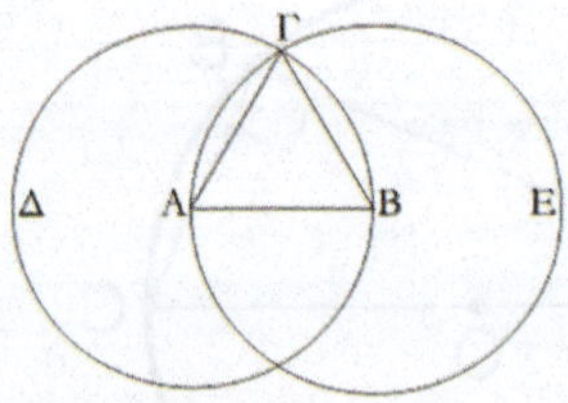

Κέντρῳ μὲν τῷ Α διαστήματι δὲ τῷ ΑΒ κύκλος γεγράφθω ὁ ΒΓΔ, καὶ πάλιν κέντρῳ μὲν τῷ Β διαστήματι δὲ τῷ ΒΑ κύκλος γεγράφθω ὁ ΑΓΕ, καὶ ἀπὸ τοῦ Γ σημείου, καθ᾽ ὃ τέμνουσιν ἀλλήλους οἱ κύκλοι, ἐπὶ τὰ Α, Β σημεῖα ἐπεζεύχθωσαν εὐθεῖαι αἱ ΓΑ, ΓΒ.

Καὶ ἐπεὶ τὸ Α σημεῖον κέντρον ἐστὶ τοῦ ΓΔΒ κύκλου, ἴση ἐστὶν ἡ ΑΓ τῇ ΑΒ· πάλιν, ἐπεὶ τὸ Β σημεῖον κέντρον ἐστὶ τοῦ ΓΑΕ κύκλου, ἴση ἐστὶν ἡ ΒΓ τῇ ΒΑ. ἐδείχθη δὲ καὶ ἡ ΓΑ τῇ ΑΒ ἴση· ἑκατέρα ἄρα τῶν ΓΑ, ΓΒ τῇ ΑΒ ἐστιν ἴση. τὰ δὲ τῷ αὐτῷ ἴσα καὶ ἀλλήλοις ἐστὶν ἴσα· καὶ ἡ ΓΑ ἄρα τῇ ΓΒ ἐστιν ἴση· αἱ τρεῖς ἄρα αἱ ΓΑ, ΑΒ, ΒΓ ἴσαι ἀλλήλαις εἰσίν.

Ἰσόπλευρον ἄρα ἐστὶ τὸ ΑΒΓ τρίγωνον. καὶ συνέσταται ἐπὶ τῆς δοθείσης εὐθείας πεπερασμένης τῆς ΑΒ. ὅπερ ἔδει ποιῆσαι.

Fig. 2.7 Euclid of Alexandria was active during the reign of Ptolemy I (323–283 BC). His treatise the Elements ($\Sigma TOIKEI\Omega N$ in Greek) is probably the most famous mathematical textbook of all times. The Greek original was transmitted through the edition cured by Theon from Alexandria in the IVth century AD. In 1808 François Peyrard discovered in Vatican a manuscript of the Elements coming from a byzantine workshop of the X century that was not based on Theon's edition. The first latin translation appears to have been produced by Boethius in the VI century AD but then the Elements disappeared in Western Europe, until the English monk Adelard of Bath produced a Latin translation of an Arabic version. The Arabs received the Elements from the Byzantines approximately around 760; this version was translated into Arabic under Harun al Rashid c. 800 and became the source of Adelard. Theon's Greek edition was recovered in 1533. In the picture we see Euclid as imagined by Raffaello in his School of Athens and on the side an example of a proof from the original Greek version of the Elements

The elaborate historical development in the course of the Vth and IVth century BC of the *Elements of Geometry*, that were finally systematized in Euclid's book with the same title (see Fig. 2.7), are masterfully reviewed in Heath's book and we do not dwell on them. What is important for us to stress is the axiomatic crystallization of the science of geometrical figures, *points, triangles, circles, polygons and polyhedra* that was the end-point of this process and that lasted for about two thousand year as a back-bone of mathematics, but also as a very severe bias on philosophical thinking.

The axioms, or postulates of euclidian geometry were analyzed for almost twenty centuries by mathematicians and philosophers, who gave special attention to the Vth postulate (see Fig. 2.8):

If a line segment intersects two straight lines forming two interior angles on the same side that sum to less than two right angles, then the two lines, if extended indefinitely, meet on that side on which the angles sum to less than two right angles.

Numberless efforts were devoted to prove the above proposition that invariably ended up into a failure. Indeed, as we now firmly know, the Vth postulate is logically independent from the others and can be substituted with different ones that lead to various *non euclidian geometries*.

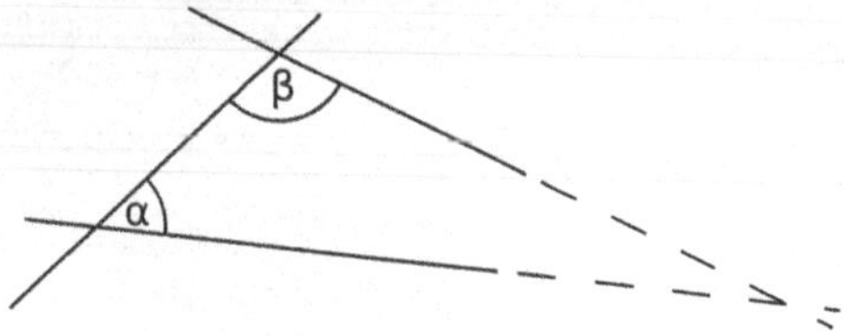

Fig. 2.8 Euclid's Vth postulate

The main justification for this stubborn determination of the philosophers of all times to erect Euclidian Geometry into an unavoidable logical scheme is largely psychological: Euclidian Geometry is the axiomatic codification of the properties of a plane surface without curvature, such as a portion of the surface of the Earth appears to be at the human scale. The temptation to identify Euclidian Geometry with Reality is very strong and, in line with what we said above, we have to rationalize Reality in order to make it satisfactorily understandable to us. Immanuel Kant (see Fig. 2.9) boosted this way of arguing to its extreme consequences stating that:

Space is not an empirical concept which has been derived from outer experiences. On the contrary: it is the subjective condition of sensibility, under which alone outer intuition is possible for us.

By space he meant Euclidian Space as conveyed to us by the *Elements*. On this we can remark that, had the human race developed on the surface of a small asteroid, where curvature effects are strongly perceivable, the Vth postulate probably would not have been introduced by anyone. Yet it was and it took almost 2000 years to overcome the prejudice that it is unavoidable.

Fig. 2.9 Portrait of Immanuel Kant (1724–1804). Kant is considered the central figure of modern philosophy. He spent almost all of his life in Königsberg, Eastern Prussia, now Kaliningrad in Russia

Fig. 2.10 Portrait of Nicolai Ivanovich Lobachevsky (1793–1856). Born in Nizhny Novgorod in a poor family he was raised with many difficulties by his mother, the widow of a small public officer. He studied in the newly founded University of Kazan' of which he later became Rector. His new conceptions of Geometry were originally presented to his Kazan' colleagues in some seminars based on lecture notes. His paper was refused publication in Russia and he succeeded to publish it in French only in 1837 with the title "Geometrie Imaginaire". A summary of his results was later written in German and published in Berlin in 1840 with the title "Geometrische Untersuchungen zur Theorie der Parallelinien". Removed from his position by an "ukas" of the Tzar in 1846, Lobachevsky died in poverty and afflicted by complete blindness in 1856. One year before death he composed a summary of his entire geometrical conception in a book entitled "Pangeometria"

Although there are some evidences that Gauss considered non euclidian geometries with positive curvatures almost at the same time,[1] it is a historically established fact that non-euclidian geometry, in its negative curvature variant, was introduced by the Russian mathematician Lobachevsky (see Fig. 2.10) in 1826. Explicit realizations of Lobachevsky's geometries on curved surfaces immersed in three-dimensional Euclidian space were constructed in the 1860s by the Italian mathematician Eugenio Beltrami (see Fig. 2.11). Lobachevsky geometry was brought under full analytic command by Poincaré and constitutes a paradigmatic simple example of a curved manifold whose geodesics can be analytically determined.

[1]See Chap. 7 and Sect. 7.3.1 for more details on this.

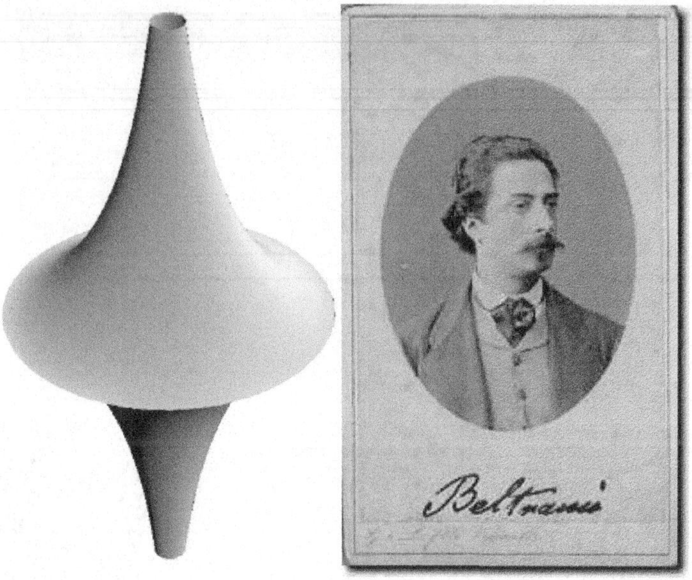

Fig. 2.11 Eugenio Beltrami (1836–1900) who was Professor in Bologna, Pisa, Rome and Pavia invented the pseudo-sphere, a surface with constant negative curvature where geodesics realize Lobachevsky's geometry

Relevant to us at this point of our introductory discussion is not the issue of non-euclidian geometry for its own sake, rather the very conception of geometry, as we have outlined it here from pre-euclidian time to the XIXth century. The objects of study for the whole span of this very long time have been idealized entities (the points, the segments, the lines) that certainly had symmetries, in the classical Greek sense of fixed and possibly harmonious proportions in their sizes and angles, and obeyed interesting relations among themselves, yet they were not operated on by means of any prescribed algorithms. Indeed, as we plan to explain in more detail in Chap. 8, the very word algorithm is of Arabic origin, from the name of the IXth century scholar *al-Khwarizmi*, who shares with Diophantus the title of Father of Algebra, also a word of Arabic descent.

The conception of operations to be performed on abstract mathematical objects and forming together with their own targets a single structure to be analyzed for internal consistency is the very heart of Algebra, but it was never attained by the Ancient Greeks. On the other hand, in order to answer questions such as the one posed at the end of the last section, namely *how many patterns do exist such as those represented on the Alhambra walls*, this conception was precisely the necessary viewpoint.

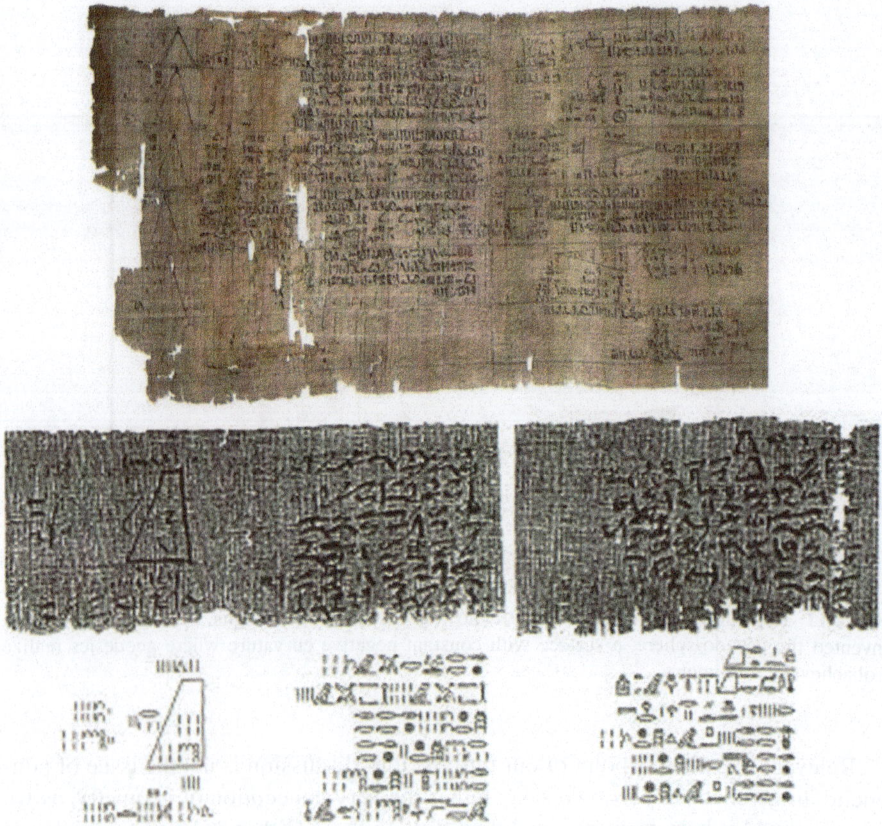

Fig. 2.12 On top the Rhind Papyrus, conserved in the British Museum that dates to around 1650 BC. In the lower part of the picture the Moscow Papyrus conserved in the Pushkin Museum in Moscow and slightly older than the first. These are the two known to us mathematical papiri of ancient Egypt

Quoting once again Sir Thomas Heath:

In algebra, as in geometry, the Greeks learnt the beginnings from the Egyptians. Familiarity on the part of the Greeks with the Egyptian methods of calculation is well attested. These methods are found in operation in the Heronian writings and collections.

From the Egyptians the Greeks learnt the *hau*-calculations which essentially amounted to the solution of first order equations, just as the following one

$$\frac{2}{3}x + \frac{1}{2}x + \frac{1}{7}x + x = 33 \qquad (2.3.1)$$

from the Papyrus Rhind, where *hau*, the heap, is the name given to the unknown x.

The goal of Algebra remained up to the beginning of the XIXth century the solution of algebraic equations[2] but, while pursuing this goal, many new mathematical conceptions were developed. First of all, similarly to what already happened at Pythagorean times with the radicals, it was understood that the field of numbers had to be enlarged with the inclusion of transcendental numbers like π or the Euler number e and then with the imaginary unit $i = \sqrt{-1}$ which leads to complex numbers,[3] secondly, when it was established, by means of the Ruffini-Abel theorem, that the algebraic equation of the 5th degree is not generically solvable by radicals, the notion of *Group* made its appearance in Mathematics through the work of *Évariste Galois* and a new season began in which the question posed at the end of last section could be rephrased and answered in the proper conceptual framework.

2.4 Galois and the Advent of Group Theory

Everything is exceptional about *Évariste Galois* (see Fig. 2.13), both his mathematical achievements and his short unlucky personal life. No more romantic and tragic cradle for the Theory of Groups might have been invented by the capricious destiny. Furthermore just as full of contradictory aspects was his human career and his relation with the other humans that came across it, so his first class mathematical results present two quite contrasting faces. The theory named after him and the theorem which, within such a theory, constitutes Galois' major result, are rather difficult both at the level of the definitions and of the proofs: in addition one can honestly say that Galois theory of the solubility of algebraic equations is a rather specialized topic which, nowadays, finds relevant applications eminently in number theory and associated topics, but not too many in geometry at large and in physics. On the contrary the weapon that Galois developed to obtain his own results, namely the Theory of Groups, has proved of extraordinary conceptual relevance and fertility, being the starting point for an entirely new vision of Mathematics and in particular of Symmetry.

Évariste Galois was born October 25th of 1811 in the small town of Bourg-le-Reine. He died at the dawn of May 31st 1832 from the wounds received the day before in a duel. During the 21 years of his life he suffered all types of misfortunes and blows, mainly caused by the incomprehension and stiff stupidity of his teachers, by the political turmoils of the time and by his naiveness. Both his mother and his father were highly educated persons, committed to Revolutionary Ideals and fierily opposed to the Restoration. Galois' father acted as Mayor of Bourg-le-Reine and in 1827 he fell victim of a clerical conspiracy organized by a priest who circulated a false poem, full of obscenities, that he pretended written by the Mayor; Galois father, full of rage and shame, escaped to Paris and committed suicide in a hotel

[2]See Sect. 8.1 for many more details.

[3]See once again sect. 8.1 for more details.

Fig. 2.13 Evariste Galois (1811–1832)

room. During the funeral Évariste suffered the aggravating sorrow to see his father's coffin at the center of a violent brawl between clericals and liberals.

In the Lyceum Luis-le-Grand where he was studying, Galois met quite stiff and stupid teachers, who did not understand his exceptional talents for mathematics and treated him as an idiot. Twice he was rejected at the entrance exams to the *École Polytechnique* where he ardently desired to enroll, not only for the excellent tuition there available, but also for the democratic ideals that inflamed all of the Polytechnic students. Notwithstanding these adversities, Galois studied mathematics by himself, directly reading books and articles by Legendre, Fourier, Abel and Gauss and, at the age of 17, he was already well advanced on the development of his own theory of algebraic equations. He wrote his results in a paper that he wanted to submit to the Academy and, for that purpose, he managed to give it to Cauchy who promised to support its publication. Unfortunately Cauchy lost Galois's manuscript.

In 1830 Galois tried once again to publish his own results by giving a new paper to the scientific secretary of the Academy. This latter brought home Galois' manuscript to read it, but the very same night he unexpectedly died; Galois' work was once again lost. Disappointed and disgusted by life, Galois entered the political agon, just at the eve of the July Revolution, supporting the Republicans.

In the last two years of his life Galois was twice arrested as a subversive, spent some months in prison, was released, participated to other political quarrels, had a love affair with a girl of vulgar personality, who disgusted him also on that front, finally was involved in a stupid debate with a political exponent of opposite views, that ended up in the duel which caused his death. Perfectly aware of being confronted with almost sure death, the night before the duel, Évariste wrote an exposition of all his mathematical results that he gave to his loyal friend Auguste Chevalier. Fortunately, this latter did not loose the sixty pages received from Galois and in 1846 Galois main theorem was finally published [98] on the *Journal de Mathématiques Pures et Appliquées*, with the praising comments of its main editor, namely *Joseph Liouville* (see Figs. 2.14, 2.15, 2.16).

2.5 A New Conception of Symmetry

What had Galois actually done? He had established a criterion for the solubility of algebraic equations in terms of radicals. Permutations had already been used by Ruffini and Lagrange, but he grasped their essence, which is that of being elements of a group and he started building on that fact. Let us outline his arguments.

Consider an algebraic equation of degree n with rational coefficients $u_i \in \mathbb{Q}$.

$$\mathscr{P}_n(x) = x^n + u_1 x^{n-1} + \cdots + u_n = 0\,, \tag{2.5.1}$$

The set of rational numbers $\mathbb{Q}$ fulfils the axioms of a field. Using a more abstract view point we can say that the field $\mathbb{F} = \mathbb{Q}$ to which the coefficients belong is provided by the set of rational functions with rational coefficients of the coefficients u_i. Since a rational function of a rational number is a rational number this definition seems rather tautological, yet it is convenient to envisage the next step that is the field extension.

Let α_i $(i = 1, \ldots n)$ be the n roots of the considered polynomial Eq. (2.5.1). Define:

$$\mathbb{K} \equiv \mathbb{F}\,(\alpha_1, \ldots \alpha_n) \tag{2.5.2}$$

the field obtained by adjoining to $\mathbb{F}$ the roots. This means that every element $a \in \mathbb{K}$ can be written as a rational function, with coefficients in $\mathbb{F}$, of the roots: $a = \mathfrak{r}(\alpha_1, \ldots, \alpha_n)$. Consider a permutation of the roots:

$$P\,:\,(\alpha_1, \ldots, \alpha_n) \mapsto (P\alpha_1, \ldots, P\alpha_n). \tag{2.5.3}$$

The action of the permutation P can be defined on the field $\mathbb{K}$, by setting:

$$P\,:\,a = \mathfrak{r}(\alpha_1, \ldots \alpha_n) \mapsto Pa \equiv \mathfrak{r}(P\alpha_1, \ldots, P\alpha_n)\,, \tag{2.5.4}$$

A given element of $\mathbb{K}$ has different representations in terms rational functions of the roots. If $a = \mathfrak{r}_1(\alpha_1, \ldots) = \mathfrak{r}_2(\alpha_1, \ldots)$ are two such representations then Eq. (2.5.4) defines a consistent action of the permutation on the field $\mathbb{K}$ only if $\mathfrak{r}_1(P\alpha_1, \ldots) = \mathfrak{r}_2(P\alpha_1, \ldots)$. This amounts to requiring that the two internal operations of the field $\mathbb{K}$ are preserved by the map P:

$$\begin{aligned} \forall a, b \in \mathbb{K} \quad &: \quad P(a+b) = P(a) + P(b)\,, \\ \forall a, b \in \mathbb{K} \quad &: \quad P(ab) = P(a)P(b) \end{aligned} \tag{2.5.5}$$

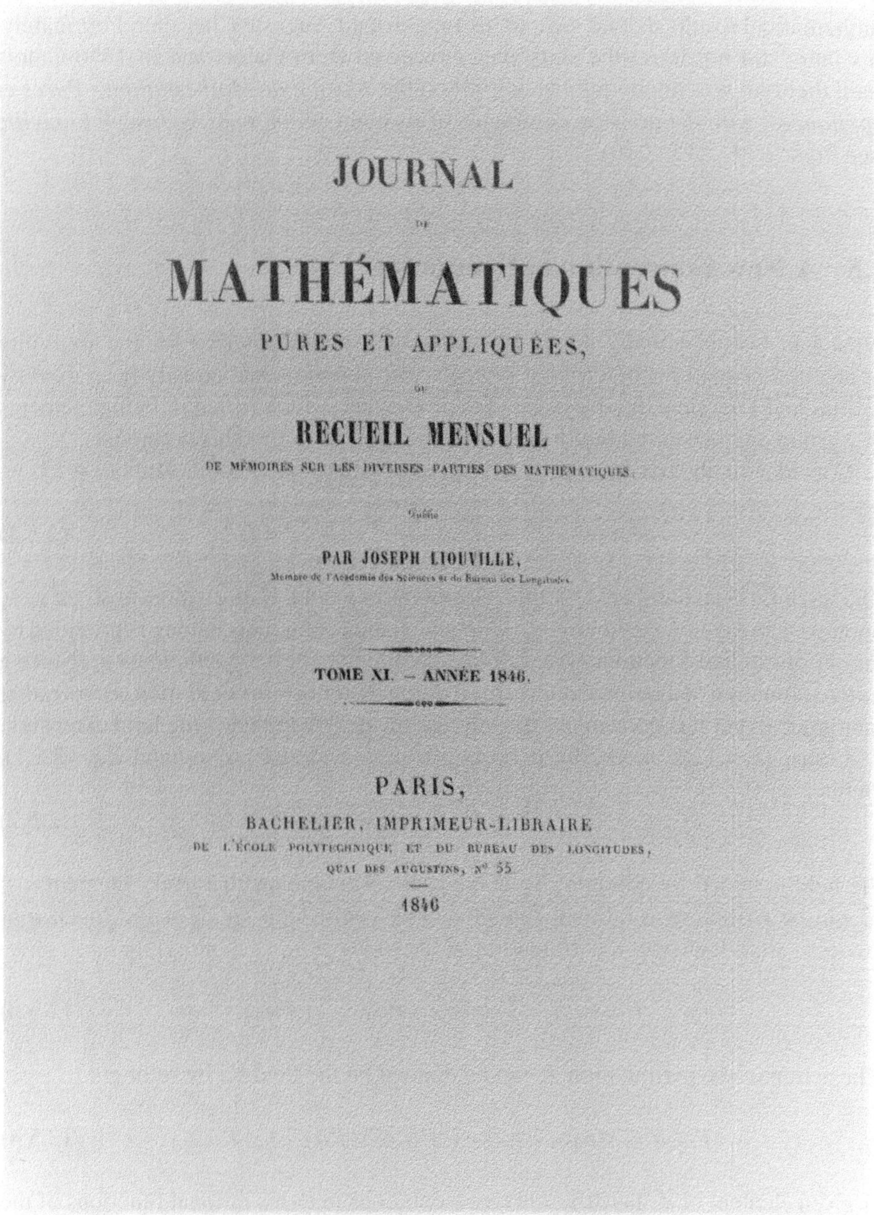
JOURNAL

DE

MATHÉMATIQUES

PURES ET APPLIQUÉES,

OU

RECUEIL MENSUEL

DE MÉMOIRES SUR LES DIVERSES PARTIES DES MATHÉMATIQUES;

Publié

PAR JOSEPH LIOUVILLE,

Membre de l'Académie des Sciences et du Bureau des Longitudes.

TOME XI. – ANNÉE 1846.

PARIS,

BACHELIER, IMPRIMEUR-LIBRAIRE

DE L'ÉCOLE POLYTECHNIQUE ET DU BUREAU DES LONGITUDES,

QUAI DES AUGUSTINS, N° 55

1846

Fig. 2.14 The issue of the *Journal de Mathématiques Pures et Appliquées* containing the paper of Galois, published posthumous by Liouville. Courtesy of the *Biblioteca-Peano of the Dipartimento di Matematica* of Torino University

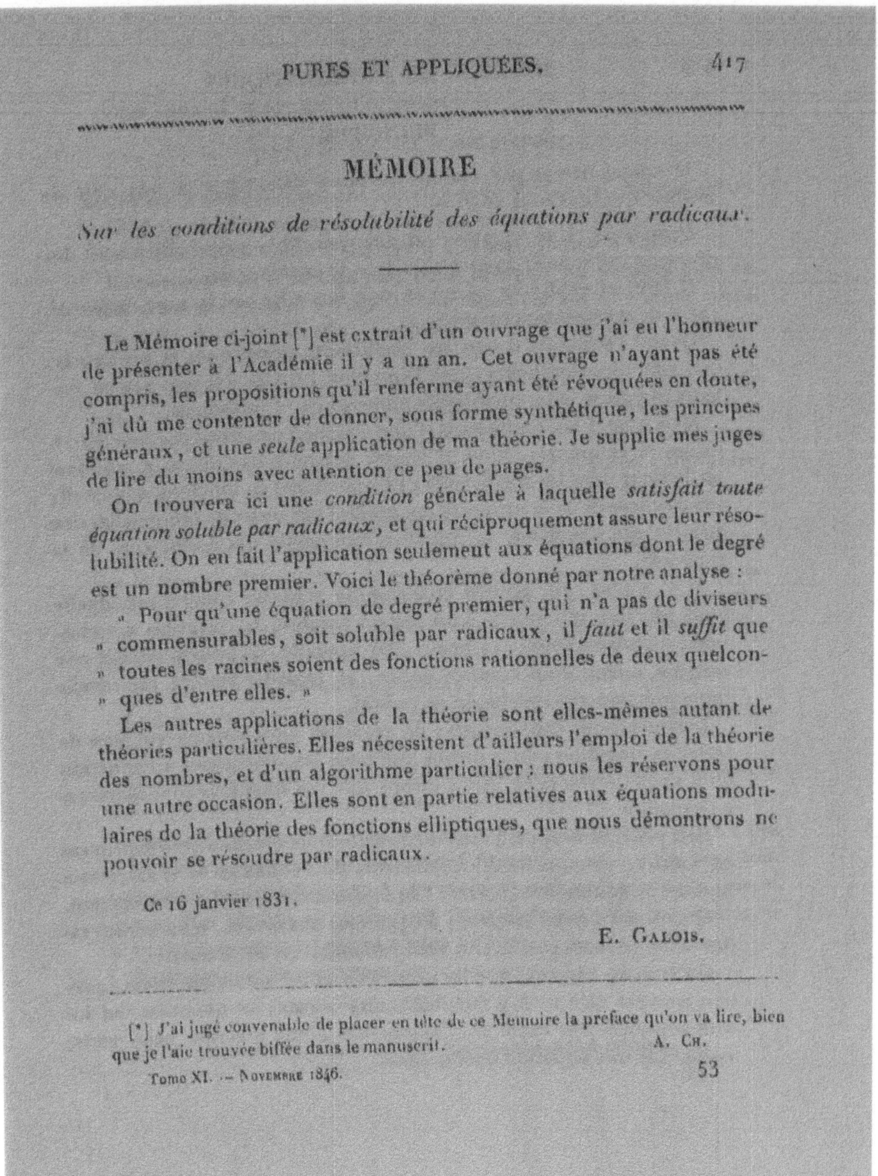

PURES ET APPLIQUÉES. 417

MÉMOIRE

Sur les conditions de résolubilité des équations par radicaux.

Le Mémoire ci-joint [*] est extrait d'un ouvrage que j'ai eu l'honneur de présenter à l'Académie il y a un an. Cet ouvrage n'ayant pas été compris, les propositions qu'il renferme ayant été révoquées en doute, j'ai dû me contenter de donner, sous forme synthétique, les principes généraux, et une *seule* application de ma théorie. Je supplie mes juges de lire du moins avec attention ce peu de pages.

On trouvera ici une *condition* générale à laquelle *satisfait toute équation soluble par radicaux*, et qui réciproquement assure leur résolubilité. On en fait l'application seulement aux équations dont le degré est un nombre premier. Voici le théorème donné par notre analyse :

« Pour qu'une équation de degré premier, qui n'a pas de diviseurs » commensurables, soit soluble par radicaux, il *faut* et il *suffit* que » toutes les racines soient des fonctions rationnelles de deux quelcon» ques d'entre elles. »

Les autres applications de la théorie sont elles-mêmes autant de théories particulières. Elles nécessitent d'ailleurs l'emploi de la théorie des nombres, et d'un algorithme particulier : nous les réservons pour une autre occasion. Elles sont en partie relatives aux équations modulaires de la théorie des fonctions elliptiques, que nous démontrons ne pouvoir se résoudre par radicaux.

Ce 16 janvier 1831.

E. GALOIS.

[*] J'ai jugé convenable de placer en tête de ce Mémoire la préface qu'on va lire, bien que je l'aie trouvée biffée dans le manuscrit. A. CH.

Tome XI. — NOVEMBRE 1846. 53

Fig. 2.15 Galois' introduction to his own Memoire. Courtesy of the *Biblioteca-Peano of the Dipartimento di Matematica* of Torino University

430 JOURNAL DE MATHÉMATIQUES

nique, XVIIe cahier.) Ainsi l'avant-dernier groupe sera

$$(G)\quad \left\{\begin{array}{llllll} x_0, & x_1, & x_2, & x_3,\ldots, & x_{n-3}, & x_{n-2}, \quad x_{n-1},\\ x_1, & x_2, & x_3, & x_4,\ldots, & x_{n-2}, & x_{n-1}, \quad x_0,\\ x_2, & x_3,\ldots\ldots & & \ldots, & x_{n-1}, & x_0, \quad\;\; x_1,\\ \ldots & \ldots & \ldots & \ldots & \ldots & \ldots\\ x_{n-1}, & x_0, & x_1,\ldots\ldots, & & x_{n-4}, & x_{n-3}, \quad x_{n-2}, \end{array}\right.$$

$x_0, x_1, x_2, \ldots, x_{n-1}$ étant les racines.

Maintenant, le groupe qui précédera immédiatement celui-ci dans l'ordre des décompositions devra se composer d'un certain nombre de groupes ayant tous les mêmes substitutions que celui-ci. Or j'observe que ces substitutions peuvent s'exprimer ainsi : (Faisons en général $x_n = x_0$, $x_{n+1} = x_1, \ldots$, il est clair que chacune des substitutions du groupe (G) s'obtient en mettant partout à la place de x_k, x_{k+c}, c étant une constante.)

Considérons l'un quelconque des groupes semblables au groupe (G). D'après le théorème II, il devra s'obtenir en opérant partout dans ce groupe une même substitution; par exemple, en mettant partout dans le groupe (G), à la place de x_k, $x_{f(k)}$, f étant une certaine fonction.

Les substitutions de ces nouveaux groupes devant être les mêmes que celles du groupe (G), on devra avoir

$$f(k+c) = f(k) + C,$$

C étant indépendant de k.

Donc

$$f(k+2c) = f(k) + 2C,$$
$$\ldots\ldots\ldots\ldots\ldots\ldots\ldots\ldots\ldots$$
$$f(k+mc) = f(k) + mC.$$

Si $c = 1$, $k = 0$, on trouvera

$$f(m) = am + b,$$

ou bien

$$f(k) = ak + b,$$

a et b étant des constantes.

Fig. 2.16 A sample page of Galois' posthumous Memoire published in 1846. Courtesy of the *Biblioteca-Peano of the Dipartimento di Matematica* of Torino University

Maps P satisfying Eq. (2.5.5) are named *automorphisms* of the extend field $\mathbb{K}$ *relative to the base field* $\mathbb{F}$. They form a group $\mathfrak{G}(\mathbb{K})$ (the *Galois group* of the original polynomial) since the product of any two elements $P_1, P_2 \in \mathfrak{G}(\mathbb{K})$ is still an element in the same set which contains also an identity element and the inverse of any element in the set. By product of two permutations we mean here the result of performing P_1 first and then P_2.

A solvable group, by definition is one that possesses a chain of normal subgroups ending in the trivial group composed only by the identity. Referring to such a notion we can state Galois theorem. He showed that an algebraic equation such as that in (2.5.1) is *solvable by radicals* if and only if *its Galois group is solvable*.

2.5.1 *Conceptual Analysis of Galois Results*

Let us analyse the revolutionary content of what Galois did in the field of Mathematics.

(1) First of all he changed perspective and, rather than analyze the properties of static mathematical objects, he emphasized the relevance of the transformations one can operate on them.
(2) Secondly he put to the forefront the notion of group: the transformations one can consider are combined together by an operation, the product and with respect to that operation they form an algebraic structure, the group G.
(3) Thirdly he put into evidence that once a group G is introduced, its action, originally defined on something (the roots in this case) can be extended to something else (the field extension $\mathbb{K}$). This was the beginning of representation theory which will concern us a lot in the sequel.
(4) Next, by means of the concept of automorphism he showed that given a group G which acts on some *space* **V** (the extended field $\mathbb{K}$ in this case), the most important things to study are the properties of objects contained in **V** that are left unchanged by G-transformations. This is the beginning of the theory of invariants that was central to mathematics for the whole XIXth century and still is essential in contemporary scientific thought.
(5) Last but not least, by means of his very theorem he showed that the key information about the structure of a mathematical object (in this case the algebraic equation) operated on by a group of transformations G (in this case the Galois group acting on the roots) resides in the algebraic structure of G. In this case the equation is solvable if G is solvable.

2.5.2 *Symmetry After Galois*

What we summarized above was going to be the new mathematical conception of Symmetry that would be slowly developed through the XIXth century, eventually producing the scientific revolutions of the XXth century, forming, up to the present day, the backbone of our understanding of the Fundamental Forces of Nature.

After Galois, symmetry is no longer some adequate proportion of the parts of something, as it was for the Classical Greeks, symmetry is just the group G. Indeed we started talking about *symmetry groups*. The old notion of symmetry is not entirely lost: it remains, to some extent, in the notion of *invariant*. Something is symmetric, with respect to a group of transformations G, if it is left unchanged by such transformations.

We can come back to the Alhambra tilings and ask ourselves the question, what are the transformations in a plane that leave one of those tilings invariant? Obviously they are made of some rotations with prescribed angles $\theta = \frac{\pi}{2}$ or $\theta = \frac{2\pi}{3}$, just to mention realized examples, and of some translations with prescribed directions and lengths. The set of all such transformations necessarily form a group G. Groups of this type that combine together in a consistent way discrete rotations and discrete translations are named *Wall Paper Groups*. The many time repeated question *how many different Alhambra patterns are conceivable?* is reformulated into the algebraic question *How many Wall Paper Groups do exist?* The answer is 17. The same answer was formulated at the end of the XIXth century in 3-dimensional space and its answer, obtained by the Russian Mathematician and Geologist Fyodorov is a list of 230, so named, *Space Groups*. This list forms the basis of *Crystallography* and of *Molecular Chemistry*.

Point (4) in our analysis of Galois results corresponds to one of the most fertile implications of his work. It had, in the long run, an enormous influence on our conception of *Geometry* and eventually it is even at the origin of *Special and General Relativity*.

Once the notion of a transformation group G is introduced, the notion of *equivalence classes* naturally arises. A set of objects acted on by G can be rationalized by dividing it into stocks, each of which contains all those that are mapped one into the other by some transformation of the group. In some sense all the objects that happen to be in the same stock are different realizations of the same entity which is none of them, but just the entire equivalence class. As we explain in detail in Chap. 5, directly influenced by Galois' ideas that came to them through Darboux and Jordan, Sophus Lie and Felix Klein started rethinking classical geometry from a new viewpoint. In particular Klein realized that Euclid axiomatic definitions of what is an equilateral triangle, a rectangular triangle and so on, can be recast into the notion of equivalence classes. There are many triangles that one can draw in a plane but two triangles that can be mapped one into other by means of a rotation or a translation, namely an element of what we shall name the Euclidian Group $\mathbf{E}_2$, have to be identified and considered just the same triangle. Hence the objects of study in Euclidian Geometry are just the equivalence classes with respect to $\mathbf{E}_2$. It follows immediately that all the propositions of Euclidian Geometry are just statements on properties and relations that are invariant with respect to $\mathbf{E}_2$ or in three-space with respect to $\mathbf{E}_3$. In this way Klein came to conceive the momentous *Erlangen Programme* (see Sect. 5.2.2). Since there are other groups different from the Euclidian Group, you can conceive other geometries, among which the non-euclidian one introduced by Lobachevsky. Actually you can classify geometries according to the group G with respect to which the relations considered in that geometry are invariant.

This way of thinking shifted the interest from the mathematical objects that are transformed to the group of transformations that operate on the objects. Through several conceptual steps, the main issue became the classification of Groups, both the discrete finite ones, whose theory was the first to be fully developed and the continuous, Lie ones, introduced in the last quarter of the XIXth century. Essentially this is the full-fledged development of point (5) in the above list.

2.6 Symmetry, Geometry and Space

Indeed the XIXth century witnessed a powerful development of Group Theory, this, after Galois, being the proper new name for Symmetry, but also of the mathematical conception of Geometries, the plural number now replacing the singular one, referred to Euclidian Geometry, which had dominated Philosophy and Mathematics for 2000 years.

One fundamental aspect of the groups, shared by the discrete and by the continuous ones, is that their intrinsic algebraic structure determines also their possible realizations as linear transformations in some vector space. This very fact is a new component of the mathematical and physical *episteme* having revolutionary consequences.

Reality, namely what exists in *actuality* is no longer investigated to learn about its symmetries (=properties) rather it is *a priori* defined as one of the available *states*, which realize the *a priori* known symmetries and exist in *potentiality*.

In this new vision the *episteme*, namely knowledge, mainly consists in a *principle of choice* which selects a symmetry (i.e. a group G) and one of its realizations (i.e. a representation $D(\mathrm{G})$) attributing to them *actuality*.

As it was emphasized in the Preface, this encodes in a nutshell the structure of the Standard Model of the Fundamental Interactions, provided we specify within larger categories the principle of choice mentioned above.

In a ramified process which unfolded throughout the XIXth century up to the middle of the XXth century, Geometry, defined as the science of *Space* was deeply influenced by the developments of Group Theory.

On one side, starting with the pioneering work of Gauss and Riemann (see Chap. 7), Differential Geometry was born, able to describe curved spaces and this eventually led to Einstein General Relativity. The new concept of *Space* is encoded in the notion of a *Differentiable Manifold* formalized by Whitney in 1936 [110] but already implicit in Riemann's Habilitation thesis.

On the other side, the discovery of Lie Groups, of their associated Lie algebras and the classification of the latter (a story which is the main theme of Chap. 5) led to the idea of groups acting by isometries on Riemannian manifolds and to the classification of symmetric spaces G/H, realized by Cartan. Slowly a new concept of Space emerged in the first half of the XXth century, corresponding to the notion of fibre-bundle and in particular of principal fibre-bundle $P(\mathrm{G}, \mathscr{M})$. The language of fibre-bundles is that of modern geometry and it is the one which underlies modern gauge theories and substantiate the present *episteme* of Theoretical Physics (the conceptual history of these geometrical developments is told in Sect. 9.1).

The complicated frontier line of contemporary researches in geometry is heavily marked by the influence of Supergravity/Superstrings and involves three new conceptual ingredients, contributing new categories in the *episteme*:

(A) Consideration of compatible Complex Structures defined over the tangent bundle $T\mathscr{M}$ which qualify the geometry of the considered manifold $\mathscr{M}$ as Complex, Quaternionic or even Octonionic.
(B) The *geometry of geometries*, namely the consideration of metrics and other structures defined on those manifolds (*moduli spaces*) whose points parameterize the available geometries of other manifolds.
(C) Manifolds $\mathscr{M}$ featuring *Special Geometries*, typically characterized by assumptions made on the structure and the characteristic classes of certain fibre-bundles constructed over $\mathscr{M}$.

The conceptual historical analysis of these developments is contained in Chaps. 8 and 9. Section 9.2 and then Chap. 10 outline instead the applications of special geometries with particular attention to symmetric spaces. The last Chap. 11 recalls how new mechanisms were found able to produce new non trivial geometries from trivial ones, by means of Kähler and HyperKähler quotients which have a very important field theoretical interpretation.

We do not dwell on the content of the last four chapters since, in order to understand the addressed issues, it is necessary for the reader to have already well digested the theory whose historical development concerns us in the previous chapters. We remark instead that the most relevant for the *episteme* is item (B) of the above list. The geometry of geometries is indeed a new philosophical category in some sense approaching, at least methodologically, the issue of the *principle of choice*. To say it bluntly it is like introducing the idea of a theory of theories. One has defined a collection, finite or infinite of options, each of which describes a particular theory. Now these options become the degrees of freedom of a new theory which has a dynamics able to determine a smaller set of options. Now it can happen that the theory of theories is also defined up to some options and one can imagine to repeat the process. One can eventually dream of a tree, at whose top everything becomes uniquely determined.

Although this idea is fairly general and can be applied in various contexts (the renormalization group, for instance), the really important thing for the *episteme* of fundamental physics is that everything appears to be geometrical at every level; indeed we are talking of the *geometry of geometries*, meaning that the mathematical structures advocated at the superior level are the same ones used one step below: fibre-bundles, riemannian manifolds, geodetics and so on. This is best exemplified by the case of black-hole solutions of supergravity (see Chap. 10). The items we are dealing with in that context are the possible geometries of a black-hole, equipped with electric and magnetic fields that describe the connection of a fibre-bundle. Each of such articulated geometries corresponds just to a geodetic of a larger manifold that can be chosen to be a suitable lorentzian symmetric space.

Chapter 3
How Group Theory Came into Being

Ancora indietro un poco ti rivolvi,
diss'io, là dove di' ch'usura offende
la divina bontade, e 'l groppo solvi.

Dante, Inferno XI, 94

3.1 The Essentials of Group Theory in Modern Parlance

The essentials of group theory can be summarized in few mathematical definitions that admit a description in relatively simple words.

A group G is first of all a set of elements. There are three cases:

1. The set G contains a finite number r of elements $\{\gamma_1, \gamma_2 \ldots, \gamma_r\}$. In this case G is a finite group and the number r, usually denoted $|G|$ is named the order of the group G (see Fig. 3.1).
2. The set G contains an infinite number of elements, but it is denumerable, namely we can count the elements as $\{\gamma_1, \gamma_2, \gamma_3, \ldots, \gamma_\infty\}$. In this case the group is infinite but discrete.
3. The set G is a continuous space as it is for instance the plane, or a sphere or some higher dimensional variety. In this case the group G is named a *continuous group* (see Fig. 3.2) and when additional properties of analyticity are satisfied it is a *Lie group*.

The feature that promotes a set G (falling in one of the above specified cases) to the status of a group is the existence of a binary operation:

$$p \quad : \quad G \times G \longrightarrow G \tag{3.1.1}$$

Modern mathematics has at its centre the notion of *map*. In simple words a map φ is a correspondence between two sets A and B:

$$\varphi \quad : \quad A \longrightarrow B \tag{3.1.2}$$

P. G. Fré, *A Conceptual History of Space and Symmetry*,
https://doi.org/10.1007/978-3-319-98023-2_3

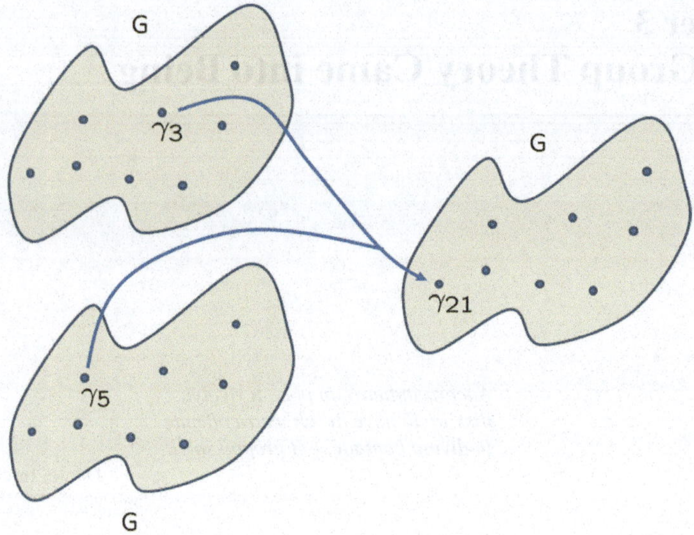

Fig. 3.1 A finite group G is a set with a finite number of elements and an internal binary operation named the product. In the above picture we imagine a finite group where the product of the element γ_3 with the element γ_5 produces the element γ_{21}

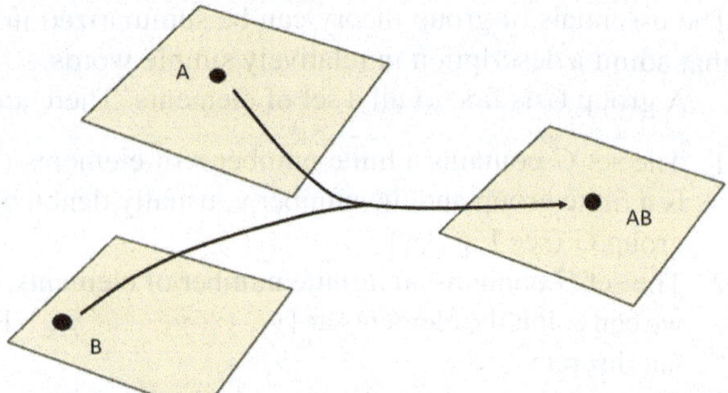

Fig. 3.2 A continuous group G is a continuous (topological) space like the plane or some other higher dimensional manifold, whose points can be labeled by coordinates and which is endowed with an internal binary operation named the product

that to each element $a \in A$ of the first set associates an element $\varphi(a) \in B$ of the second set. The element $\varphi(a)$ is named the image of a in B. On the other hand any element $a \in A$ whose image is a given element $b \in B$ is said to be in the preimage $\varphi^{-1}(b)$. In general the preimage $\varphi^{-1}(b)$ can contain more than one element.

The binary product p of a group G is a map from the set of ordered pairs $\{a, b\}$, where $a, b \in G$ are elements of the group, to the group G. The image of the pair:

$$p(a, b) \equiv a \cdot b \in G \tag{3.1.3}$$

is an element of the same set G and it is named the product of a with b.

In order for G to deserve the name of *group*, the *product* should have the following necessary properties:

(a) The set G must include a specific element $\mathbf{e} \in G$, named the *identity*, which multiplied either on the left or on the right with any element $x \in G$ reproduces the latter, i.e.:

$$x \cdot \mathbf{e} = \mathbf{e} \cdot x = x \tag{3.1.4}$$

(b) Chosen any element x belonging to the group G, the latter must contain also a unique element x^{-1}, named the inverse of x such that:

$$x \cdot x^{-1} = x^{-1} \cdot x = \mathbf{e} \tag{3.1.5}$$

where $\mathbf{e}$ is the previously introduced identity element.

3.1.1 Examples

A familiar example of infinite, but discrete group is provided by the integer relative numbers $\mathbb{Z}$. In that case the product is simply the sum $a + b$, the identity element is the number zero 0 and the inverse of any element a is just $-a$. A very simple example of continuous group is provided by the complex numbers deprived of the number 0, namely $\mathbb{C}^\star = \mathbb{C} - \{0\}$. In this case the product is the ordinary product, the identity element is the number 1 and the inverse of $z \in \mathbb{C}^\star$ is the reciprocal $\frac{1}{z}$ which always exists, since we have excluded $z = 0$. The simplest example of finite group is $\mathbb{Z}_2$ formed by the two-element set $\{1, -1\}$. The product is the ordinary one, the identity element is 1 and the inverse of -1 is just the same element, since $(-1) \times (-1) = 1$.

3.1.2 Groups as Transformation Groups

In all the examples quoted above the product operation is commutative, namely the product $a \cdot b$ of the element a with the element b yields the same result as the product $b \cdot a$ taken in the reverse order. This is not the general case and the groups that possess such a property form the subclass of *abelian groups*. The generic case is that of *non abelian groups*.

To understand how the apparently unfamiliar situation $a \cdot b \neq b \cdot a$ enters the stage we have to think of the groups not as sets of numbers rather as sets of *transformations* that act on another set S, which can be either finite, or infinite discrete, or continuous. In other words every element $\gamma \in G$ of a given group G is viewed as a map:

$$\gamma \; : \; S \longrightarrow S \tag{3.1.6}$$

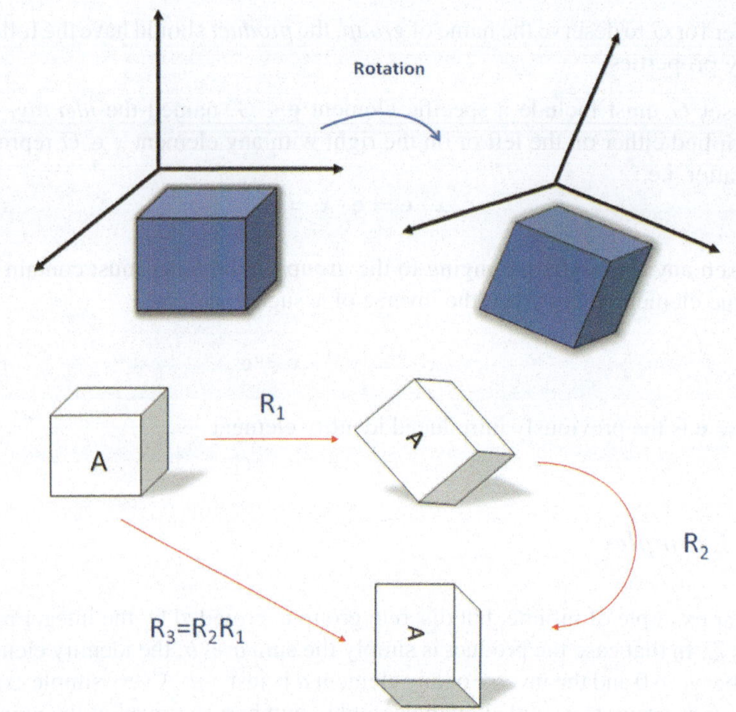

Fig. 3.3 In this picture we illustrate the notion of group product with the example of familiar three-dimensional rotations. A finite rotation of a three-dimensional solid object is effected around some axis for the extension of some angle θ. After performing a rotation R_1, we can perform a second rotation R_2. The net result of the sequence of the two transformations is a new rotation R_3 around some new axis and for the extension of some new angle

that associates an image $\gamma(a) \in S$ in the same set to every element a of the set S. The product $\gamma_2 \cdot \gamma_1$ of two group elements is just the transformation of the set S into itself that is obtained by applying first the transformation γ_1 and then the transformation γ_2 in the specified sequence. This fundamental idea is illustrated in Fig. 3.3 with the example of the rotations in three-dimensional space. The set of such rotations is the rotation group that has the mathematical name SO(3). It is evident from familiar experience that once thought in this way the group–product can be non commutative. The result on any three-dimensional object of performing first a rotation around the x-axis and then a rotation along the y-axis is typically different from the result obtained by performing the same rotations in reversed order.

It is precisely in the capacity of sets of transformations that groups became the pivot item in the modern conception of symmetry originating from the fundamental work of Galois (see Sect. 2.5.2). The above summarized concept of group came into being through a rather long historical process which we plan to outline in the present chapter.

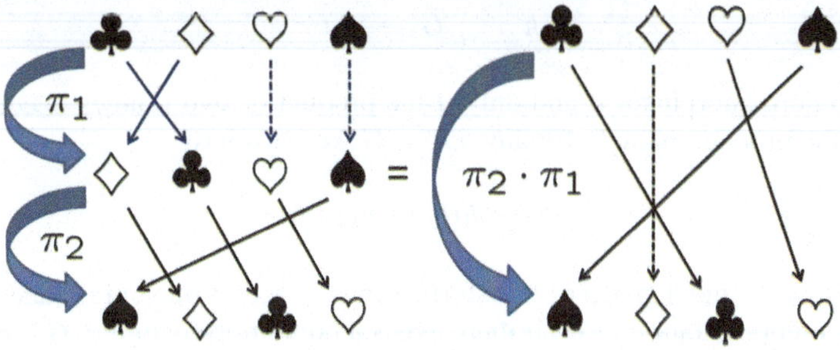

Fig. 3.4 Focusing on the case of four objects in the above figure we exemplify the product law within the permutation group

The first group to be considered, which is also at the basis of Galois' work and which actually encompasses all the other finite groups as subgroups, was the *permutation group* of n-objects. The elements of this latter, denoted S_n and named the *nth symmetric group* are the permutations of an array of n objects into a different order like, for instance:

$$\pi \quad : \quad \{\spadesuit, \heartsuit, \diamondsuit, \clubsuit\} \longrightarrow \{\heartsuit, \spadesuit, \clubsuit, \diamondsuit\} \tag{3.1.7}$$

which is an element of S_4. As it is probably known to most readers, the total number of permutations of n objects is $n! = n \times (n-1) \times (n-2) \times \cdots \times 2 \times 1$ which is the order $|S_n|$ of the symmetric group S_n. In the case of four objects, like the playing card suits, the number of permutations is just 24 and such is the order of the corresponding symmetric group.

An example of product of permutations is provided in Fig. 3.4.

3.1.3 Representations of a Group

Once the idea of transformation is absorbed, it becomes evident that every group G acts as a transformation group on itself, since each of its elements $g \in G$ acts, via the product, on all the group elements $\gamma \in G$ (the same g included) and maps them in other elements of G. Furthermore it is evident that the same group can operate as a transformation–group on different spaces. Each of these incarnations of G is named one of its *representations* and a central issue of group–theory is the classification of all possible representations of each G.

To be more precise we ought to rely on the notion of *homomorphism*. A homomorphism is a map from one group G to another group Γ that respects the product law of the two groups. Let denote by $\cdot$ the product in the first group G and by $\diamond$ the product in the second group Γ. A map:

$$h \ : \ G \longrightarrow \Gamma \tag{3.1.8}$$

is a group homomorphism if and only if the product of two images is equal to the image of the product, namely for any $a, b \in G$ we must have:

$$h(a) \diamond h(b) = h(a \cdot b) \tag{3.1.9}$$

Given a group G and a group of transformations $\mathscr{T}$ acting on some space S we say that $\mathscr{T}$ is a representation of G, if there exists a homomorphism $h \, : \, G \longrightarrow \mathscr{T}$.

Among the possible representations a distinctive privileged role is played by the linear ones. What do we mean by this? To answer such a question we need the notion of *vector space*. This is the generalization to arbitrary dimension of the familiar notion of three-dimensional vectors.

VECTOR SPACES

Let us consider Fig. 3.5 which displays two vectors **v** and **w** in the ordinary three–dimensional space $\mathbb{R}^3$. The basic property of the space of vectors is that vectors can be summed and the sum is a vector in the same space. For instance the new vector $\mathbf{v} + \mathbf{w}$ is displayed in the figure. A vector **v** can be multiplied also by real numbers $\lambda \in \mathbb{R}$ obtaining a vector $\lambda\,\mathbf{v}$ that has the same direction if $\lambda > 0$ but length $\lambda \times |\mathbf{v}|$ where the latter symbol denotes the length of **v**. In the case $\lambda < 0$ the vector $\lambda\,\mathbf{v}$ has direction opposite to the direction of **v** and length $-\lambda\,|\mathbf{v}|$.

Actually the entire vector space V_3 of three dimensional vectors can be viewed as the set of all possible linear combinations of three linear independent vectors $\mathbf{e}_{1,2,3}$ such as the orthonormal versors displayed in Fig. 3.5:

$$V_3 = \left\{ \bigoplus_{i=1}^{3} \lambda^i \, \mathbf{e}_i \mid \lambda_i \in \mathbb{R} \right\} \tag{3.1.10}$$

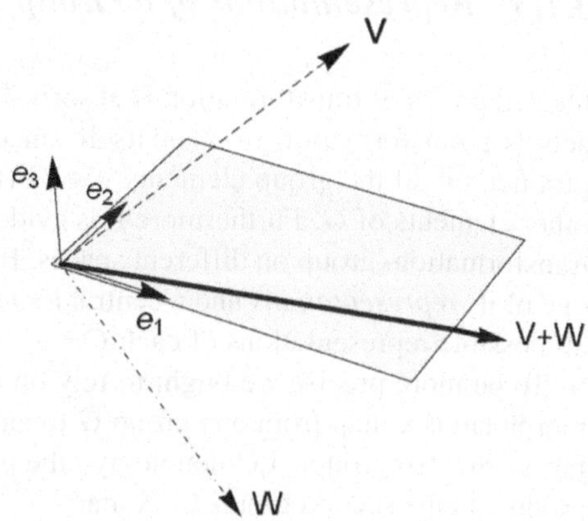

Fig. 3.5 In this picture we display two vectors **v** and **w** in the ordinary three-dimensional vector space $V_3 \simeq \mathbb{R}^3$ and we show their sum. For reference we display also the three unit vectors $\mathbf{e}_{1,2,3}$ respectively aligned with the x, y and z axis. Every vector in V_3 is a linear combination of the basis vectors $\mathbf{e}_{1,2,3}$

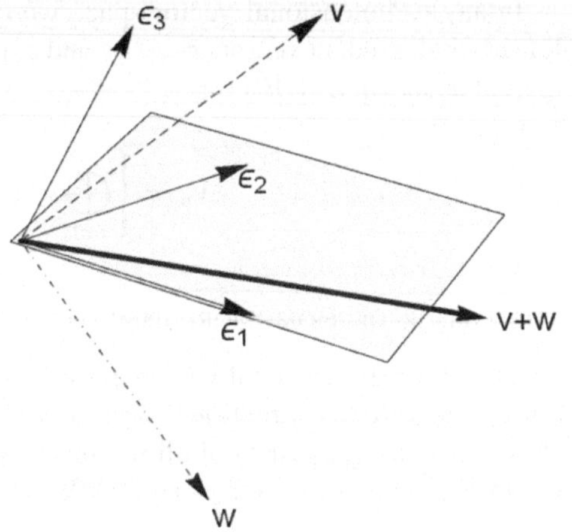

Fig. 3.6 The vectors **v** and **w** (and their sum) displayed in Fig. 3.5 can be expressed as linear combinations with real coefficients also of the non orthonormal basis vectors $\boldsymbol{\varepsilon}_{1,2,3}$ displayed in this figure

The essential point is that the basis of a given vector space is not uniquely defined. In the case of V_3 any other triplet $\boldsymbol{\varepsilon}_{1,2,3}$ of three vectors that do not lie in the same plane (this is the concept of linear independence) provides an equally good basis as the orthonormal set $\mathbf{e}_{1,2,3}$. This is illustrated in Fig. 3.6 which displays the same vectors **v** and **w** already displayed in Fig. 3.5 but emphasizes that they, as any other vector in the same vector space, can be expressed also as linear combinations of the non orthonormal triplet $\boldsymbol{\varepsilon}_{1,2,3}$.

The general concept of vector space emerges therefore from such a discussion. A vector space V is first of all a commutative group with respect to an operation that we can name the vector addition. The identity element is the **0**-vector and the inverse of any element (vector) **v** is just $-\mathbf{v}$. In addition the vector space has a second operation that is a map:

$$\mathfrak{s} : \mathbb{K} \times V \longrightarrow V \tag{3.1.11}$$

where $\mathbb{K}$ is a field typically $\mathbb{R}$ - in this case V is a real vector space - or $\mathbb{C}$ - in this case V is a complex vector space. The vector space is of finite dimension $n < \infty$ if the maximal number of vectors $\mathbf{v}_{i=1,\dots,n}$ that can be linearly indipendent is n. Linear independence of r vectors $\mathbf{v}_{i=1,\dots,r}$ means that the equation:

$$\sum_{i=1}^{r} \lambda^i \, \mathbf{v}_i = 0 \tag{3.1.12}$$

has the unique solution $\lambda_1 = \lambda_2 = \cdots = \lambda_r$. This is the generalization to higher dimension of the condition applying to the $n = 3$ case that three linear indipendent vectors cannot lie in the same plane. Surprisingly it took about a century to arrive at the above four definitions which appear at first sight extremely simple and natural.

In any n-dimensional vector space we can always choose a basis formed by n linearly indipendent vectors $\boldsymbol{\varepsilon}_{i=1,\dots,n}$ and express all the others as linear combination thereof as in Eq. (3.1.10):

$$V_{\mathbf{n}} = \left\{ \bigoplus_{i=1}^{n} \lambda^i \boldsymbol{\varepsilon}_i \mid \lambda_i \in \mathbb{K} \right\} \tag{3.1.13}$$

THE GROUP OF HOMOMORPHISMS OF A VECTOR SPACE V INTO ITSELF

Given a finite dimensional vector space $V_{\mathbf{n}}$ there is always a very interesting continuous group of transformations that is associated with it. Mathematically it is named Hom(V,V) and it consists of all the invertible linear homomorphisms of V into V. In practice an element of $h \in$ Hom(V,V) is a map:

$$h \ : \ V \longrightarrow V \tag{3.1.14}$$

that to each vector $\mathbf{v} \in V$ associates and image vector $h(\mathbf{v}) \in V$. Linearity of the map is the property that the image of a linear combination must be the linear combination of the images, namely:

$$h\left(\bigoplus_{i=1}^{n} \lambda^i \mathbf{v}_i\right) = \bigoplus_{i=1}^{n} \lambda^i h(\mathbf{v}_i) \tag{3.1.15}$$

Finally the map is invertible if the preimage $h^{-1}(\mathbf{w})$ of any vector $\mathbf{w}$ in the image of the map exists and it is unique.

THE LINEAR REPRESENTATIONS OF AN ABSTRACT GROUP G

Given the notion of vector space a linear representation D of dimension n of a group G is a homomorphism:

$$D \ : \ G \longrightarrow \text{Hom(V,V)} \tag{3.1.16}$$

Choosing a basis of V, each homomorphism $h \in$ Hom(V,V) is translated into an $n \times n$ matrix h_{ij}, so that the representation $D(\gamma)$ of every group element $\gamma \in G$ is provided, at the end of the day, by a suitable matrix D_{ij}. Hence the theory of groups is intimately related with the theory of matrices that was a crucial focus of attention in the middle of the XIXth century. In the next sections we trace the historical development finally leading to the concepts and definitions briefly reviewed in the present section.

3.2 From Cayley and Sylvester's Matrices to Vector Spaces and Groups: A Long Gestation

Any introductory text on group theory makes extensive use of matrices; from the point of view of contemporary students in *Math and Phys* we rightly consider such material as simple and elementary, yet this fact, which is true, should not induce us into the erroneous assumption that what nowadays we call *linear algebra* is something naturally obvious for the human mind. Historically it took a rather long time before the fundamental concepts of linear algebra were consolidated and settled down to the apparently simple shape used in current textbooks and lecture courses. The same is true for the abstract notion of a group.

Reconsidering the conceptual history of these ideas is very useful in order to fully appreciate the degree of abstraction which is tacitly involved in our current way of thinking, both in elementary mathematics and in elementary physics, a degree of abstraction which has percolated down the generation–tree and currently makes part of the educational process. As a result, in the average, the mentality of XXth–XXIst century students, already incorporates such categorical structures as part of their logical thinking, which is a non trivial advance.

3.2.1 Cayley and Sylvester: A Short Account of Their Lives

The main figures in the early history of *linear algebra* are Arthur Cayley and Joseph Sylvester who became life long friends and whose lives often intersected. Hence we start our historic outline with a summary of their biographies.

Cayley was first educated at the King's College School in London and then entered Cambridge University where he studied mathematics. At the beginning he could not continue an academic career in Cambridge since he refused to take the minor orders of the Church of England. Having turned to Law, and having worked for 14 years as an attorney in the City of London, during which he never stopped doing research in mathematics, at the age of 42 Cayley was elected Sadleirian Professor of Mathematics in Cambridge, a position that he occupied until his death. He has been one of the most prolific mathematicians of history giving extensive contributions to different provinces of algebra, geometry and analysis.

James Joseph Sylvester studied mathematics at St John's College, Cambridge. He was not awarded a Cambridge degree since, to that purpose, he had to renounce his Jewish religion and accept the Thirty-Nine Articles of the Church of England, which he refused to do. After holding for some time a teaching position in London and then obtaining a degree from Trinity College in Dublin he became professor in the United States in Virginia. He stayed there briefly and came back to London where he studied Law; the following 10 years he worked in an Insurance Company. In London he met with Cayley and a life-long interaction between the two mathematicians started which was very fruitful for both. Then he crossed once again the Atlantic and for several years he was professor of mathematics at the John Hopkins University in Maryland. In 1883 he returned to England where he was appointed Savilian Professor of Geometry at Oxford University.

Cayley and Sylvester have been nicknamed the *Invariant Twins* for their extensive and outstanding contributions to the theory of invariants.

3.2.2 Matrices in the Middle 1850s

Probably the most important paper in the history of linear algebra was written by Arthur Cayley (see Fig. 3.7) in 1857–1858 and was published in 1859 on the *Philosophical Transactions of the Royal Society of London* with the title *A Memoir on the Theory of Matrices* [39].

At beginning of his Memoir, Cayley says:

The term matrix might be used in a more general sense, but in the present memoir I consider only square and rectangular matrices and the term used without qualification is to be understood as meaning a square matrix; in this restricted sense, a set of quantities arranged in the form of a square, e.g.

$$\begin{pmatrix} a, & b, & c \\ a', & b', & c' \\ a'', & b'', & c'' \end{pmatrix}$$

is said to be a matrix.

Fig. 3.7 Arthur Cayley (Richmond 1821–Cambridge 1895)–James Sylvester (London 1814–London 1897)

What originally a matrix meant for Cayley is immediately stated by him. He says that the notion of a matrix arises as an abbreviation for a system of linear equations:

$$\begin{aligned} X &= ax + by + cz, \\ Y &= a'x + b'y + c'z, \\ Z &= a''x + b''y + c''z, \end{aligned}$$

which *may be more simply represented by*:

$$(X, Y, Z) = \left(\begin{array}{|ccc|} a, & b, & c \\ a', & b', & c' \\ a'', & b'', & c'' \end{array} \right)(x, y, z),$$

The rather curious and obsolete notation used by Cayley reveals its graphical motivation in the above equation. The round brackets embracing each other make a sort of times symbol $\times$ which is probably what he liked.

So matrices were born in the field of algebraic equations as many other mathematical items of the XIXth century.

Notwithstanding this equation–based start point, Cayley's article contains, spelled out for the first time, most of the concepts that elevate the set of matrices to an algebra. He introduced the multiplication of matrices $\mathbf{L}.\mathbf{M}$, the addition of matrices $\mathbf{L}+\mathbf{M}$, the $\mathbf{0}$-matrix and the identity matrix $\mathbf{1}$. He also defined the operation of transposition and gave rules for the calculation of determinants. He went as far as defining rational functions of matrices but there is no indication in this paper that Cayley had developed the abstract notion of a vector space V and that he identified the matrix as an explicit representation of a linear map $\mu : V \to V$ in a basis of V.

The most important result of Cayley's 1858 Memoir is what was named by posterity the *Cayley–Hamilton theorem*, namely the statement that the eigenvalues λ of a matrix $\mathbf{M}$ are roots of the characteristic polynomial:

$$0 = \mathfrak{P}(\lambda) \equiv \det(\mathbf{M} - \lambda\,\mathbf{1}) \tag{3.2.1}$$

After verifying the statement for 2×2 matrices, Cayley wrote: *I have verified the theorem in the next simplest case, of a matrix of order 3, but I have not thought it necessary to undertake the labour of a formal proof of the theorem in the general case of a matrix of any degree.*

In his marvelous thesis which has become the main reference for the early history of linear algebra [48], Crilly says:

Cayley's Memoir, which could have been a useful starting point for further developments, went largely ignored...His habit of instant publication and not waiting for maturation had the effect of making the idea available even if it was effectively shelved.

Crilly also wrote: *As it is well known, it was Sylvester who introduced the word* ***matrix*** *into mathematical language in 1850, but he then meant an array of numbers* *from which determinants could be formed. Sylvester's concern was with determinants, as his consideration for the multiplication rule for determinants shows. A letter written in 1852 to Cayley by Sylvester shows a matrix being multiplied by another matrix. It is quite natural for Sylvester to be using* ***row by row*** *multiplication to obtain the resulting matrix as he was interested in determinants.*

The multiplication of matrices row by row introduced by Sylvester (see Fig. 3.7) is the clearest evidence how far the notion still was in the middle 1850s of a matrix as an explicit representation of a linear map of a vector space V into itself.

Almost thirty years later in 1882, Cayley visited his friend Sylvester in Baltimore (Maryland) for the spring semester. There Sylvester told Cayley of his own new discovery of a theory of matrices, only to be reminded by Cayley of the 1858 memoir. Nevertheless Sylvester explained his rediscovery, in his typical prolix and flourishing style:

Much as I owe in the way of fruitful suggestion to Cayley's immortal memoir, the idea of subjecting matrices to the additive process and of their consequent amenability to the laws of functional operation was not taken from it, but occurred to me independently before I had seen the memoir or was acquainted with its contents; and indeed forced itself upon my attention as a means of giving simplicity and generality to my formula for the powers or roots of matrices, published in the Comptes Rendus of the Institute for 1882 (Vol. 94, pp. 55, 396).

So it was Sylvester, more than Cayley, who took up the development of matrix theory towards the end of his time at the Johns Hopkins University. N. J. Higham wrote [115]:

We owe quite a lot of our linear algebra terminology to Sylvester, including the words ***annihilator***, ***canonical form***, ***discriminant***, ***Hessian***, ***Jacobian minor***, *and* ***nullity***. *Sylvester coined* ***latent roots***, *and after reading his explanation of the term, one may wonder why* ***eigenvalue*** *has supplanted it:*

It will be convenient to introduce here a notion (which plays a conspicuous part in my new theory of multiple algebra), namely that of the latent roots of a matrix ...latent in a somewhat similar sense as vapour may be said to be latent in water or smoke in a tobacco-leaf.

3.2.3 Cayley and Sylvester: The Invariant Twins

Before we continue tracing back the conceptual history of the notions of a group and of linear algebra, it is appropriate that we pause to consider in some more detail the human figures of Cayley and Sylvester.

Let us begin with a vivid picture of Cayley drawn in 1859 by his friend Hirst, as reported in Crilly's thesis [48]:

This evening, Friday Dec. 23rd 1859, I called upon Cayley and we had a very interesting hour's talk on Curves of the Third Order a propos of Möbius, ...
What a wonderful head he has, not merely round but spheroidal with the largest diameter parallel to his eyes, or rather to the line joining his ears. He never sits

upright on his chair but his posterior on the very edge he leans one elbow on the seat of the chair and throws the other arm over the back. Yet he is a keen sighted and extraordinary man, gentle I think by nature and at once timid, modest and reticent. Often when he speaks he shuts his eyes and talks as if he were reading from an unseen book, and talks well too that one has to sharper one's own wits to follow him.

Crilly comments on that as follows:

Yet for this exceptionally gifted man there was no suitable academic position available and at one point he considered taking private pupils.
...employment prospects for research scientists in general were bleak in mid century Britain. Most research was done outside the Universities and those already in teaching posts and whose main interest was primarily in teaching had a little aptitude for research.

Indeed this was still the time when Cayley, notwithstanding his extraordinary mathematical production, which proceeded at the pace of six-seven articles per year and reached, once, that of thirty in one year, was obliged to make a living for himself working as a conveyancing barrister in the City of London.

A political comment is unavoidable at this point. With respect to this issue, the difference existing at this time between France and the United Kingdom is striking. In 1851 Britain was fast becoming the most industrialized nation of the world. The British Empire was the strongest and largest Dominion on the Planet, the British Industrial Revolution, based on the use of Technologies originating from Science was coming to a second stage of sustained development which was influencing society and the world market to a degree never seen before in history, yet, as Crilly remarks, in 1857 the only University Institutions with University Status in England and Wales were Oxford and Cambridge, Durham, King's College and University College in London, Owen's College at Manchester and Lampeter in Wales. Furthermore Oxford and Cambridge, the main intellectual centres of the Kingdom, were still immobilized under the obsolete and stiff cape of the Church of England.

On the contrary, in France the *École Polytechnique,* and the *École Normale Superiéure,* both supported by the State, had already been shaped into the main *formation and research centres* of a modern, laic, industrial nation. French mathematicians, if they were talented, had a typically much easier career then their British colleagues and obtained much greater respect and honour from the various governments that succeeded one the other during the complicated French history of the XIXth century. This was the result of the French Revolution and of the epochal modernizations, never too much estimated, that had been introduced by the napoleonic administration. It is not by chance that revolutionary and napoleonic times have been so immensely prolific in mathematics in France with Laplace, Lagrange, Legendre, Fourier, Monge, Poncelet, Delambre, Carnot and several others.

As it is well known, the French Revolution had been prepared by the philosophical remaking of the European Civilization conducted during the *Age of the Enlightenment* by the *philosophes* of France, Britain and Germany. With Napoleon the Revolution swept the whole of Europe and, although Napoleon was eventually defeated and his time was followed by the gloomy Restoration, administrated by clerical Austria, yet, wherever napoleonic administration had once been, even for a short time, nothing was really again as it used to be before. Also in Prussia, one of the main defeaters of Napoleon, the legacy of the French type of Enlightenment found its original way of continuing its own development path through the German Idealism of Kant, Hegel and finally Marx. The hegelian conception of the State certainly helped the development of the great, state–owned German Universities of the XIXth century, Göttingen, Berlin, Königsberg, Bonn, Leipzig and highly contributed to the high prestige and comfortable life of german professors. This is certainly an important factor in the impressive development of German mathematics in the XIXth century. Only the British Islands had not been swept by the Napoleonic Armies and for twenty years His Majesty's Governments had patiently prepared Napoleon's final annihilation. This made the difference in the early lives of two of the greatest British mathematicians of the same century: Cayley and Sylvester.

In the course of time they came to know each other very well, they became close friends, they shared very similar mathematical interests and gave important contributions to the same provinces of mathematics, namely linear algebra and the theory of invariants. At the beginning of their career, without knowing each other they stumbled, for the same motivation, in the same obstacle that, in my opinion, is a shame for Victorian England.

Arthur Cayley was born 1821 in England in a well-to-do family of English merchants boasting descent from the Norman Conquerors. He spent his early years in Sankt Peterburg, capital of the Russian Empire where his father had established his business. There are not sure evidences, but it seems that Cayley's mother was of Russian origin. He made his secondary studies in London in King's College School where he immediately showed many talents, both in mathematics and in languages. In his adult life he conserved the ability to read Greek and Latin classics in the original and furthermore he spoke French as well as English and he was also proficient in German, language in which he later wrote some scientific papers, as well as he did in French. At age of 17 he enrolled at Cambridge University, brilliantly passing the entrance examinations. He mainly studied mathematics and was extremely successful in all the tests. In the first three years after graduation Cayley had already published twenty-five mathematical research papers and he could easily continue a Cambridge academic career. It was only required that he entered the religious orders of the Church of England. Cayley was a christian believer and all of his life he piously followed the practices of the Church of England, yet he could not accept the idea that, in order to make an academic career, he should unwillingly become a minister of the cult. Hence he refused and in 1846 he left Cambridge.

James Joseph Sylvester was born in London seven years before Cayley in a Jewish family. It seems that the surname Sylvester was invented by James' senior brother who had emigrated to the United States and that surname was adopted by the entire family. Similarly to Cayley, also Sylvester had a brilliant start in secondary studies, first in London, then in Liverpool. Just as his future friend, he learnt classical languages, Latin and Greek to such a high degree of perfection that in his late years he could amuse himself translating *Horace poems*, writing an essay on *Classical Metrics* and the like. Throughout all of his life he used to replenish his scientific papers with classical quotations that demonstrate his huge classical culture. He also studied and became absolutely proficient in French, German and Italian. His interest in mathematics developed early and similarly to Cayley he enrolled at the University of Cambridge at the age of 17, entering St. John's College in 1831. Cayley could not continue an academic career in Cambridge because he did not want to become a minister of the cult. Sylvester, by religion being a Jew, could not even obtain his doctoral degree. The necessary condition to obtain the diploma was that of renouncing his Jewish faith and subscribe to the Thirty-Nine Articles of the Church of England. The University of Cambridge, *honoris causa*, awarded Sylvester the delated doctor degree in 1871, after that it had been freed from Church surveillance by an Act of Parliament. At that time Sylvester was already on pension from the Royal Military Academy of Woolwich where he had been professor for 16 years.

Before telling that part of the story let us go back to the years 1840s when both Cayley and Sylvester had left, for similar reasons Cambridge University.

In 1841 Sylvester went to the United States where he was Professor at the University of Virginia for only one year. Then, having a quarrel with the administration of the university, he resigned and came back to England where in 1846 he entered the Inner Temple and studied Law. The same did Cayley entering instead Lincoln's Inn also in the City of London. Cayley was called to the bar on May 3rd 1849 at the age of twenty seven and for the next 14 years he practiced his profession as a conveyancing barrister. Sylvester instead worked as an Actuary for the *Equity and Law Life Assurance* up to 1856 and the two friends who, by that time, had become acquainted, had many opportunities to meet and discuss mathematics in the yards around Lincoln's Inn. They were both members of the Royal Society and had already written an impressive number of scientific papers on closely related topics although they never signed a joint paper.

In 1863 a new chair was created in Cambridge, the Saidlerian chair of Mathematics and Cayley was elected to it with the duty *to explain and teach the principles of pure mathematics and to apply himself to the advancement of that science*. In other words it was a teaching and research position at the same time, which, as Crilly clearly explains, was a sort of new gear for Cambridge. Cayley settled down in Cambridge, married and continued an intense research activity up to his death in 1895.

In 1876 at the age of 62, already a pensioneer, Sylvester was appointed on a full chair of mathematics by the John's Hopkins University in Baltimore and worked there until 1883, becoming also one of the founders of the *American Journal of Mathematics.*

In 1883, leaving with regret his American friends, Sylvester made return to England being appointed on the Chair of Geometry of Oxford University which he occupied until his death in 1897.

3.2.4 The Calculus of Operations: Cayley and George Boole

As we emphasized above, the notion of a vector space was completely absent in the society of 1850s mathematicians although, as we recall in the next section, a revolutionary book had already been published in 1844 by an obscure German school teacher, which, for a long time, no one wanted either to read or to understand.

The lack of this fundamental notion is what prevented Sylvester and Cayley to properly understand the *essence* of the matrices with which they were playing a lot of successful games. Yet Cayley came quite close to the right modern conception. The first key point was to identify a matrix with an *operator* or, in the language of that time, with an *operation*. The next two obligatory logical points were:

(a) To realize that operations can be combined together and are the objects of an algebraic calculus. Indeed they are elements of an *Algebra*.
(b) To inquire what are the mathematical entities on which the action of these operators is defined. Assessing the definition of such objects means to introduce the *space* of which they are elements and there you are: you have discovered the concept of linear realization of an algebra (or of a group).

Cayley had come to grasp point (a) already four years before publishing the *Memoir on the Theory of Matrices* when he wrote another article [38] which can be viewed as the first paper on abstract group theory. In that publication he started precisely from the notion of an *operation*. We shortly come back to this 1854 work by Cayley. Here we note that, quite curiously and probably quite significantly, the notion of operation is not mentioned in the 1858 *Memoir*. However someone else detected this notion in Cayley's paper. Crilly discovered that one of the referees of Cayley's paper was George Boole, the founder of modern *Symbolic Logic* (see Fig. 3.8). In his report on the paper dated March 29th 1858, Boole stated:

This memoir is an application of what has recently been termed ***Calculus of Operations****, to a particular branch of the Calculus of Functions. A matrix is a complex symbol denoting an operation by which from any set of quantities* $\{x, y, z\}$, *we form a set of linear functions of these quantities, e.g.*

$$a\,x + b\,y + c\,z\,,\; a'\,x + b'\,y + c'\,z\,,\; \&c$$

the number of such functions being in the class of Matrices chiefly considered by author, equal to the number of the subject quantities. As operations such as the above may be performed in succession, as the results to which they lead are capable of addiction and subtraction, as also, here as elsewhere, a direct operation supposes the existence of a corresponding inverse operation - the inquiry is suggested what are the distinctive laws of this class of operations, and to what special forms of Calculus, included under the more general calculus of operations, they give birth. This inquiry forms the business of the memoir, and its results are developed with clearness and ability.
In certain general feature they resemble, and necessarily so, the results of all other special developments of the Calculus of Operations and they are certainly of an interesting character ...

The founder of symbolic logic had clearly spotted the key point of the paper and in modern language, if he had it at his own disposal, he could give himself the answer to the question *to what special forms of Calculus, included under the more general*

Fig. 3.8 George Boole (Lincoln 1815–Ballintemple 1864)–Boole was born in Lincoln, Lincolnshire, England, the son of John Boole, a shoemaker and Mary Ann Joyce. He had a primary school education, and received lessons from his father, but because of the poverty of his family he could not receive any academic education. William Brooke, a bookseller in Lincoln, may have helped him with Latin. He was self-taught in modern languages. At age 16 Boole became the bread-winner for his family, taking up a junior teaching position. He taught briefly in Liverpool. Later he made a living running a boarding school. From 1838 onwards Boole was in touch with sympathetic British academic mathematicians and enlarged his own culture widely. He studied algebra in the form of symbolic methods and published research papers. In 1849 he was appointed the first professor of mathematics at Queen's College, Cork in Ireland. He died from pneumonia in 1864 as a consequence of his walking under a heavy rain. In 1841 Boole published an influential paper in early invariant theory. He receive a Medal in 1844 from the Royal Society for another paper on the same subject. In 1847 he published *The Mathematical Analysis of Logic* which founded a new field: symbolic logic, named after him *Boolean Algebra*

calculus of operations, they (the matrices) give birth. The answer is: an **Algebra**. Subsets singled out by proper restrictions can be **Lie Agebras** and subsets singled out by other proper restrictions can be **Groups**. However such an answer in 1858 was out of reach, even for the founder of symbolic logic, since the three quoted definitions did not yet exist. The necessary degree of abstraction had not yet been reached.

Yet at least for the last definition, which is actually the first in the hierarchy of algebraic complexity, namely that of a group, Cayley had come quite close to it in his paper of 1854 [38]. An interesting analysis of this paper from the viewpoint of a XXIst century student is provided in a note by David Pengelley [144]. Let us follow it closely.

At the time Cayley was writing, the only well known group was the permutation group and to prominence had come its applications in the issue of algebraic equations introduced by Galois. Yet linear *substitutions* were considered in the framework of quadratic forms and of their invariants (the favorite theme of Cayley and Sylvester) and also in relation with differential equations. So a general idea of *transformation* as an *operation* was emerging and Cayley tried to capture some general features of the *Calculus of these operations* that would correspond to the notion of a group.

Pengelley selected some extracts from Cayley's paper that are quite useful in order to appreciate the slow development of fundamental ideas.

At the beginning Cayley says: *Let* θ *be a symbol of operation, which may, if we please, have for its operand, not a single quantity* x*, but a system* $(x, y, \dots)$*, so that:*

$$\theta(x, y, \dots) = (x', y', \dots)$$

where $x', y', \dots$ *are any functions whatever of* $x, y, \dots$*, it is not even necessary that* $x', y', \dots$ *should be the same in number with* $x, y, \dots$*. In particular,* x', y'*, &c. may represent a permutation of* x, y*, &c.,* θ *is in this case what is termed a substitution; and if, instead of a set* $x, y \dots$*, the operand is a single quantity* x*, so that* $\theta\, x = x' = f(x)$*,* θ *is an ordinary functional symbol.*

Pengelley comments at this point: *It is delightfully unclear just how Cayley's initial general notion of operation really differs from that of function, and this makes good classroom discussion.*

Indeed this sentence reveals that the notion of function was at that time something more concrete and analytic then the simple abstract notion of a *map* from some *space* to another *space*.

Next Cayley expresses very clearly the idea that the symbol $\theta\phi$ denotes *the compound operation, the performance of which is equivalent to the performance, first of the operation* ϕ*, and then of the operation* θ*;* $\theta\phi$ *is of course in general different from* $\phi\theta$. This shows that he had captured the abstract notion of product which, in general, is not commutative, yet as he says next, it should be associative since $\theta \cdot \phi\chi = \theta\phi \cdot \chi$.

Few lines below Cayley arrives at his closest approach to the axiomatic definition of a group saying:

A set of symbols, $1, \alpha, \beta, \dots$ *all of them different, and such that the product of any two of them (no matter what order), or the product of any one of them into itself, belongs to the set, is said to be a group.*[1] *It follows that if the entire group is multiplied by any one of the symbols, either as further or nearer factor, the effect is simply to reproduce the group.*

Is this the complete axiomatic definition of a group? Let us translate Cayley's words into logical symbols. He makes three statements:

(1) $\forall a, b \in G \;:\; a \cdot b \in G.$
(2) $\forall a, b \in G \;:\; \exists b' \in G \setminus a \cdot b' = b$
(3) $\forall a, b \in G \;:\; \exists b' \in G \setminus b' \cdot a = b$

---

[1] [Cayley's footnote]: The idea of a group as applied to permutations or substitutions is due to Galois, and the introduction of it may be considered as marking an epoch in the progress of the theory of algebraic equations.

If we know that the set G contains also the neutral element e such that $\forall a \in G$, $a \cdot e = e \cdot a = a$, then the above three statements imply also the existence of the inverse a^{-1} for each element a: indeed it suffices to choose $b = e$ in the property (2) or (3). Yet if we do not know that such element exists the properties (2) and (3) are not sufficient to imply its existence. They imply the existence of some element that mulitplied by a reproduces a but we are not guaranteed that it is the same for all the elements $a \in G$.

Cayley did not explicitly define the neutral element but he implicitly assumed its existence since his list of symbols begins with the number 1. The full fledged axiomatic definition of a group was given by Camille Jordan in 1870 in [118], yet it is fair to say that Cayley had essentially put it forward in his 1854 paper. Notwithstanding this, Cayley's 1858 memoir does not make a clear connection between matrices and the abstract notion of a group. The main reason is probably that the notion of a vector space was absent. The x, y, z-operand was seen as a set of quantities and not yet as the description of another abstract object like a vector $v \in V$.

3.2.5 Grassmann, Peano and the Birth of Vector Spaces

The life and the achievements in mathematics of Hermann Günther Grassmann (see Fig. 3.9) present similarities with those of another great German scientist, Wilhelm Killing about whom we will say a lot later on. Both Grassmann and Killing did not make an academic career and became teachers in secondary schools of Eastern Prussia; both, notwithstanding their isolation from the main currents of academic mathematical research, contributed extremely original breakthroughs. Furthermore for both of them the recognition of their achievements came only later. In the case of Killing, his work was resumed and brought to perfection by Cartan during Killing's life-time. The case of Grassmann was worse. No one considered seriously his masterpiece, namely the book *Die Lineale Ausdehnungslehre, ein neuer Zweig der Mathematik* (The Theory of Linear Extension, a New Branch of Mathematics) that he published in 1844 (see Fig. 3.10). His abstract approach which was much ahead of his time, was not appreciated even by such great mathematicians like Kummer or Möbius, who dismissed it as obscure, criticizing the lack in his work of intuitive examples. The second revised edition of 1862 had no better fortune. Yet a contemporary mathematician, Fearnley Sander, in 1979 described Grassmann's foundation of linear algebra as follows:

The definition of a linear space (vector space)... became widely known around 1920, when Hermann Weyl and others published formal definitions. In fact, such a definition had been given thirty years previously by Peano, who was thoroughly acquainted with Grassmann's mathematical work. Grassmann did not put down a formal definition — the language was not available — but there is no doubt that he had the concept. Beginning with a collection of ***units*** $\mathbf{e}_1, \mathbf{e}_2, \mathbf{e}_3, \ldots$, *he effectively defines the free linear space which they generate; that is to say, he considers formal linear combinations* $a_1\,\mathbf{e}_1 + a_2\,\mathbf{e}_2 + a_3\,\mathbf{e}_3 + \ldots$ *where the* a_j *are real numbers, defines*

Fig. 3.9 On the left Hermann Günther Grassman (Stettin 1809–Stettin 1877) - On the right Giuseppe Peano (Cuneo 1858–Torino 1932). Grassman has been both a mathematician and a linguist. His aspirations to become an academic researcher were frustrated and he reached only the level of teacher first in lower secondary school and later of higher secondary school. During Grassmann life-time, his book which essentially founded modern linear algebra was completely ignored almost by everyone both in its first and its second edition. It was only after Grassmann's death that the great value of its conception started being appreciated. Born in a village near Cuneo, Peano studied at the Liceo Classico Cavour of Torino, and then at the University of Torino. He was full Professor of *Infinitesimal Calculus* at Torino University since 1888 till 1930 when his chair was renamed *Complementary mathematics*. In 1890 he produced his famous curve, the first example of a fractal. Strongly admired by Bertrand Russel, Peano has been the founder of Modern Mathematical Logic after Boole. He has contributed also in the fields of differential equations and vector calculus. He invented a new language, *Latino sine flexione*, a drastic grammatical simplification of Latin that he wanted to propose as the international language for mathematics and science. He taught until the last day of his life in 1932

addition and multiplication by real numbers and formally proves the linear space properties for these operations. ... He then develops the theory of linear independence in a way which is astonishingly similar to the presentation one finds in modern linear algebra texts. He defines the notions of subspace, linear independence, span, dimension, join and meet of subspaces, and projections of elements onto subspaces. ...few have come closer than Hermann Grassmann to creating, single-handedly, a new subject.

Indeed it was only in 1888 that Giuseppe Peano (see Figs. 3.9, 3.11 and 3.12) finally provided the axiomatic definition of a vector space which, from our contemporary viewpoint, is so simple and fundamental. Quite significant is the subtitle of Peano's book: *Preceded by the Operations of Deductive Logic*. Apparently it was necessary, in order to arrive at this fundamental concept that opened up the world of

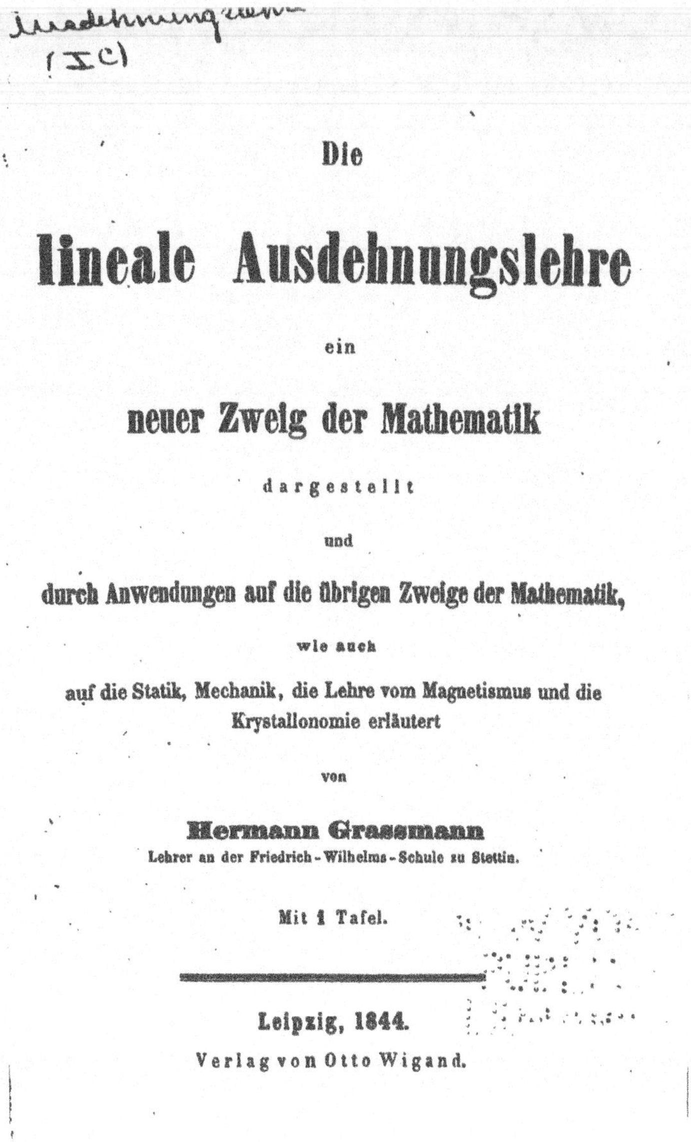

Die

lineale Ausdehnungslehre

ein

neuer Zweig der Mathematik

dargestellt

und

durch Anwendungen auf die übrigen Zweige der Mathematik,

wie auch

auf die Statik, Mechanik, die Lehre vom Magnetismus und die Krystallonomie erläutert

von

Hermann Grassmann

Lehrer an der Friedrich-Wilhelms-Schule zu Stettin.

Mit 1 Tafel.

Leipzig, 1844.

Verlag von Otto Wigand.

Fig. 3.10 On the left the frontispiece of the original book *Die Lineale Ausdehnungslehre, ein neuer Zweig der Mathematik* of 1844, authored by Grassmann

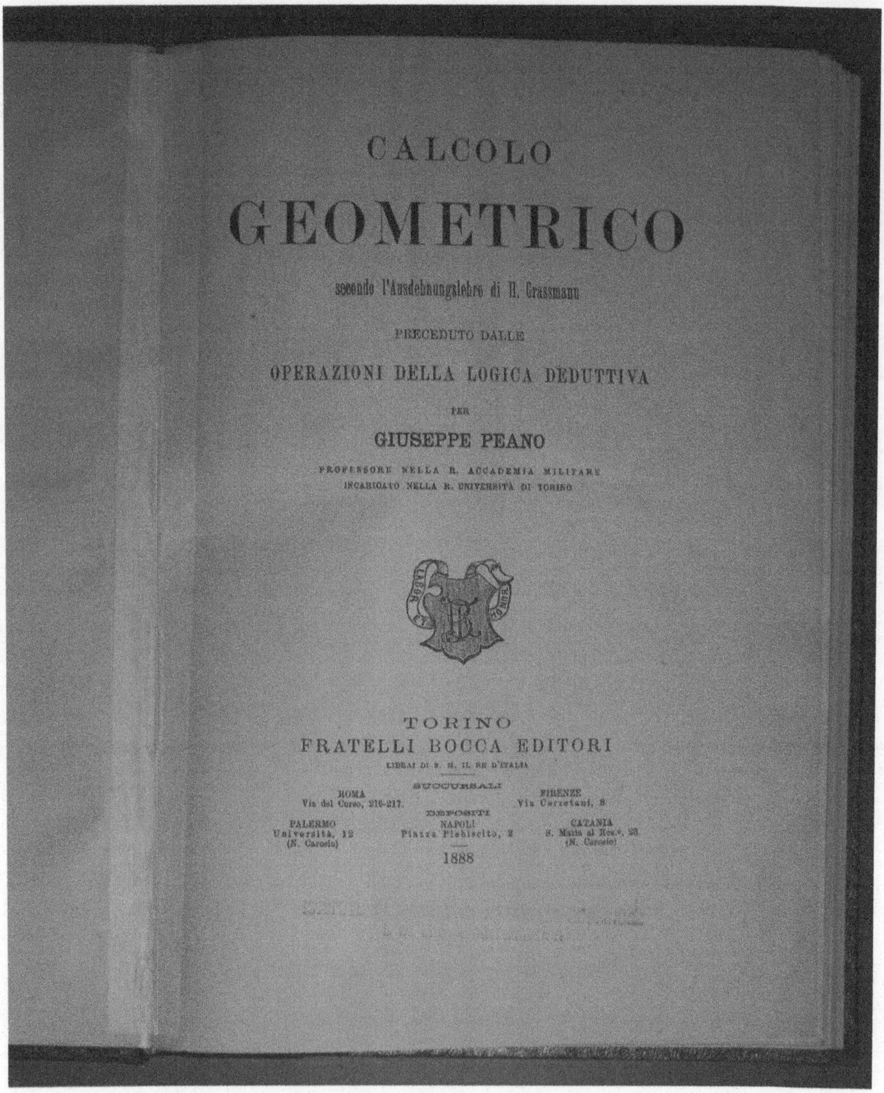
CALCOLO
GEOMETRICO
secondo l'Ausdehnungslehre di H. Grassmann
PRECEDUTO DALLE
OPERAZIONI DELLA LOGICA DEDUTTIVA
PER
GIUSEPPE PEANO
PROFESSORE NELLA R. ACCADEMIA MILITARE
INCARICATO NELLA R. UNIVERSITÀ DI TORINO
TORINO
FRATELLI BOCCA EDITORI
1888

Fig. 3.11 The 1888 book by Peano *Calcolo Geometrico secondo l'Ausdehnungslehre di Grasmann* that contains the formal definition of a vector space. Frontespice. Courtesy of the *Bilioteca-Peano - Dipartimento di Matematica - Universitá di Torino*

representation theory, to make few step further on the path of abstraction. The concept of space, even of the familiar flat one, had to be *dematerialized* and *axiomatized*.

As later Hermann Weyl would say in his *the mathematical way of thinking* (see Sect. 6.1.2), also coordinates have to become mere symbols and the mathematician should forget what the symbols stand for and concentrate only on the operations one can make on them.

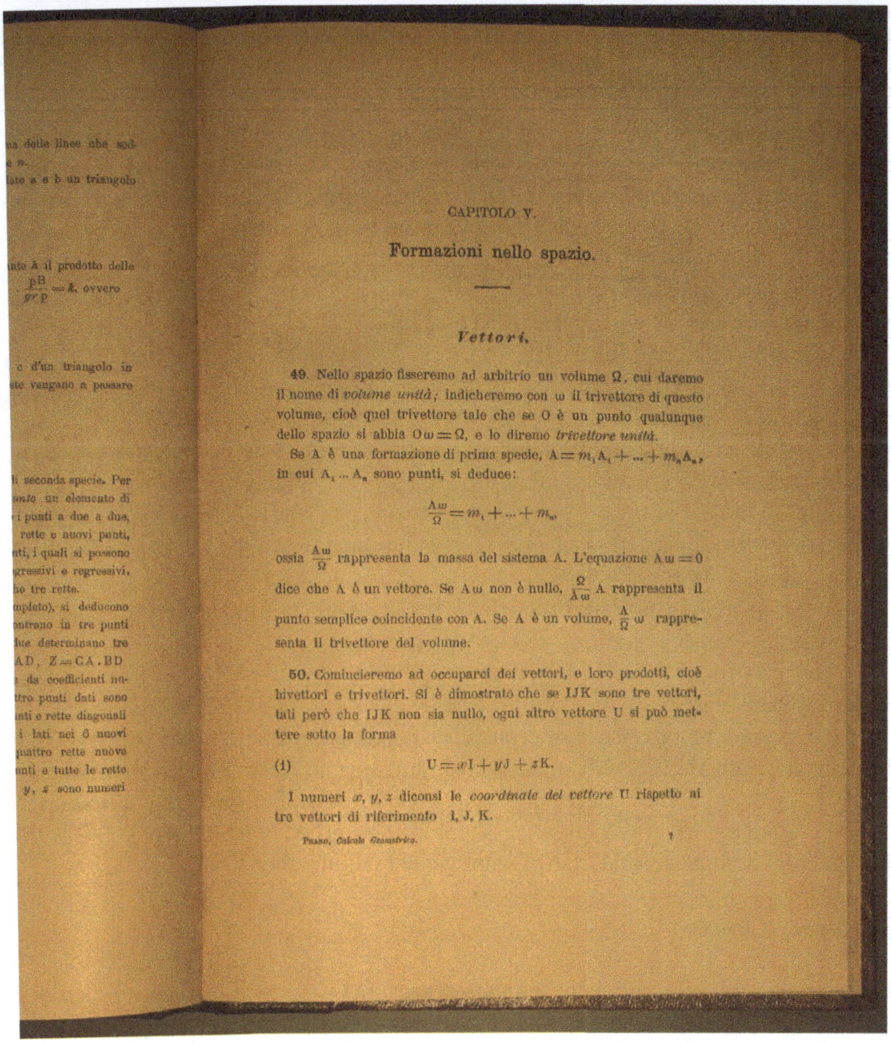

CAPITOLO V.

Formazioni nello spazio.

Vettori.

49. Nello spazio fisseremo ad arbitrio un volume Ω, cui daremo il nome di *volume unità*; indicheremo con ω il trivettore di questo volume, cioè quel trivettore tale che se O è un punto qualunque dello spazio si abbia $O\omega = \Omega$, e lo diremo *trivettore unità*.

Se A è una formazione di prima specie, $A = m_1 A_1 + \ldots + m_n A_n$, in cui $A_1 \ldots A_n$ sono punti, si deduce:

$$\frac{A\omega}{\Omega} = m_1 + \ldots + m_n,$$

ossia $\frac{A\omega}{\Omega}$ rappresenta la massa del sistema A. L'equazione $A\omega = 0$ dice che A è un vettore. Se $A\omega$ non è nullo, $\frac{\Omega}{A\omega} A$ rappresenta il punto semplice coincidente con A. Se A è un volume, $\frac{A}{\Omega}\omega$ rappresenta il trivettore del volume.

50. Comincieremo ad occuparci dei vettori, e loro prodotti, cioè bivettori e trivettori. Si è dimostrato che se IJK sono tre vettori, tali però che IJK non sia nullo, ogni altro vettore U si può mettere sotto la forma

$$U = xI + yJ + zK. \qquad (1)$$

I numeri x, y, z diconsi le *coordinate del vettore* U rispetto ai tre vettori di riferimento I, J, K.

PEANO, *Calcolo Geometrico.* 7

Fig. 3.12 The 1888 book by Peano *Calcolo Geometrico secondo l'Ausdehnungslehre di Grasmann* that contains the formal definition of a vector space. Some interior pages. Courtesy of the *Bilioteca-Peano - Dipartimento di Matematica - Universitá di Torino*

3.3 Classification of Finite Groups

In the previous sections we traced the slow development through the XIXth century of the abstract definition of a group as an algebraic structure.

Once the conception of groups as abstract mathematical objects is completely integrated into the fabrics of Mathematics and Physics, the next two natural questions are:

(a) The classification issue. Which are the possible groups G?
(b) The representation issue. Which are the possible explicit realizations of each given abstract group G.

Let us comment on the first issue. If the *classification* of groups were accomplished, given a specific realization of a group arising e.g. in some physical system, one might just identify its isomorphic class in the general classification and know a priori, via the isomorphism, all the relevant symmetry properties of the considered physical systems. For finite groups, this means the classification of all possible distinct (non-isomorphic) groups of a given finite order n. Such a goal is too ambitious and cannot be realized, yet there is a logical way to proceed.

One is able to single out certain types of groups (the so-called *simple groups*) which are the *hard core* of the possible different group structures. The complete classification of simple finite groups is one notable achievement of modern mathematics in the seventies and eighties of the XXth century which became possible thanks to massive computer calculations.

Assuming the list of simple groups as given one can study the possible *extensions* which allow the construction of new groups having the simple groups as building blocks.

Also for groups of infinite order there are some general results in the line of a classification, mainly regarding abelian groups.

For Lie groups the quest of classification follows a pattern very similar to the case of finite groups, involving the definition of *simple Lie groups* to be classified first.

In order to address the issue of group classification one ought to introduce several concepts and general theorems related with the inner structure of a given group; for instance, essential is the concept of *conjugacy classes* and of *invariant* subgroups.

Although the present is a history essay it is appropriate to recall here some of the fundamental definitions that clarify what we are talking about.

Conjugacy Classes

The conjugacy relation between elements of a group G, ($g' \sim g \Leftrightarrow \exists h \in G$ such that $g' = h^{-1}gh$) is an equivalence relation. Therefore we can consider the quotient of G (as a set) by means of this equivalence relation. The elements of the quotient set are named the *conjugacy classes*. Any group element g defines a conjugacy class $[g]$:

$$[g] \equiv \{g' \in G \text{ such that } g' \sim g\} = \{h^{-1}gh\,,\ \text{for } h \in G\}\,. \tag{3.3.1}$$

Basically, conjugation is the implementation of an inner automorphism of the group; we may think of it as a *change of basis* in the group (it is indeed so for matrix groups).

Conjugate Subgroups

Let H be a subgroup of a group G. Let us consider

$$H_g \equiv \left\{h_g \in G \; : \; h_g = g^{-1}hg \, , \; \text{for } h \in H\right\} \tag{3.3.2}$$

which we simply write as $H_g = g^{-1}Hg$. It is esay to see that H_g is a subgroup. The subgroups H_g are called *conjugate subgroups* to H.

Invariant Subgroups

A subgroup N of a group G is called an *invariant* (or *normal*) subgroup if it coincides with all its conjugate subgroups: $\forall g \in G, \; N_g = N$. When a subgroup is normal the coset G/N, namely the set of equivalence classes with respect to the relation that identifies two G elements if they differ by multiplication on the right by an H-element is itself a group, namely the *factor group*.

Simple, Semi-simple, Solvable Groups

In general a group G admits a chain of invariant subgroups, called its *subnormal series*[2]:

$$G = G_r \rhd G_{r-1} \rhd G_{r-2} \rhd \ldots \rhd G_1 \rhd \{e\} \, , \tag{3.3.3}$$

where every G_i is a normal subgroup.

G is a *simple* group if it has *no proper invariant subgroup*. For simple groups, the subnormal series is minimal:

$$G \rhd \{e\} \, . \tag{3.3.4}$$

Simple groups are the *hard core* of possible group structures. There is no factor group G/H smaller than G out of which the group G could be obtained by some *extension*, because there is no normal subgroup H other than the trivial one $\{e\}$ or G itself.

G is a *semi-simple* group if it has *no proper invariant subgroup* which is *abelian*.

A group G is *solvable* if it admits a subnormal series Eq. (3.3.3) such that all the factor groups $G/G_1, G_1/G_2, \ldots, G_{k-1}/G_k, \ldots$ are *abelian*.

A fundamental property of groups which has a profound influence on the contemporary *episteme* is that their intrinsic structure determines their possible linear representations.

---

[2]Following a convention widely utilized in finite group theory we make a distinction between subgroups and normal subgroups. The notation $G \supset H$ simply means that H is a subgroup of G, not necessarily an invariant one. On the other hand $G \rhd N$ means that N is a normal (invariant) subgroup of G.

Fig. 3.13 Issai Schur 1875 Mogilev, (Russian Empire) - 1941 in Tel Aviv (Palestine)

3.4 Group Representation and the Unhappy Life of the Man of Two Lemmas

Coming to the second issue, a fundamental property of groups which has a profound influence on the contemporary *episteme* is that their intrinsic structure determines their possible linear representations. This is true both for finite and for continuous groups. Any representation can be split into a finite number of so named *irreducible representations* which constitute the building blocks, usually nicknamed *irreps*. It is the set of these blocks that the intrinsic group structure uniquely determines. The only difference between finite groups and continuous groups resides in that for finite groups the set of irreps is also finite, while for continuous groups it is infinite, but denumerable and classifiable. In the case of finite groups, irreps are in one-to-one correspondence with conjugacy classes of G and they are as many as these latter are.

The profound implications of such basic mathematical properties of groups for the *episteme* is easily explained. In the thirties of the XXth century, after the advent of Quantum Mechanics, particularly under the influence of the fundamental book *Gruppentheorie und Quantenmechanik* by Hermann Weyl, the labels identifying an irreducible representation started being renamed *quantum numbers* and were associated with the possible eigenvalues of suitable observable operators. In this way the spectrum of available physical states appeared to be partially or fully determined in terms of group theory. Objects of reality are just representations of a symmetry and the primary entity is, in a fully neo-platonic mode, the abstract algebraic structure, namely the group. This viewpoint underpins the five points (A)–(E) in which the contemporary episteme was articulated in Chap. 1. Indeed the possible matter fields are just sections of vector bundles corresponding to specific representations of the *gauge group*, whose intrinsic structure determines the possible types of matter.

These considerations demonstrate the fundamental relevance of group representation theory.

In the last part of the XIXth century representation theory for finite groups was developed independently by William Burnside in Britain and by Georg Frobenius in Germany. About Frobenius we talk more extensively in Chap. 8 in connection with his fundamental result on division algebras (see Fig. 8.7).

In its present form the theory of finite group representations owes a lot to Issai Schur who was Frobenius' student in Berlin. In particular Schur's lemmas, the first and the second, are the main instrument to work out the conclusion that irreducible representations are just in equal number to the conjugacy classes.

Schur was born in a Jewish family in Mogilev, at the time part of the Russian Empire, today a city of Belaruș. In his late childhood he went to Latvia and since then he always attended German schools and all of his life he considered himself a German. Indeed he spoke the German language without any accent. He entered Berlin University in 1894 and there he became Frobenius student obtaining his doctorate in 1901. From Frobenius, with whom he collaborated, he took up the field of group representations that he brought to perfection. He made advances also in other fields of mathematics, in particular in number theory. He invented *ante litteram* the second cohomology group of a manifold, utilizing for it a different name. For a short time he was professor in Bonn. In 1919 he was appointed full professor in Berlin where he built a famous school. In 1922 Schur was elected to the Prussian Academy, on Planck's proposal.

When the Nazis came to power, Schur, as a Jew, was first dismissed from his University chair, then forced to resign from the Academy and finally in 1939 to emigrate to Palestine where two years later he died in Tel Aviv. He stubbornly refused appointments in Britain and the USA, unable to understand why a German could not be Professor in Germany.

Chapter 4
From Crystals to Plato

χαλεπὰ, τὰ καλά
Nothing beautiful without struggle.
Plato

4.1 Mathematics and Crystallography

On the basis of finite group theory, that by the end of the XIXth century was reaching a firm state of ripeness, the question raised in Sect. 2.2, *how many of the Alhambra patterns are possible?* could be answered. The man who found the answer, establishing that they are exactly 17, as many as those realized in the decorations of the XIIIth century arabic palace, was the Russian geologist, crystallographer and mathematician Evgar Stepanovich Fyodorov (see Fig. 4.1).

He published his result about the classification of the symmetry groups of two-dimensional regular lattices (*wallpaper groups*) in a paper of 1891 [94] that was just a warming up exercise for the more ambitious task accomplished by the same author in the same year, namely the classification of the symmetry groups of three dimensional lattices and figures, presently dubbed space-groups. Fyodorov found the list of 230 space groups [93] which constitutes to the present day the back-bone of crystallography and plays a fundamental rule in all aspects of chemical-physics.

We shortly dwell on the biography of this extraordinary scientist before analyzing from a mathematical point of view what was his problem, that encodes some of the most fundamental questions about symmetry and group-theory.

FYODOROV

Fyodorov was born in Orenburg in the Southern Ural region of the Russian Empire in 1853. Son of a military engineer who left the Urals for the northern capital, Evgraf Stepanovich graduated from the Sankt Peterburg Military Engineering School in 1872. In 1874, after a brief service in a military engineering unit at Belaya Tzerkov, he followed courses at the Military Medical Academy of the imperial capital. While being a student he entered the secret organization *Zemlya i Volya* (Land and Will), developing tight contacts with the German Worker Movement. In 1877 Evgraf mar-

P. G. Fré, *A Conceptual History of Space and Symmetry*,
https://doi.org/10.1007/978-3-319-98023-2_4

Fig. 4.1 Evgraf Stepanovich Fyodorov (Orenburg 1853–Sankt Peterburg 1919). The man who classified the 17 wallpaper groups in $d = 2$ and the 230 space-groups in $d = 3$

ried Lyudmila Vasilievna Panyutina (1851–1936). In their flat, the new married couple organized the publication of the two revolutionary journals, *Nachalo* (the Beginning) and *Zemlya i Volya*.

In 1881, at the eve of the *Narodnaya Volya* (People's Will)[1] revolutionary movement's defeat after the assassination of Alexander II, the 26 year old Fyodorov developed a strong interest in crystallography and enrolled at the *Gorny Institut* (Mining Institut) of Sankt Peterburg graduating in 1883.

Many years later, at the time of the first Russian Revolution of 1905, elections were conducted to appoint the Director of that Institute and the chosen scientist was just Evgraf Stepanopvich.

In 1885 Fyodorov joined the staff of the Geological Committee and carried out geological research in the Northern Urals from 1885 to 1890. In 1894 he was a mining engineer at Turinskie Rudniki in the Urals. In 1895 he was appointed professor at the Moscow Agricultural Institute. After the revolutionary events of 1905, Fyodorov became, as we already said, the first elected director of the Mining Institute in St. Petersburg. His reelection in 1910 was nullified by the government, which feared the development of revolutionary sentiments among the students and believed that Fyodorov promoted such development. He was elected a member of the Bavarian Academy of Sciences in 1896 and an adjunct of the Imperial St. Petersburg Academy of Sciences in 1901. He resigned from the Imperial St. Petersburg Academy of Sciences in 1905 after failing to obtain support for the establishment of a mineralogical institute.

Fyodorov began writing his first major work, *Principles of the Theory of Figures*, when he was quite young. This fundamental paper published in 1885 contained the ideas of most of Fyodorov's subsequent discoveries in geometry and crystallography. In particular, this work introduced parallelohedrons, that is, the convex polyhedrons

[1]This movement was an offspring of *Zemlya i Volya*.

upon which Fyodorov based his theory of crystal structure. From 1885 to 1890 he wrote a series of papers on the structure and symmetry of crystals, culminating in the classic work *The Symmetry of Regular Systems of Figures* [93]. This work presented the first derivation of the 230 space groups known as Fyodorov groups. The groups were derived at almost the same time by the German mathematician A. Schoenflies. An epistolar correspondence between Fyodorov and Schoenflies provided mutual consultations on the derivation of the space groups, and Schoenflies later published a letter in which he confirmed that Fyodorov's derivation was the first.

The 1917 Bolshevik Revolution found in Fyodorov an enthusiastic supporter who believed that it would bring a brilliant future to Russia and to Russian Science. Unfortunately the material conditions of living in Sankt Peterburg were not so brilliant, in the early Bolshevik years, as Fyodorov hoped for: nutrition and heating were quite scan. Evgraf Stepanovich fell ill of pneumonia in the spring of 1919 and died on May 21st of that year.

4.1.1 Crystallographic Groups

Most chemical compounds that exist at room temperature in a solid state have a crystalline structure and crystals are known to mankind since remote ages. The macroscopically observable structure of crystals corresponds, at the atomic level, to a periodic disposition of the atoms, ions or molecules that form the material. This results in symmetric shapes of macroscopic samples of the compound that are invariant under rotations and translations forming a group (see for instance Fig. 4.2).

Fig. 4.2 Apatite crystal from Cerro de Mercado Mine, Victoria de Durango, Cerro de los Remedios, Durango, Mexico. Matteo Chinellato - ChinellatoPhoto/Getty Images

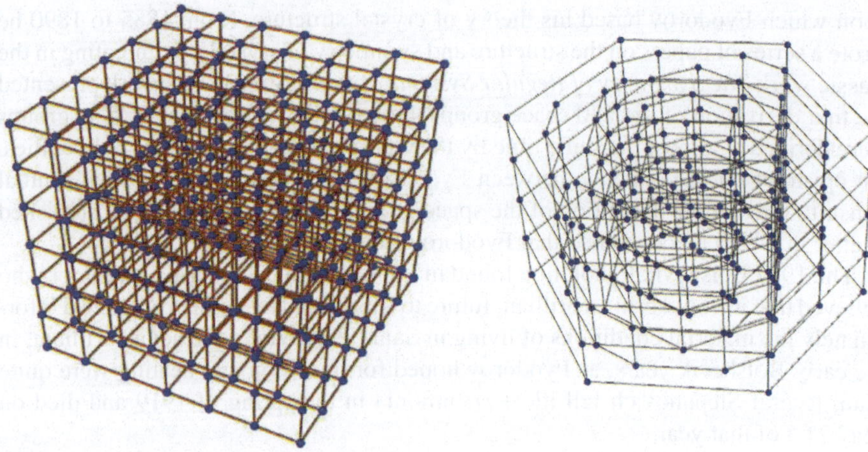

Fig. 4.3 A view of the self-dual cubic lattice on the left and of the hexagonal lattice on the right

Mathematically we describe the structure of a crystal saying that the elementary constituents (ions, atoms or molecules) are disposed on the vertices of a lattice such as those depicted in Fig. 4.3, which are images of the cubic and of the hexagonal lattice, respectively.

4.1.1.1 Lattices

The notion of lattice is relatively simple, once the notion of vector space is granted. Let us begin with the planar case which allows for simple visualizations. According to what we explained in Sect. 3.1.3 the plane $\mathbb{R}^2$ can be viewed as a two-dimensional vector space and each of its points $\{x, y\}$ can be identified with a vector $\mathbf{v}$, namely the oriented segment which reaches it from the origin $\{0, 0\}$.

As we emphasized in the quoted section, there are infinitely many possible bases of the same n-dimensional vector space V. It suffices to choose an n-tuple of linearly independent vectors $\mathbf{w}_1, \ldots, \mathbf{w}_n$. In the planar case it suffices to choose a pair $\mathbf{w}_1, \mathbf{w}_2$. Hence let us choose a specific basis, for instance the orthonormal basis $\mathbf{w}_1 = \mathbf{e}_1 = \{1, 0\}, \mathbf{w}_2 = \mathbf{e}_2 = \{0, 1\}$. Once the basis is chosen we can do something weird. Instead of considering all possible vectors, namely the entire vector space V, we can consider the infinite subset $\Lambda_{\mathbf{w}} \subset V$ formed by all those vectors that in the given basis have integer valued components:

$$\Lambda_{\mathbf{w}} = \left\{ \bigoplus_{i=1}^{n} m^i \, \mathbf{w}_i \;\mid\; m^i \in \mathbb{Z} \right\} \tag{4.1.1}$$

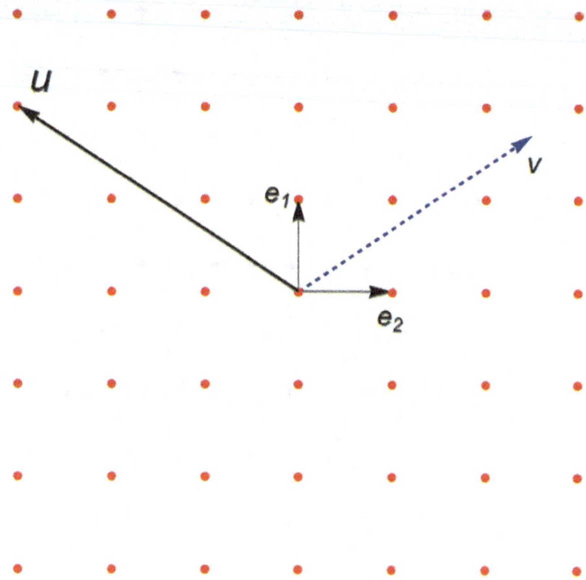

Fig. 4.4 In this picture we display an image of the square lattice generated by the orthonormal basis $\mathbf{e}_1, \mathbf{e}_2$. The points of the lattice are marked as small circles. The vector $\mathbf{v}$ shown in the figure belongs to the vector space $V = \mathbb{R}^2$ generated by $\mathbf{e}_1, \mathbf{e}_2$, but not to the corresponding lattice, since its components with respect to the chosen basis are not integer numbers. On the other hand the vector $\mathbf{u}$, having integer components in this basis, belongs to the square lattice

The set $\Lambda_{\mathbf{w}}$ defined in Eq. (4.1.1) is what we name a *lattice* and its elements form an abelian group with respect to the addition (the sum of two vectors in the lattice belong to the lattice), the identity element being the 0-vector. As an abstract group, the lattice is just the tensor product of n copies of the group $\mathbb{Z}$.

If we apply this recipe to the planar case with the above chosen orthonormal basis we obtain the square lattice shown in Fig. 4.4.

One can choose another basis, for instance:

$$\mathbf{w}_1 = \{1, 0\} \quad ; \quad \mathbf{w}_2 = \left\{-\frac{1}{2}, \frac{\sqrt{3}}{2}\right\} \tag{4.1.2}$$

and applying the definition (4.1.1) one obtains another lattice, in this case the hexagonal lattice which is shown in Fig. 4.5.

In higher dimensional vector spaces everything is analogous. Once a basis is chosen, the corresponding lattice is given. In three dimensions, for instance, the cubic lattice displayed in Fig. 4.3 is generated by the orthonormal basis $\mathbf{e}_1 = \{1, 0, 0\}$, $\mathbf{e}_2 = \{0, 1, 0\}$, $\mathbf{e}_3 = \{0, 0, 1\}$. The hexagonal lattice shown in the same figure is instead generated by the basis vectors $\mathbf{w}_1 = \{1, 0, 0\}$, $\mathbf{w}_2 = \left\{-\frac{1}{2}, \frac{\sqrt{3}}{2}, 0\right\}$, $\mathbf{w}_3 = \{0, 0, 1\}$.

4.1.1.2 Crystallographic Groups and the Bravais Lattices for $n = 3$

The continuous rotation group O(n), transforms the vector space $V \simeq \mathbb{R}^n$ into itself. Given a lattice Λ, the question is whether there is a non trivial subgroup $G_\Lambda \subset \mathrm{O(n)}$

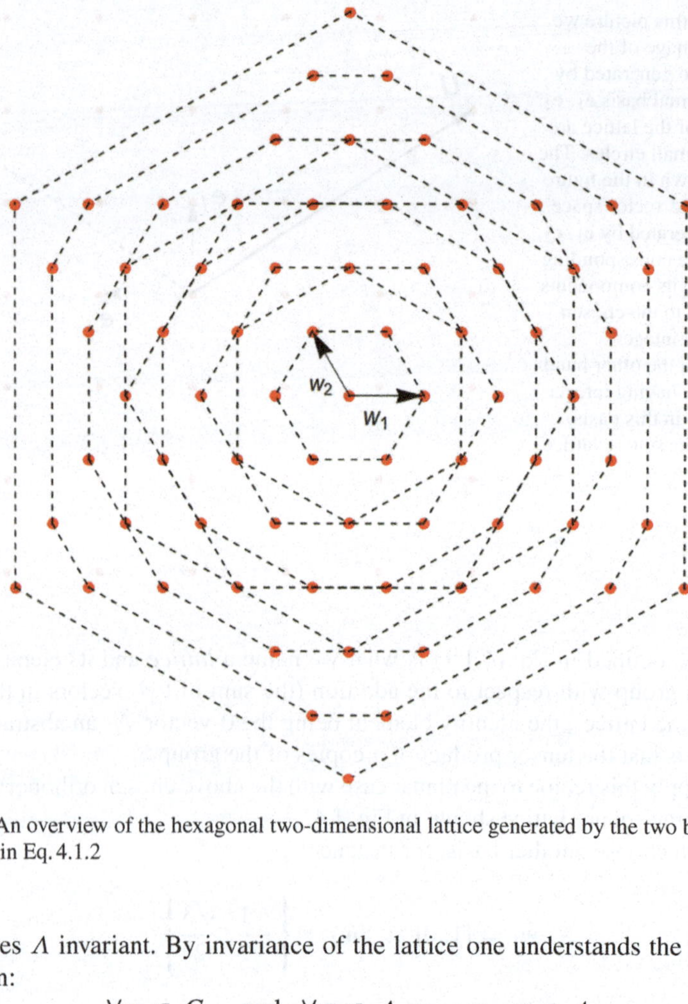

Fig. 4.5 An overview of the hexagonal two-dimensional lattice generated by the two base vectors displayed in Eq. 4.1.2

that leaves Λ invariant. By invariance of the lattice one understands the following condition:

$$\forall \gamma \in G_\Lambda \quad \text{and} \quad \forall \mathbf{q} \in \Lambda \; : \qquad \gamma \cdot \mathbf{q} \in \Lambda \tag{4.1.3}$$

namely the image of any lattice point created by any element of the group is another lattice point. A generic lattice Λ is not invariant with respect to any proper subgroup of the rotation group $\mathrm{G} \subset \mathrm{SO(n)}$, namely the corresponding symmetry group is just the identity element.

For $n = 3$ lattices that have a non trivial symmetry group $\mathrm{G} \subset \mathrm{O(3)}$ are those relevant to Solid State Physics and Crystallography. There are 14 of them grouped in 7 classes that were already classified in the XIXth century by Bravais. The symmetry group G of each of these Bravais lattices Λ_B is necessarily one of the well known finite subgroups of the three-dimensional rotation group O(3). In the language universally adopted by Chemistry and Crystallography for each Bravais lattice Λ_B the corresponding invariance group $\mathrm{G_B}$ is named the *Point Group*.

According to a standard nomenclature the 7 classes of Bravais lattices are respectively named *Triclinic, Monoclinic, Orthorombic, Tetragonal, Rhombohedral, Hexagonal and Cubic*. Such classes are specified by giving the lengths of the basis vectors $\mathbf{w}_\mu$ and the three angles between them.

An abstract group Γ is named crystallographic in n-dimensions if there exists an n-dimensional lattice Λ_n with basis vectors $\mathbf{w}_\mu$ such that there is a isomorphism:

$$\omega : \Gamma \to \mathrm{H} \subset \mathrm{O}(n) \tag{4.1.4}$$

where $\mathrm{O}(n)$ is the n-dimensional rotation group and the subgroup H leaves the lattice Λ_n invariant. Obviously in the basis $\mathbf{w}_\mu$ all group elements of $\mathrm{H} \sim \Gamma$ are integer valued matrices.

When a group Γ is crystallographic with respect to a given n-dimensional lattice Λ_n we say that it is the *Point Group* of Λ_n.

4.1.1.3 The Proper Point Groups

Restricting one's attention to $n = 3$, it was shown in the classical crystallographic literature that the proper point groups that appear in the 7 lattice classes are either the cyclic groups $\mathbb{Z}_h$ with $h = 2, 3, 4, 6$ or the dihedral groups Dih_k with $k = 3, 4, 6$ or the tetrahedral group T_{12} or the octahedral group O_{24}. Indeed the $n = 3$ crystallographic point groups are, by definition, finite subgroups of the rotation group, hence they must fall in the classification of these latter. Yet not every finite rotation group is crystallographic. For instance there is no lattice that is invariant under the icosahedral group and in general in a $n = 3$ point group there are no elements with orders different from $1, 2, 3, 4, 6$.

4.1.2 Platonic Groups

We arrive in this way at an important conclusion. The physical structure of reality, for what attains such a relevant subclass of materials as those that have a crystalline form, is a priori mathematically determined by available and classifiable symmetries, namely by available groups of a certain class i.e. **crystallographic groups** Γ_{crys} in n-dimensions which are a subclass of the **finite subgroups** of the rotation group in the same dimension:

$$\Gamma_{finite} \subset \mathrm{O(n)} \quad ; \quad |\Gamma_{finite}| < \infty \tag{4.1.5}$$

We have used the generic notation n instead of the value $n = 3$ in order to stress that the problem can be formulated in any dimension and consists of two steps, first the derivation of all the finite subgroups $\Gamma_{finite} \subset \mathrm{O(n)}$, next the determination among them of the crystallographic ones, which entrains the construction of the corresponding invariant lattices.

For $n = 3$ the finite subgroups of the rotation group are the *Platonic Groups* since they correspond to the symmetry groups of the regular solids or of the regular polygons.

Hence the seeds of crystallography are contained in the Platonic conceptions of 2500 years ago! Even more fascinating is the isomorphism of the Platonic Group classification with the classification of simple Lie algebras, to be discussed in Sect. 5.6.

Let us analyze Plato's philosophy and mathematics in relation with the regular solids.

4.2 Plato and the Regular Solids

The first mention in history of the five regular solids of Euclidian Geometry in three-dimensions, namely the Tetrahedron, the Cube, the Octahedron, the Icosahedron and the Dodecahedron, occurs in Plato's dialogue Timaeus [44], which was probably composed about the year 360 B.C. (see Fig. 4.6).

In our contemporary understanding of mathematics we have a powerful technique to prove that the possible regular solids are just five, starting from the notion of symmetry, namely from finite group theory. Each regular solid singles out a finite

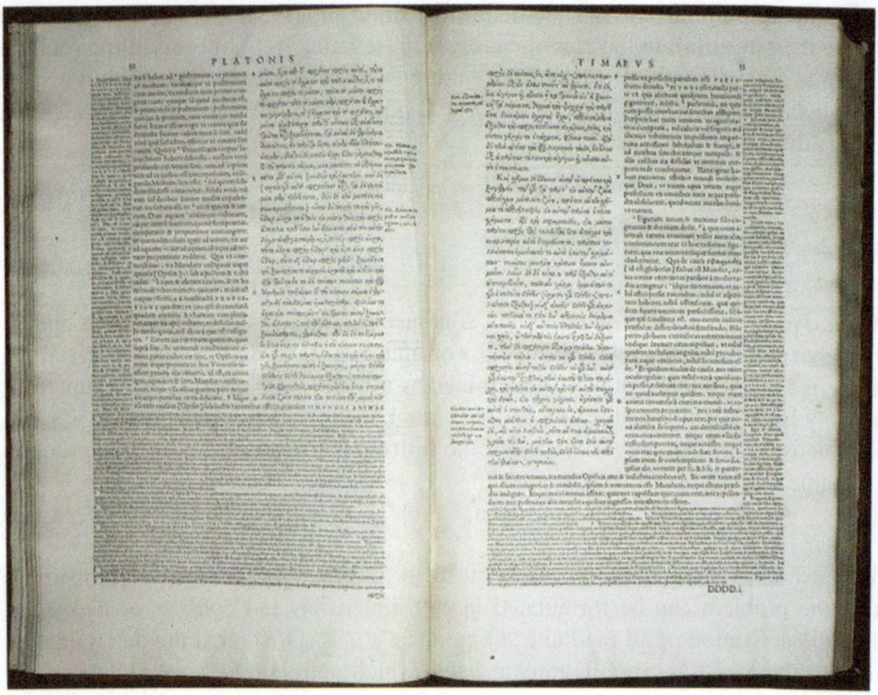

Fig. 4.6 The 1578 Stephanus edition of Plato's works

Fig. 4.7 The first of the five platonic regular solids is the tetrahedron that has 4 faces that are equilateral triangles, 4 vertices and 6 edges. The discrete subgroup $T_{12} \subset SO(3)$ of the rotation group each element of which maps the regular tetrahedron into itself is a finite group or order 12

subgroup of the rotation group containing those transformations that map the solid into itself. If we are able to classify the possible symmetry groups, a classification of the solids follows from that.

There is just one loophole that we have to take into account in this reasoning. There might be two solids that have the same symmetry group. Under a rotation, vertices are mapped into vertices, faces into faces and edges into edges. Hence if two solids are dual to each other in the sense that one obtains one from the other by interchanging vertices with faces (this is done by taking as vertices of the new solid the central points in the faces of the old one), then it is clear that those rotations that map the first solid into itself do the same with the second one.

It is a very remarkable thing that by a simple argument that we sketch in the next section, the classification of finite discrete subgroups of the rotation group is reduced to the enumeration of solutions of a certain Diophantine equation, namely an algebraic equation with integer coefficients for integer valued unknowns. It is even more remarkable that the very same Diophantine equation emerges in the classification of simple Lie algebras, as we are going to see in Sect. 5.6, and even in the classification of singularity types as we briefly mention further on.

Such Diophantine equation admits two infinite families of solutions and three exceptional sporadic solutions. In the case of finite subgroups of SO(3) the two infinite families of solutions of the Diophantine equation correspond to the cyclic groups $\mathbb{Z}_n$ that we have already introduced and to the dihedral groups D_n that we are going to introduce in next section. In the case of simple Lie algebras the two infinite families of diophantine solutions correspond, in the same order, to the Lie algebras $\mathfrak{a}_n$ associated with the matrix groups $SL(n+1, \mathbb{C})$ and to the Lie algebras $\mathfrak{d}_n$ associated with the matrix groups $SO(2n, \mathbb{C})$. The three sporadic solutions, correspond instead to the three symmetry groups T_{12}, O_{24} and I_{60} of the Tetrahedron (see Fig. 4.7), of the Octahedron (see Fig. 4.8) and of the Icosahedron (see Fig. 4.9).

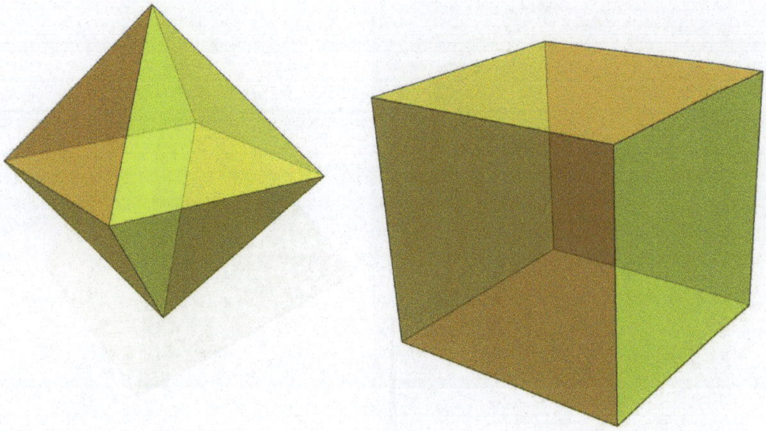

Fig. 4.8 The next two of the five platonic regular solids are the regular octahedron and the cube. The octahedron has 8 faces that are equilateral triangles, 6 vertices and 12 edges. The cube instead has 6 faces that are squares, 8 vertices and 12 edges. These two solids are dual to each other since one obtains one from the other by exchanging faces with vertices. If we take as vertices of a new solid the central points of the six faces of a cube and we join them with edges we obtain an octahedron. This means the discrete subgroup $O_{24} \subset SO(3)$ made by those rotations that map a cube into itself has the same property with respect to the octahedron. This unique discrete subgroup has order 24 and it is named the octahedral group

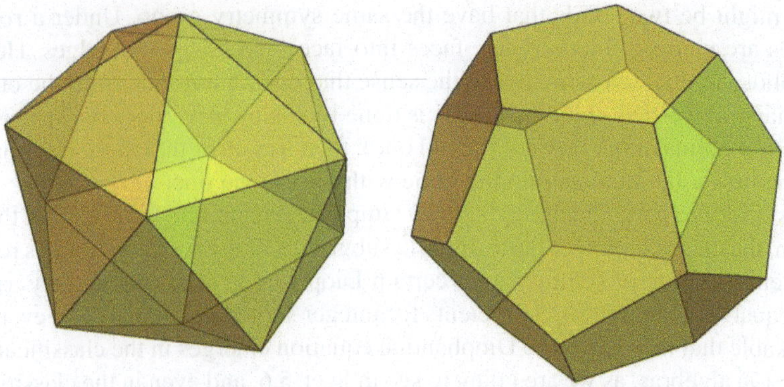

Fig. 4.9 The last two of the five platonic regular solids are the regular icosahedron and the cube. The icosahedron has 20 faces that are equilateral triangles, 12 vertices and 30 edges. The regular dodecahedron has 12 faces that are regular pentagons, 20 vertices and 30 edges. These two solids are dual to each other in the same way as the octahedron and the cube. Taking as vertices of a new solid the central point in the 12 faces of the dodecahedron and joining them with edges one obtains the icosahedron. It follows that also in this case there is a unique subgroup of the rotation group named $I_{60} \subset SO(3)$ that is the symmetry group of both the icosahedron and the dodecahedron. This group is named the icosahedral group and has order 60

In the Lie algebra case the three sporadic solutions are respectively associated with the exceptional Lie algebras $\mathfrak{e}_6$, $\mathfrak{e}_7$ and $\mathfrak{e}_8$. In a language that will become clear while considering the constructions of Chap. 5, the simple algebras $\mathfrak{a}_n$, $\mathfrak{d}_n$, $\mathfrak{e}_{6,7,8}$ form the set of *simply laced Lie algebras*. Hence the classification of finite subgroups of the rotation group is in one-to-one correspondence with the classification of simply laced Lie algebras and this classification is named the ADE-classification.

We can now understand the enumeration of platonic solids. There is a pair (a solid S and its dual $S^\star$) for each of the exceptional groups T_{12}, O_{24} and I_{60}. The result would be six but it is 5 since the first solid in the list, the tetrahedron is self-dual. Indeed there are four faces and four vertices in this solid and constructing the dual of a tetrahedron we obtain another tetrahedron.

In view of the fundamental relevance of the mentioned Diophantine equation it is mandatory to recall the little we know about Diophantus.

DIOPHANTUS

We have very scarce information about the life of Diophantus from Alexandria (see Fig. 4.10) and even the dates of his birth and death are unknown. Various philological arguments and a few certain historical references place his life somewhere in a rather wide time-range of about five centuries from 150 BC to AD 415. Accepting as true a story related by the great XI century byzantine historian Michael Psellus in one of his letters that was discovered by the French scholar Paul Tannery (1843–1904) in the Escurial Library in Spain, we can pin down the life span of Diophantus to the third century AD. According to Psellus, Anatolius of Alexandria, a philosopher who was Bishop of Laodicea in the decade 270/280 AD, dedicated a treatise on arithmetic composed by himself to the great mathematician Diophantus who, on the basis of this, was presumably his contemporary [155].

What is certain is that the main work of Diophantus *Arithmetica* (Ἀριθμητικά) written, according to tradition, in thirteen books, six of which have come down to us in the original Greek version, has played an outstanding role in the history of Modern Mathematics. Translated from Greek into Latin in 1621 by the French Mathematician Claude Gaspard Bachet de Méziriac [9] was studied by many scholars and won to Diophantus the nick-name of *Father of Algebra*. A copy of this published Latin version was in possession of Pierre de Fermat and on the margin of one of its pages, the French genius noted with his pen his famous third theorem, whose proof he omitted stating that it was very simple. As it is well known, after many efforts that lasted three centuries, the proof was finally established by Andrew Wiles in 1995. As we stressed, only six of the thirteen books of the Arithmetica are extant in the Greek original. The remaining books were believed to be lost, until the discovery of a medieval Arabic translation of four of the remaining books in a manuscript that was kept in the Shrine Library in Meshed in Iran. The manuscript was discovered in 1968 by F. Sezgin [158].

Let us now return to the analysis of Plato's dialogue and of his philosophical-mathematical conception.

Fig. 4.10 Diophantus and its book translated into Latin by Gaspard Bachet de Méziriac

Plato's Timaeus and the Regular Solids

All sporadic solutions of a given problem have always duely excited the imagination of any mathematician or theoretical physicist throughout the history of science. The property of being sporadic generates the suspect that the considered sporadic mathematical object is endowed with special significance, having some hidden fundamental property which promotes its candidacy to be a fundamental brick in Nature's Architecture. Therefore it is not too much surprising that the five sporadic regular solids strongly impressed Plato and excited his philosophical creativity, leading him to associate them with the fundamental architecture of the physical world.

As any serious philosopher or scientist should do at any time, Plato had to encompass, within the framework of his new *theory of the world*, all the laws of Nature as they were known by his time. In the fourth century B.C. the prevailing conception of the physical world was that originated from Empedocle's theory, which envisaged four fundamental elements to be the ultimate constituents of matter: *Earth, Air, Fire* and *Water*. Plato produced a mathematization of Empedocle's theory associating a regular solid to each of the four elements. He associated *Fire* with the *Tetrahedron*, *Air* with the *Octahedron*, *Water* with the *Icosahedron* and he left apart the *Dodecahedron* as an extra *Decoration of the World* for which his pupil Aristotle found later a dignified association with the Empyreal Aether (see Fig. 4.11).

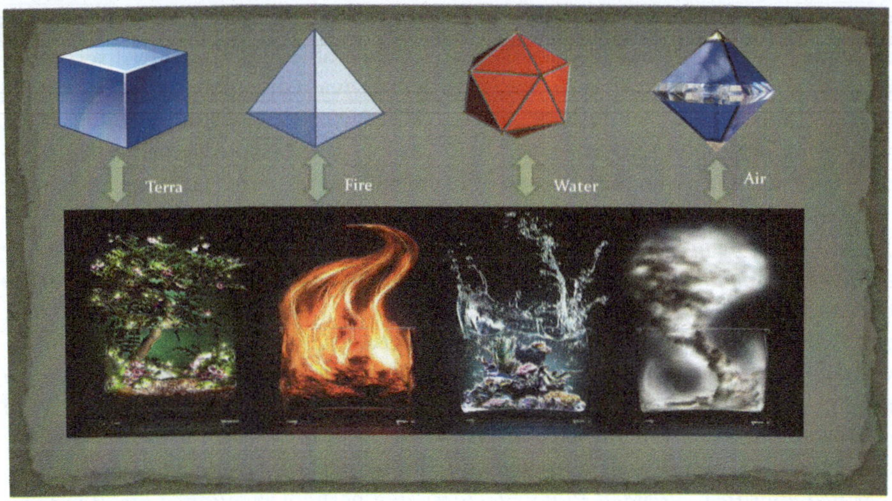

Fig. 4.11 The four elements of ancient physics, Earth, Fire, Water and Air are identified with four of the five regular solids in Plato's Timaeus

Plato's choices were not arbitrary, rather they were motivated by an underlying quite elaborated conception which is deeply rooted in the fundamentals of Platonic philosophy. Indeed *Timaeus* is one of the most important of Plato's dialogues, where the exposition of his philosophy is most extended and systematic.

To begin with, the main speaker of the dialogue, the very Timaeus of Locri, quite seemingly an invented character, is implicitly presented as a philosopher of the Pythagorean school. As I mentioned in Chap. 2, the Pythagoreans were the first to conceive a mathematical vision of the world, actually an arithmetical one. After they discovered that musical instruments that produce consonant sounds are related to one another by simple numerical ratios, the Pythagoreans sought to establish other correspondences between integer numbers and natural processes. In the already quoted words of Aristotle, they concluded that *the elements of numbers are the elements of things*.

The pythagorean dream of an *arithmetical physics* came to a sudden end when one member of the same school discovered the existence of the *incommensurables*, that is, of magnitudes which can be constructed geometrically but stand in no conceivable proportion to one another. As we know from our modern mathematical wisdom, this was the first step towards the discovery of the real numbers $\mathbb{R}$ that were properly axiomatized only in the XIXth century. Indeed what the Pythagorean discovered were the irrational numbers like $\sqrt{2}$, the trascendental numbers like the Euler e being still to come. From our current point of view, educated by differential and integral calculus, which, incidentally might have been also the viewpoint of the later Archimedes (287–212 B.C), nothing *non mathematical* was inherent to the incommensurables, yet from the philosophical stand point of the pythagoreans who viewed the essence of things in the integer numbers this was a serious inconvenience and produced a serious doubt about the existence of a *mathematical theory* of the world. In particular the existence of the incommensurables highlighted a discrepancy between *geometry* and *arithmetic* that in the pythagorean conception, probably shared by Plato, was the essence of mathematics. This raises the question whether Plato, admired by the modern founders of physics Copernicus, Kepler and Galilei, as the ancient philosopher who shared their belief in mathematics as the language of Nature, was actually committed to a mathematical theory of the world. According to Roberto Torretti [164] there are several hints that he was not. In the dialogue *The Republic* he says: *Motion presents not just one, but many forms. Someone truly wise might list them all, but there are two which are manifest to us.*[2] According to Torretti: *One is that is imperfectly illustrated by celestial motions. The other is the musical motion, studied by Pythagorean acoustics*. The same Torretti stresses that *Plato's warning to would-be astronomers, that they should not expect heavenly bodies to be excessively punctual, nor spend too much effort observing them in order to grasp their truth, was probably aimed at none other than the young Eudoxus, who, while the Republic was being written, attended Plato's lectures and perhaps mentioned his plan for a mathematical theory of planetary motions* [164].

The life span of Eudoxus of Cnidus is probably (408–355 B.C.) and it is generally believed that his work is the ground basis of Euclid's Vth Book, dealing with the exact quantitative comparison of geometrical magnitudes, irrespectively whether they are commensurable or not. He was also the first developper of Greek mathematical astronomy. He introduced models of the heavenly motions based on uniformly rotating spheres whose poles, at the extremity of the rotation axes, are pinned on other rotating spheres. Such models evolved in the Ptolemaic theory of *the cycles and epicycles* that dominated astronomy for almost two thousand years up to the Copernican revolution.

The quite successful outcome of the models conceived by his pupil Eudoxus seem to have convinced Plato, in his older age, of the feasibility of a mathematical theory of heavenly motions. According to Torretti, Plato reconciled his opposition to a mathematical theory of physical phenomena with this counter evidence from astronomy by setting apart the heavenly bodies, as intelligent entities, whose behavior is quite different from the *clumsy, unpredictable behavior of the inanimate objects that surround us* [164].

In modern parlance Plato's distinction might be rephrased as the distinction between simple few-body systems following the fundamental laws of fundamental interactions and complex systems admitting just a thermodynamical description. Nonetheless the question remains whether Plato's theory of the world is ultimately mathematical in character or not.

In my humble opinion, the answer is very simple and fully sustained by the long exposition of the Pythagorean Timaeus in the homonymous dialogue. Plato's theory of the physical world is *geometrical* as opposed to *arithmetic*, Geometry and Arithmetic being two branches of Mathematics in our present day view, yet philosophically different in Plato's and Pythagorean perception.

---

[2] Plato, *Respub* 530^d1.

Let us review the essentials of the theory presented in Timaeus.

The Demiurge, Creator of the Universe, was by definition the *Very Good One* and therefore he wanted to create things as much similar to himself as he might do. The Very Good One nothing is allowed to do that is not the *most beautiful*, since in Greek mentality the *good and the beautiful* coincide (remember the καλοκἁγαθὸς of Xenophon). Hence the Demiurge made the World beautiful, which, in line with the Canon of Polykleitos (see Sect. 2.1), means in good proportions. Indeed Timaeus says: *Now that which is created is of necessity corporeal, and also visible and tangible. And nothing is visible where there is no fire, or tangible which has no solidity, and nothing is solid without earth. Wherefore also God in the beginning of creation made the* ***body of the universe to consist of fire and earth****. But two things cannot be rightly put together without a third; there must be some bond of union between them. And the fairest bond is that which makes the most complete fusion of itself and the things which it combines; and* ***proportion is best adapted to effect such a union****. For whenever in any three numbers, whether cube or square, there is a mean, which is to the last term what the first term is to it; and again, when the mean is to the first term as the last term is to the mean-then the mean becoming first and last, and the first and last both becoming means, they will all of them of necessity come to be the same, and having become the same with one another will be all one. If the universal frame had been created a* ***surface only*** *and having no depth,* ***a single mean would have sufficed*** *to bind together itself and the other terms; but now, as the* ***world must be solid****, and solid bodies are always compacted* ***not by one mean but by two****, God placed* ***water and air in the mean between fire and earth****, and made them to have the same proportion so far as was possible (as fire is to air so is air to water, and as air is to water so is water to earth); and thus he bound and put together a visible and tangible heaven. And for these reasons, and out of such elements which are in number four, the body of the world was created, and it was harmonised by proportion, and therefore has the spirit of friendship; and having been reconciled to itself, it was indissoluble by the hand of any other than the framer.*

Translating into modern physical terms, the Demiurge created the Universe out of radiation (=fire) and matter (=earth) since what is generated has to be felt (=gravity) and it has to be seen (you need light). Yet in order to fulfil the imperative of beauty you need proportions and this implies at least one mean x as to be able to write $\frac{F}{X} = \frac{X}{E}$ where F stands for fire and E for earth (this is certainly reminiscent of Eudoxus' theory of magnitude comparisons exposed in Euclid's Vth book). However, the Universe had to be three-dimensional, rather than two-dimensional. In $d = 3$ one mean is not sufficient to write proportions of solid bodies so you need two means X, Y so to be able to write $\frac{F}{X} = \frac{X}{Y} = \frac{Y}{E}$. The X and the Y were the two additional elements *Air* and *Water*.

The argument that leads to the identification of the four elements with four out of the five regular solids is quite elaborate and based on the fundamentals of Platonic Idealism. For Plato, as it is well known, the real beings are just the Ideas, while the material world is made of imperfect copies of the true entities, the shadows on the walls of the cavern in the myth of the *Antrum Platonicum*. Yet the subdivision in two is not sufficient as Timaeus explains: *This new beginning of our discussion of the universe requires a fuller division than the former; for then we made two classes, now a third must be revealed. The two sufficed for the former discussion: one, which we assumed, was a pattern intelligible and always the same; and the second was only the imitation of the pattern, generated and visible. There is also a third kind which we did not distinguish at the time, conceiving that the two would be enough. But now the argument seems to require that we should set forth in words another kind, which is difficult of explanation and dimly seen. What nature are we to attribute to this new kind of being? We reply, that it is the* **receptacle**, *and in a manner the nurse, of all generation.*

The patterns are the *Ideas*, the imitations of the patterns are the *Physical Phenomena*, the receptacle is just *Space*, the space of Geometry.

After this momentous conceptual insertion, Timaeus embarks on two auxiliary discussions. On one side he observes the regime of continuous transformations of the four elements, their dynamical and chemical interactions that make it difficult to distinguish them in a firm way: *I must first raise questions concerning fire and the other elements, and determine what each of them is; for to say, with any probability or certitude, which of them should be called water rather than fire, and which should be called any of them rather than all or some one of them, is a difficult matter*. In modern terms Timaeus remarks that matter changes shape and status since it is probably made of subconstituents as it was claimed by the atomist Democritus (see Fig. 4.12). Secondly Timaeus calls geometry into the play by saying: *In the first place, then, as is evident to all, fire and earth and water and air are bodies. And every sort of body possesses solidity, and every solid must necessarily be contained in planes; and every plane rectilinear figure is composed of triangles; and all triangles are originally of two kinds, both of which are made up of one right and two acute angles; one of them has at either end of the base the half of a divided right angle, having equal sides, while in the other the right angle is divided into unequal parts, having unequal sides.* Thus Timaeus argues that all the elements are three-dimensional bodies that occupy a finite portion of space. Such finite volumes correspond to some solids delimited by faces that are, on their turn, finite portions of the plane. All plane figures and in particular the faces of the considered solid can be triangulated, namely they can be decomposed into triangles. All triangles, on their turn, can be decomposed into two rectangular triangles as shown in Fig. 4.13. There are two cases: either the two constituent triangles of the considered triangle, have unequal sides meeting at the right angle or they have equal sides. Timaeus states that the constituent triangles are the final subconstituents of matter and hence of the four elements: in short these are the platonic version of Democritus' atoms. Indeed Timaeus says: *These (the constituent triangles), then, proceeding by a combination of probability with demonstration, we assume to be the original elements of fire and the other bodies; but the principles which are prior to these God only knows, and he of men who is the friend God.* At this very point the supreme principle of beauty is utilized which is the Greek word for symmetry, as we emphasized several times. Timaeus continues: *Now of the two (constituent) triangles, the isosceles has one form only; the scalene or unequal-sided has an infinite number. Of the infinite forms we must select the most beautiful, if we are to proceed in due order, and any one who can point out a more beautiful form*

Fig. 4.12 Democritus (c. 460–c. 370 BC)

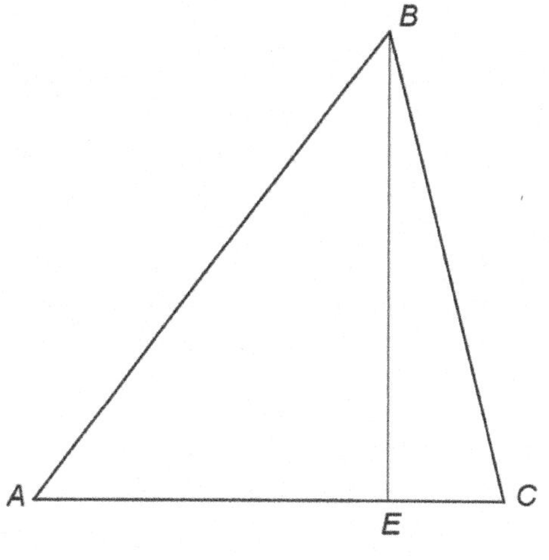

Fig. 4.13 Every triangle ABC can be decomposed in two rectangle triangles ABE and EBC. Generically the remaining two angles, of these two triangles, apart from the 90° angle are different, being their sides unequal

than ours for the construction of these bodies, shall carry off the palm, not as an enemy, but as a friend. Now, the one which we maintain to be the most beautiful of all the many triangles (and we need not speak of the others) is that of which the double forms a third triangle which is equilateral. The case where the two constituent triangles of another triangle are isosceles is unique as it is shown in Fig. 4.14. Such triangles have two 45° angles and one 90° angle.

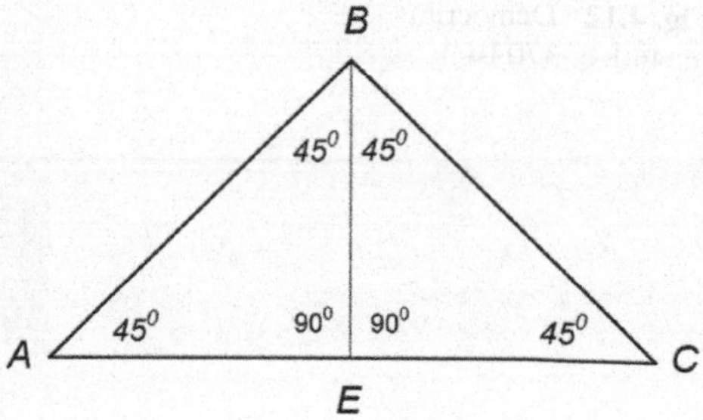

Fig. 4.14 The isoscele rectangular triangle is unique: it has two 45° angles apart from the 90° one

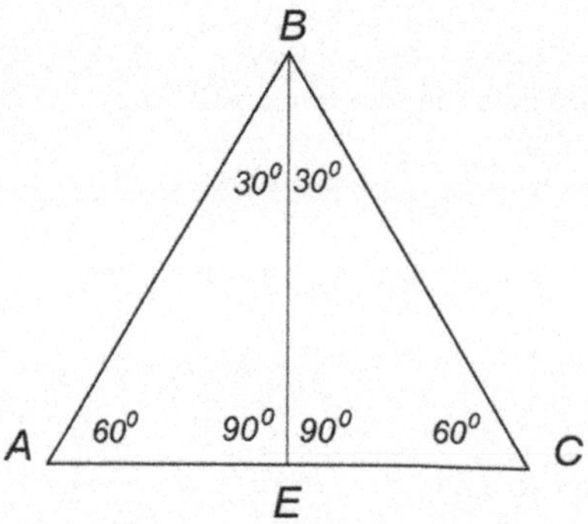

Fig. 4.15 The most beautiful scalene rectangular triangle is that one that has one 90° angle, one 60° angle and one 30° angle

It remains to choose one case among the infinite scalene instances of rectangular triangles and our choice must correspond to the most beautiful case. There is no other such triangle - asserts Timaeus - that is more beautiful than that one which, once doubled, makes an equilateral triangle. Such a triangle has one angle of 90°, one of 60° and one of 30°. One of the two sides meeting at the right angle is obviously $\frac{\sqrt{3}}{2}$ times longer than its companion (see Fig. 4.15).

In this way Plato elaborates a conception according to which there are subconstituents of the basic elements, namely triangles with quantized angles $\theta = 30°, 45°, 60°$ or $90°$ and the faces of the fundamental elements∼solids are either squares or equilateral triangles that can be constructed by adjoining two triangles of the above classes (Figs. 4.14 and 4.15).

The Tetrahedron, the Octahedron and the Icosahedron have equilateral triangular faces so that they are made of the same unique type of subconstituents. Their identification with Fire, Air and Water is somewhat arbitrary and it is made in order of complexity, assuming that Fire is thinner than Air and Air is thinner than Water. The Cube, whose square faces can be decomposed in terms of the other species of constituent triangles, is set apart and associated with Earth which looks less akin to be transformed into Fire, Air or Water than the other three are akin to transform into eachother. This conception provided also an embryonal idea of *chemical reactions* conceived as the dissociation and reassociation of the elementary constituent triangles.

Notwithstanding the self-consistency of the platonic scheme it is rather obvious that it had no other real motivation but the need to exclude one of the five solids, since the elements to be explained were four rather than five. The Dodecahedron, having pentagonal faces cannot be decomposed into the same type of subconstituent triangles as the other four platonic solids. Had Plato come to the concept of duality and had he arrived at the mathematical notion of symmetry group, which was instead introduced two thousand years after him, he would have had at his disposal three rather than five candidates for the fundamental bricks of matter. There is no doubt that Plato would have highly liked the intrinsic notion of symmetry group, which perfectly fits into his Philosophy of Ideas, the representation of a group being a magnificent example of the *material shadows* or *receptacles of becoming*, yet it is difficult to imagine how he would have solved the excess of fundamental elements to be explained: may be by identifying two of them? It is anyhow curious to note that the quantization of angles in the first quadrant ($\theta = 30°, 45°, 60°$ or $90°$) introduced by Plato is the same that occurs in the classification of root-spaces, the main basis in the classification of simple Lie algebras.

An important lesson told by the above analysis is the following. It is indeed fruitful to focus on sporadic mathematical objects in the quest for the interpretation of natural laws, yet if you have to introduce arbitrary choices among the sporadic structures a strong suspicion should arise that you are not on the right track. Logically there are three possible origins of error:

1. Notwithstanding its appeal, the considered sporadic mathematical structure has no place in Nature's Architecture.
2. The current formulation of the fundamental laws of Nature has a flaw. It needs a motivated revision, leading either to an increase or to a reduction by identification of the postulated constituents.
3. The considered features of the sporadic mathematical structure under study are not the really fundamental ones. New mathematical principles are possibly needed to revise our understanding of the mathematical objects we play with.

The first possibility cannot be excluded yet it is unlikely. Sooner or later all exceptional mathematical objects have come into play in Physics and Chemistry. The other two possibilities are not mutually exclusive. In the case of Plato's Timaeus, Empedocle's four elements had to be replaced by more adequate constituents whose number is not four. A substantial flaw occurred in the laws of Nature as seen at that time. From the theoretical point of view the relevant feature was not encoded in the geometrical appearance of the solids, rather in their group of symmetries.

In the next section we obtain the ADE classification of the finite subgroups of SU(2), which are the binary extensions of the finite rotation groups.

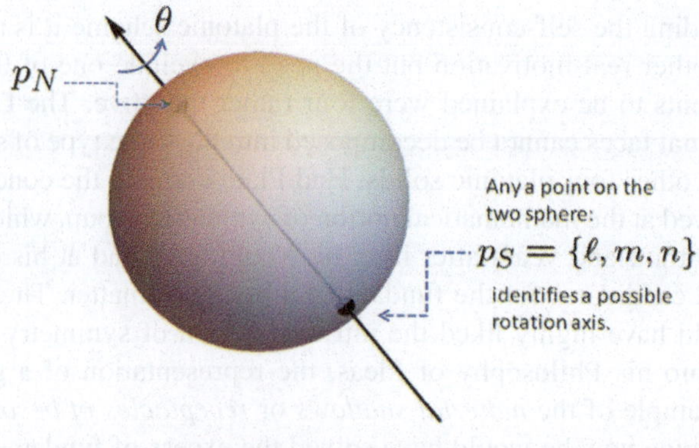

Fig. 4.16 Every element of the rotation group $\mathscr{O}_{(\ell,m,n)} \in \mathrm{SO}(3)$ corresponds to a rotation around some axis $\mathbf{a} = \{\ell, m, n\}$. On the surface of the two-sphere $\mathbb{S}^2$ this rotation has two fixed points, a North Pole and a South Pole that do not rotate to any other point. The rotation $\mathscr{O}_{(\ell,m,n)}$ is the image, under the homomorphism ω of either one of 2 × 2 - matrices $\mathscr{U}^{\pm}_{\ell,m,n}$ that, acting on the space $\mathbb{C}^2$, admit two eigenvectors $\mathbf{z}_1$ and $\mathbf{z}_2$. The one-dimensional complex spaces $p_{1,2} \equiv \lambda_{1,2}\mathbf{z}_{1,2}$ are named the two poles of the unitary rotation

4.2.1 *The Diophantine Equation that Classifies Finite Rotation Groups*

We begin by considering one parameter subgroups of SO(3). These are singled out by a rotation axis, namely by a point $p_S = \{\ell, m, n\}$ on the two–sphere $\mathbb{S}^2$. Mathematically a two-sphere is the locus in three-dimensional space of all the points $\{\ell, m, n\}$ that have the same fixed distance, say 1 in some length units, from some reference point, say $\mathbf{0} = \{0, 0, 0\}$, which is the centre of the sphere. This yields the equation:

$$\ell^2 + m^2 + n^2 = 1 \tag{4.2.1}$$

The unique infinite line that goes through the origine $\mathbf{0}$ and through the point p_S can be regarded as the rotation axis $\mathbf{a} \sim \{\ell, m, n\}$ of a one-dimensional rotation subgroup $\mathscr{O}_{(\ell,m,n)} \in \mathrm{SO}(3)$. Therefore to specify an element of the rotation group it suffices to give the axis $\mathbf{a}$ and the rotation angle θ (see Fig. 4.16). All the elements of the rotation subgroup $\mathscr{O}_{(\ell,m,n)}$ have two fixed points on the sphere, that never rotate, namely the south pole p_S and the its antipodal point $p_N = -\{\ell, m, n\}$ that we name the North Pole.

In terms of these elements any finite element of the rotation group can be represented by the following 3 × 3 matrix

$$\mathscr{O}_{(\ell,m,n),\theta} =$$
$$\begin{pmatrix} \left(m^2+n^2\right)\cos(\theta)+2 & -m\ell\cos(\theta)-n\sin(\theta)+1 & -n\ell\cos(\theta)+m\sin(\theta)+1 \\ -m\ell\cos(\theta)+n\sin(\theta)+1 & \left(n^2+\ell^2\right)\cos(\theta)+2 & -mn\cos(\theta)-\ell\sin(\theta)+1 \\ -n\ell\cos(\theta)-m\sin(\theta)+1 & -mn\cos(\theta)+\ell\sin(\theta)+1 & \left(m^2+\ell^2\right)\cos(\theta)+2 \end{pmatrix} \tag{4.2.2}$$

In Sect. 3.1.3 we discussed[3] the notion of group homomorphism. One of the most important homomorphisms in group–theory, which plays an extremely relevant role in Physics and in Quantum Mechanics, is that from the group SU(2) to the rotation group SO(3):

$$\omega : \mathrm{SU}(2) \longrightarrow \mathrm{SO}(3) \tag{4.2.3}$$

By definition the group SU(2) is composed by all those 2×2 matrices $\mathscr{U}$ that are unitary, namely satisfy the relation $\mathscr{U}\,\mathscr{U}^\dagger = \mathrm{Id}$, and, furthermore, have unit determinant, *i.e* det U = 1. The identity Id is that matrix that has all elements zero, except those on the principal diagonal that are all equal to 1.

$$\mathrm{Id} = \begin{pmatrix} 1 & 0 & \dots & \dots & \dots & 0 \\ 0 & 1 & 0 & \dots & \dots & 0 \\ 0 & 0 & 1 & 0 & \dots & \vdots \\ \vdots & \vdots & 0 & \ddots & 0 & \vdots \\ \vdots & \vdots & \vdots & 0 & 1 & 0 \\ 0 & \dots & \dots & \dots & 0 & 1 \end{pmatrix} \tag{4.2.4}$$

By definition, the hermitian conjugate $\mathscr{U}^\dagger$ of a complex matrix is the matrix obtained by interchanging the rows with the columns[4] and complex conjugating all the so obtained entries.[5]

The angular momentum **J** of macroscopic particles was known in Classical Physics since its very beginning in the XVIIth century. Mathematically, specifying the angular momentum vector $\left\{J_x, J_y, J_z\right\}$ of a particle corresponds to assigning such a particle to an irreducible representation of the rotation group SO(3). It is known from the founders of Quantum Mechanics that the third component J_z, once the reference frame is fixed in such a way that z is the rotation axis, should be quantized in either integer or half integer units. It was Wolfgang Pauli (see Fig. 4.17) the great physicist who fully understood the deep mathematical and physical meaning of such a thumb rule. He was the founder of the theory of *the spin*, namely of the

[3] See in particular Eqs. (3.1.8) and (3.1.9).

[4] The result of performing the exchange of the rows with the columns of a matrix A is a new matrix A^T, named the transpose of the previous one. The element A^T_{ij} of the transposed matrix is equal to A_{ji} of the original one.

[5] In fomulae the elements of the hermitian conjugate matrix $\mathscr{U}^\star$ are as follows $\mathscr{U}^\dagger_{ij} = \mathscr{U}^\star_{ji}$.

Fig. 4.17 Wolfgang Pauli (Vienna 1900–Zurich 1958) was born in a rich and educated family in Vienna. Wolfgang's father was a chemist coming from a prominent Jewish family based in Prague, who converted to Catholicism when he married his christian wife, Wolfgang's mother. Pauli first studied in Vienna where his diploma thesis, defended in 1918, was the basis of the first thoroughful review of Einstein's General Relativity, published one year later. Next he went for his Ph.D. to Munich where he worked under the supervision of Sommerfeld. After graduation in 1921, he had various temporary appointments in Göttingen, Copenhagen and Hamburg, where he met the most prominent physicists and mathematicians of that gold period. In 1931 he was appointed professor in Zurich ETH and then in 1940, offered a position in Princeton, he emigrated to the USA. In the meantime he obtained Swiss citizenship and after the war he made return to Zürich, where he remained for the rest of his life. Among the many fundamental contributions of Pauli to theoretical physics the most outstanding and the deepest one is *Pauli exclusion principle* stating that two fermions, namely two particles with half integer spin can not occupy the same quantum state. For this achievement he was awarded the Nobel Prize in 1945 on Einstein's nomination

intrinsic angular momentum of elementary particles. He understood that spin $\frac{1}{2}$ particles like all the basic constituents of the atoms (electrons, protons and neutrons) are assigned to the lowest dimensional representation of the rotation group which is actually two-dimensional complex. Furthermore it is, as physicists say, a double valued representation, namely to each element of SO(3) there correspond two elements in the two-dimensional representation that have the same image in the standard three-dimensional one. This is essentially the homomorphism (4.2.3).

In explicit matemathical term the homomorphism ω is described as follows. Setting

$$\lambda = \ell \sin\frac{\theta}{2} \quad ; \quad \mu = m \sin\frac{\theta}{2} \quad ; \quad \nu = n \sin\frac{\theta}{2} \quad ; \quad \rho = \cos\frac{\theta}{2} \qquad (4.2.5)$$

the two SU(2) group elements, whose image under ω is the matrix (4.2.2) are the following ones:

$$\mathscr{U}^{\pm}_{\ell,m,n} = \pm \begin{pmatrix} \rho + \mathrm{i}\nu & \mu - \mathrm{i}\lambda \\ -\mu - \mathrm{i}\lambda & \rho - \mathrm{i}\nu \end{pmatrix} \tag{4.2.6}$$

A generic $\mathscr{U} \in \mathrm{SU(2)}$ acts on a $\mathbb{C}^2$-vector $\mathbf{z} = \begin{pmatrix} z_1 \\ z_2 \end{pmatrix}$ by usual matrix multiplication $\mathscr{U}\mathbf{z}$. The matrices $\mathscr{U}^{\pm}_{\ell,m,n} \in \mathrm{SU(2)}$ have two eigenvectors:

$$\mathbf{z}_1 = \begin{pmatrix} 1 - n \\ l - im \end{pmatrix} \quad ; \quad \mathbf{z}_2 = \begin{pmatrix} -n - 1 \\ l - im \end{pmatrix} \tag{4.2.7}$$

such that

$$\begin{aligned} \mathscr{U}^{\pm}_{\ell,m,n}\,\mathbf{z}_1 &= \pm \exp\left[-\mathrm{i}\frac{\theta}{2}\right]\mathbf{z}_1 \\ \mathscr{U}^{\pm}_{\ell,m,n}\,\mathbf{z}_2 &= \pm \exp\left[\mathrm{i}\frac{\theta}{2}\right]\mathbf{z}_2 \end{aligned} \tag{4.2.8}$$

where θ is the corresponding rotation angle in three-dimension. Next consider the complex one-dimensional subspaces $\{\xi_{1,2}\,\mathbf{z}_{1,2}\}$ where $\xi_{1,2} \in \mathbb{C}$ are arbitrary complex numbers. The latter are named rays. Since $\mathbf{z}_1 \cdot \mathbf{z}_2 = \mathbf{z}_1^\dagger \mathbf{z}_2 = 0$ it follows that each element of SU(2) singles out two rays, hereafter named *poles* that are determined one from the other by the orthogonality relation. This concept of pole is the basic item in the argument leading to the classification of finite rotation groups.

Let $H \subset \mathrm{SO(3)}$ be a finite, discrete subgroup of the rotation group and let $\hat{H} \subset \mathrm{SU(2)}$ be its pre-image in SU(2) with respect to the homomorphism ω. Then the order of H is some positive integer number:

$$|H| = n \in \mathbb{N} \tag{4.2.9}$$

The total number of poles associated with H is:

$$\#\,\text{of poles} = 2n - 2 \tag{4.2.10}$$

since $n - 1$ is the number of elements in H that are different from the identity. Indeed the identity singles out no poles having the property of leaving invariant the full $\mathbb{C}^2$ space. As one sees each pair of poles is in correspondence with a rotation axis and hence, once the angle θ is specified with an element of the finite rotation subgroup.

The rest of the argument relies on the consideration that the finite group $\hat{H}$ maps the poles of each of its elements into the poles of other elements. Hence under the action of the finite group the set of poles organizes into orbits that we name Q_α. Each orbit contains an integer number of poles p_i, each of them being invariant under a subgroup $K_i \subset \hat{H}$ which is necessarily cyclic of finite order. Since the stability subgroups of the poles in the same orbit are conjugate to each other, all of them have the same order $k_\alpha \in \mathbb{N}$ which is a property of the entire orbit Q_α and it is necessary a divisor of the order of the group, namely we have $m_\alpha = \frac{n}{k_\alpha} \in \mathbb{N}$. Obviously m_α is the cardinality of the orbit $\mathcal{Q}_\alpha$, namely the number of poles it contains. Hence we assume that there are r orbits and that each orbit $\mathcal{Q}_\alpha$ contains m_α elements.

On the other hand, counting also coincidences, the total number of poles we have in the orbit $\mathcal{Q}_\alpha$ is:

$$\#\,\text{of poles in the orbit } \mathcal{Q}_\alpha = m_\alpha \,(k_\alpha - 1) \tag{4.2.11}$$

since the number of elements in each stability subgroup that are different from the identity is just $k_\alpha - 1$. So we find

$$2n - 2 = \sum_{\alpha=1}^{r} m_\alpha \,(k_\alpha - 1) \tag{4.2.12}$$

By means of a few straightforward arithmetic manipulations from Eq. (4.2.12), one obtains:

$$r + \frac{2}{n} - 2 = \sum_{\alpha=1}^{r} \frac{1}{k_\alpha} \tag{4.2.13}$$

with $k_\alpha \geq 2$ since each pole admits at least two group elements that keep it fixed, the identity and the non trivial group element that defines it by diagonalization. With few more manipulations one concludes there are only two possible cases:

$$r = 2 \quad \text{or} \quad r = 3 \tag{4.2.14}$$

Let us now consider the solutions of the diophantine equation (4.2.13) and identify the finite rotation groups and their binary extensions.

Taking into account the conclusion (4.2.14) we have two cases.

4.2.2 Case r = 2: The Infinite Series of Cyclic Groups $\mathbb{A}_n$

Choosing $r = 2$, the diophantine equation (4.2.13) reduces to:

$$\frac{2}{n} = \frac{1}{k_1} + \frac{1}{k_2} \tag{4.2.15}$$

Since we have $k_{1,2} \leq n$, the only solution of (4.2.15) is $k_1 = k_2 = n$, with n arbitrary. Since the order of the cyclic stability subgroup of the two poles coincides with the order of the full group H it follows that H itself is a cyclic subgroup of SU(2) of order n. We name it $\Gamma_b[n, n, 1]$. The two orbits are given by the two eigenvectors of the unique cyclic group generator:

$$\mathscr{A} \in \mathrm{SU}(2) \ : \ \mathscr{Z} \equiv \mathscr{A}^n \tag{4.2.16}$$

The finite subgroup of SU(2), isomorphic to the abstract group $\mathbb{Z}_{2n}$ is composed by the following $2n$ elements:

$$\mathbb{Z}_{2n} \sim \Gamma_b[n, n, 1] = \left\{\mathbf{1}, \mathscr{A}, \mathscr{A}^2, \ldots, \mathscr{A}^{n-1}, \mathscr{Z}, \mathscr{Z}\mathscr{A}, \mathscr{Z}\mathscr{A}^2, \ldots, \mathscr{Z}\mathscr{A}^{n-1}\right\} \tag{4.2.17}$$

Under the homomorphism ω, the SU(2)-element $\mathscr{Z}$ maps into the identity and both $\mathscr{A}$ and $\mathscr{Z}\mathscr{A}$ map into the same 3×3 orthogonal matrix $\mathrm{A} \in \mathrm{SO}(3)$ with the property $\mathrm{A}^n = \mathbf{1}$. Hence we have:

$$\omega\left[\Gamma_b[n, n, 1]\right] = \Gamma[n, n, 1] \sim \mathbb{Z}_n \tag{4.2.18}$$

In conclusion we can define the cyclic subgroups of SO(3) and their binary extensions in SU(2) by means of the following presentation in terms of generators and relations:

$$\mathbb{A}_n \quad \Leftrightarrow \quad \begin{cases} \Gamma_b[n, n, 1] = (\mathscr{A}, \mathscr{Z} \mid \mathscr{A}^n = \mathscr{Z} \ ; \ \mathscr{Z}^2 = \mathbf{1}) \\ \Gamma[n, n, 1] = \qquad (\mathrm{A} \mid \mathrm{A}^n = \mathbf{1}) \end{cases} \tag{4.2.19}$$

The nomenclature $\mathbb{A}_n$ introduced in the above equation is just for future comparison. As we will see, in the ADE-classification of simply laced Lie algebras the case of cyclic groups corresponds to that of $\mathbb{A}_n$ algebras.

4.2.3 Case $r = 3$ and Its Solutions

In the $r = 3$ case the Diophantine equation becomes:

$$\frac{1}{k_1} + \frac{1}{k_2} + \frac{1}{k_3} = 1 + \frac{2}{n} \tag{4.2.20}$$

In order to analyze its solutions in a unified way and inspired by the above case it is convenient to introduce the following notations:

$$\mathscr{R} = 1 + \sum_{\alpha}^{r} k_\alpha \tag{4.2.21}$$

and consider the abstract groups, that turn out to be of finite order, associated with each triple of integers $\{k_1, k_2, k_3\}$ satisfying (4.2.20) and defined by the following presentation:

$$\begin{aligned} \Gamma_b\left[k_1, k_2, k_3\right] &= \left(\mathscr{A}, \mathscr{B}, \mathscr{Z} \mid (\mathscr{A}\mathscr{B})^{k_1} = \mathscr{A}^{k_2} = \mathscr{B}^{k_3} = \mathscr{Z};\ \mathscr{Z}^2 = \mathbf{1}\right) \\ \Gamma\left[k_1, k_2, k_3\right] &= \left(\mathrm{A}, \mathrm{B} \mid (\mathrm{AB})^{k_1} = \mathrm{A}^{k_2} = \mathrm{B}^{k_3} = \mathbf{1}\right) \end{aligned} \tag{4.2.22}$$

We will see that the finite subgroups of SU(2) are indeed isomorphic to the above defined abstract groups $\Gamma_b\,[k_1, k_2, k_3]$ and that their image under the homomorphism ω are isomorphic to $\Gamma\,[k_1, k_2, k_3]$.

4.2.3.1 The Solution $(k, 2, 2)$ and the Dihedral Groups Dih_k

One infinite class of solutions of the diophantine equation (4.2.20) is given by

$$\{k_1, k_2, k_3\} = \{k, 2, 2\} \quad ; \quad 2 < k \in \mathbb{Z} \tag{4.2.23}$$

The corresponding subgroups of SU(2) and SO(3) are:

$$\mathrm{Dih}_k \quad \Leftrightarrow \quad \begin{cases} \Gamma_b[k, 2, 2] = \left(\mathscr{A}, \mathscr{B}, \mathscr{Z} \mid (\mathscr{A}\mathscr{B})^k = \mathscr{A}^2 = \mathscr{B}^2 = \mathscr{Z}\,; \; \mathscr{Z}^2 = \mathbf{1}\right) \\ \Gamma[k, 2, 2] = \quad (\mathrm{A}, \mathrm{B} \mid (\mathrm{AB})^k = \mathrm{A}^2 = \mathrm{B}^2 = \mathbf{1}) \end{cases} \tag{4.2.24}$$

whose structure we illustrate next.

$\Gamma_b[k, 2, 2] \simeq \mathrm{Dih}_k^b$ is the binary dihedral subgroup. Its order is

$$|\,\mathrm{Dih}_k^b\,| = 4\,k \tag{4.2.25}$$

and it contains a cyclic subgroup of order k that we name K. Its index in Dih_k^b is two. The elements of Dih_k^b that are not in K are of period equal to two since $k_2 = k_3 = 2$.

4.2.3.2 The Three Isolated Solutions Corresponding to the Tetrahedral, Octahedral and Icosahedral Groups

There remain three isolated solutions of the Diophantine equation (4.2.20), namely:

$$\{k_1, k_2, k_3\} = \{3, 3, 2\} \tag{4.2.26}$$
$$\{k_1, k_2, k_3\} = \{4, 3, 2\} \tag{4.2.27}$$
$$\{k_1, k_2, k_3\} = \{5, 3, 2\} \tag{4.2.28}$$

They respectively correspond to the tetrahedral T_{12}, octahedral O_{24} and icosahedral I_{60} groups and to their binary extensions, namely:

$$\Gamma[3, 3, 2] \simeq \mathrm{T}_{12} \tag{4.2.29}$$
$$\Gamma[4, 3, 2] \simeq \mathrm{O}_{24} \tag{4.2.30}$$
$$\Gamma[5, 3, 2] \simeq \mathrm{I}_{60} \tag{4.2.31}$$

As their name reveals these three groups have, 12,24 and 60 elements, respectively. The corresponding binary extensions have 24,48 and 120 elements respectively.

4.3 A Provisional Conclusion for a Tale Two Thousand Year Long

As their names reveal, the groups $\Gamma[3,3,2]$, $\Gamma[4,3,2]$, $\Gamma[5,3,2]$ are the automorphism groups of the five platonic solids, $\Gamma[5,3,2]$ being the symmetry of the icosahedron and of its dual, the dodecahedron, $\Gamma[4,3,2]$ being the symmetry of the octahedron and of its dual, the cube, $\Gamma[3,3,2]$ being the symmetry of the self dual tetrahedron. The algebraic structure of these groups is encoded in the triplet of three numbers that label them and satisfy a Diophantine inequality of deep significance also for Lie Algebra Theory.

We see in this a paradigmatic example of the gigantic progress in abstract thought provided by the conceptual revolution of group theory.

The Greeks admired symmetry and identified it with the beautiful. Plato went one step further and identified symmetry and the beautiful with the inner structure of the world, namely with the fundamental laws of Nature. From the point of view of a contemporary theoretical physicist he was essentially right, yet he still missed a good intrinsic definition of what symmetry is and that was the source of his troubles with the apparently excessive number of solids. Plato was rightly impressed by the sporadic solution of the problem encoded in the classification of the five regular solids, yet he did not know what the formulation of such a problem was.

Today we exactly know the correct formulation of this problem and we realize that the essential point is the three-dimensional extension of space (at constant time). Indeed the triplet of numbers denoting the element-orders contained in the group Γ is a characteristic feature of finite subgroups of SO(3) and of its binary extension SU(2). Finite subgroups of higher rotation groups can have more than three of such element–orders and lead to more complicated solutions of the Platonic problem for $n > 3$. Hence the relevant issue for the *episteme* is to find an a priori reason why the effective dimensions of our world are $3 \oplus 1$. The fascination of the ADE correspondence between Platonic Groups and semi–simple Lie algebras, here anticipated and discussed in Sect. 5.6, relies precisely in the triplet of integer number structure satisfying the same constraint. The classification of Lie Algebras makes no reference to three dimensions and promotes the classification of Platonic groups to a higher level of abstraction. The full fledged mathematical and philosophical consequences of this view point have still to be worked out, showing that the present two thousand year tale is still an open chapter.

4.4 Further Comments About Crystallography

At the beginning of this chapter we looked at group-theory from the point of view of crystallography emphasizing that being crystallographic is a further restriction which in any dimension $d = n$ selects, among the available Platonic Groups, the subclass of candidate Point Groups for lattices to be constructed in the same dimensions.

Let us summarize for the $n = 3$ case some group–theoretical features that follow from the ADE classification, combined with the crystallographic constraint:

(a) The Point Group $\mathfrak{P}$ must be a finite rotation group in $d = 3$ hence it must belong to the list:

$$\mathfrak{P} \in \{\mathbb{Z}_k\,,\, \mathrm{Dih}_k\,,\, \mathrm{T}_{12}\,,\, \mathrm{O}_{24}\,,\, \mathrm{I}_{60}\} \tag{4.4.1}$$

(b) The order of any element $\gamma \in \mathfrak{P}$ belonging to the Point Group must be in the range 2, 3, 4, 6

The intersection of these two conditions leads to the conclusion that:

$$\mathfrak{P} \in \left\{\mathbb{Z}_{2,3,4,6}\,,\, \mathrm{Dih}_{3,4,6}\,,\, \mathrm{T}_{12}\,,\, \mathrm{O}_{24}\right\} \tag{4.4.2}$$

The classification of Bravais lattices, which is responsible for so many chemical-physical properties of matter, is essentially encoded in Eq. (4.4.2). In this list of candidate Point Groups there is no simple one which is non abelian. They are all either solvable or abelian and this implies that their irreducible representations can be constructed by means of an induction algorithm starting from the one-dimensional representations of cyclic groups. A simple group which occurs in the ADE classification is the icosahedral group I_{60} which is isomorphic to the simple alternating group A_5 (the even permutations of 5 objects). It is barred out by the crystallographic condition because it contains elements of order 5.

Under many respects this is the analogue of what happens with algebraic equations. The algebraic equations of order 2, 3, 4 are always solvable by radicals since their Galois group is solvable. In degree $d \geq 5$ the generic equation is not solvable because the Galois group is generically not solvable.

A natural question arises at this point. Is the condition b) on the possible orders of the Point Group elements intrinsic to the crystallographic constraint in any dimension or it is a specific feature of $d = 3$?

The correct answer to the above question is the second option and there exists, for instance, a counterexample of a crystallographic group in 7-dimensions that has group elements of order 7. Not only that. The simple group of order 168 is an example of a *simple non abelian crystallographic point group*!

It is quite remarkable that the analogue of the ADE classification of finite rotation groups in $d > 5$ is so far non existing up to the knowledge of this author. Even less is known about higher dimensional crystallographic groups.

It is philosophically quite challenging to imagine what Chemistry, Geology and even Molecular Biology and Genetics might be in a world where the point group is a simple non abelian group!

Chapter 5
The Long Tale of Lie Groups

The analysts try in vain to conceal the fact that they do not deduce: they combine, they compose ... when they do arrive at the truth they stumble over it after groping their way along.
Evariste Galois

5.1 From Discrete to Continuous Groups

So far we outlined the conceptual development of group theory paying particular attention to *finite groups*. It is historically correct to do so, since finite groups were the first to be considered and studied. Indeed the very notion of group is to be credited to Galois and, by definition, Galois groups are finite.

Next we turn our attention to *continuous groups* the tale of whose birth and development is the topic of the present chapter.

For some time, as we are going to outline, the construction of this new field of mathematics went on in parallel with the conceptual evolution which finally led to the establishment of modern differential geometry, another exciting story we plan to unfold in Chap. 7.

Yet it was implicit in their very logical structure that *differential geometry* and *Lie group theory* should merge, as they eventually did, in particular because of the enormously influential ideas and monumental work of Élie Cartan.

From its very start Lie group theory was associated with a deep revision of the conception of geometry, specially promoted by Felix Klein. The mathematical idea of *Space* and that of *Symmetry* were indeed destined to compenetrate each other in an essential way and in the long run the two tales of differential geometry and Lie Groups led to both *General Relativity* and *Gauge Theories*, the two pillars of the present day *episteme*.

As we did in the case of finite groups, let us first summarize, in modern mathematical language what Lie groups are.

These latter, arise from the consistent merging of two structures:

P. G. Fré, *A Conceptual History of Space and Symmetry*,
https://doi.org/10.1007/978-3-319-98023-2_5

1. an *algebraic structure*, since the elements of a Lie group G can be composed via an internal binary operation, generically called product, that obeys the axioms of a *group*,
2. a *differential geometric structure* since G is an *analytic differentiable manifold* and the group operation are infinitely differentiable in such a topology.

General Relativity is founded on the concept of *differentiable manifolds*. The mathematical model of *space–time* that we adopt is given by a pair $(\mathcal{M}, g)$ where $\mathcal{M}$ is a differentiable manifold of dimension $D = 4$ and g is a *metric*, that is a rule to calculate the length of curves connecting points of $\mathcal{M}$. In physical terms the points of $\mathcal{M}$ take the name of *events* while every physical process is a continuous succession of events. In particular the motion of a *point–like particle* is represented by a *world–line*, namely a curve in $\mathcal{M}$ while the motion of an *extended object* of dimension p is given by a $d = p + 1$ dimensional *world–volume* obtained as a continuous succession of p–dimensional hypersurfaces $\Sigma_p \subset \mathcal{M}$.

Therefore, the discussion of such *physical concepts* is necessarily based on a collection of *geometrical concepts* that constitute the backbone of differential geometry. The latter is at the basis not only of General Relativity but of all Gauge Theories by means of which XXth century Physics obtained a consistent and experimentally verified description of all Fundamental Interactions.

The central notions are those which fix the geometric environment:

- Differentiable Manifolds
- Fibre-Bundles

and those which endow such environment with structures accounting for the measure of lengths and for the rules of parallel transport, namely:

- Metrics
- Connections

Once the geometric environments are properly mathematically defined, the metrics and connections one can introduce over them turn out to be the structures which encode the Fundamental Forces of Nature.

The above remarks clearly demonstrate that **differential geometry** and **Lie group theory**

- are **intimately and inextricably related** and
- have a **much wider range of applications** in all branches of physics and of other sciences.

since that of a manifold is the appropriate mathematical concept of a continuous space whose points can have the most disparate interpretations and that of a group is the appropriate mathematical framework to deal with symmetry operations acting on that space.

We postpone the discussion of the modern definition of differentiable manifolds and fibre-bundles to Chap. 7 were we outline the history of differential geometry and we begin our historical outline of how Lie group theory came into being.

5.2 Sophus Lie and Felix Klein

Who was Lie? Let us begin with the words dedicated to his memory, spoken on the occasion of Lie birth centennial by another giant of modern mathematics, namely by Élie Cartan (see Fig. 5.9):

Sophus Lie was of tall stature and had the classic Nordic appearance. A full blond beard framed his face and his gray-blue eyes sparkled behind his eyeglasses. One always immediately felt at ease with him, certain beforehand of his sincerity and his loyalty. He was not afraid to admit his ignorance of branches of mathematics unfamiliar to him, which nevertheless did not keep him from being aware of his own worth... Posterity will see him the genius who created the theory of transformation groups, and we French shall never be able to forget the ties which bind us to him and which make his memory dear to us.

The *theory of transformation groups*, mentioned by Cartan and to whose accomplishment Cartan himself gave the largest and deepest contributions, was born in dramatic times of war for Europe, and the history of an extremely deep conceptual revolution, whose ramifications have some of the furthest reaching consequences for modern Mathematics and Physics, is tightly linked with a quite interesting and very emotional story of a life-time friendship, eventually turned into scientific rivalry with touches of bitterness. The two main actors in such a historical drama are Sophus Lie and Felix Klein (see Fig. 5.1).

FELIX KLEIN

Felix Klein was born in Dusseldorf in 1849. After his secondary school education in Dusseldorf, he enrolled as a student at Bonn University where he became student of

Fig. 5.1 **Sophus Lie** (1842–1899) on the left and **Felix Klein** (1849–1925) on the right

Plücker. He graduated in 1868 with a thesis on Plücker's formulation of Geometry applied to Mechanics. From 1868 to 1870 he was travelling to Berlin, Paris and Göttingen. In 1871 he was nominated *dozent* in Göttingen. In 1872 he was offered a full chair in *Erlangen* where he stayed for three years. In 1875 he was appointed on a new chair at the *Technische Hochschule* in *Munich*. The same year he married with Anne Hegel the grand-daughter of the famous philosopher. In 1880, after five years of intensive work in Munich Felix Klein was appointed on the chair of Geometry at the University of Leipzig. Finally in 1886 he got an offer from *Göttingen University* where he worked until his retirement in 1913. He greatly contributed to make Göttingen the mathematical center of the world, in particular promoting the appointment of *David Hilbert* and funding the journal *Mathematische Annalen*. During World-War-One he privately taught mathematics in his home. He died in Berlin in 1925.

SOPHUS LIE

Marius Sophus Lie was born in 1842 in a small town in Norway, son of a Lutheran minister. He was the youngest of six brothers in a family with restricted economic possibilities. In 1857 he entered Nisses's Private Latin School in Christiania (the name of Oslo at that time). He entered Christiania University at beginning of the 60s and in 1862 he had the chance of attending a course on finite group theory and Galois theory given by Ludwig Sylow. He received his first diploma in Mathematics in 1865. In 1869 he published his first research paper and was granted a fellowship that allowed him to travel to Berlin and Göttingen (where he met Klein) and finally to Paris. In 1871 he obtained his Ph.D. from the University of Christiania with a thesis entitled *On a class of geometric transformations*. In 1874 he married with the 20 year old Ann Birch from whom he had three children. In 1884 Friedrich Engel, a former student of Klein, was dispatched to him by his advisor with the purpose of helping him writing his monumental books on transformation groups. In 1886 when Klein left for Göttingen, Lie was appointed on Klein's chair of geometry at Leipzig University. There he continued his work with Engel until 1898 when the *anaemia perniciosa* from which he was suffering since long progressed quite seriously and he decided to retire to his home country. Next year (1899) he died in Christiania.

5.2.1 *The Spring of 1870 in Paris*

The two decades of the XIXth century after the continental outbreak of the 1848 revolutions, mostly suppressed by the force of the arms, had witnessed a steady political evolution towards the creation of coherent Nation-States, whose gluing principle was self-identification around one national language, one national culture and one currency. This process had certainly many positive aspects in promoting the removal of the last remaining traces of feudalism, in boosting the diffusion of education among the popular masses and in strongly facilitating industrial development. Yet the negative aspects of XIXth century nationalism were the raising tensions among the emerging Nation-States and their competition for dominance both in Europe and in the rest of the World. The Great Power that was steadily declining was the multi-ethnic, catholic oriented and strongly conservative Austrian Empire. With the initial help of the French Empire, at the expenses of Austria and of the obsolete absolute monarchies that were its clients and occupied the largest part of the italian peninsula, 1861 witnessed the birth of the liberal, anti-clerically oriented United Kingdom of Italy. The next historical problem to be solved was the unification of Germany.

This unification occurred in a different way, with *iron and fire* under the leadership of the strongly military oriented Prussian Monarchy and by means of the astute diplomacy of its chancelor Otto von Bismarck. Firstly, in 1866, Austria was defeated by Prussia in the famous seven week war and Italy used that opportunity to complete its unification process, freeing Venice and its interland from the Austrians. Prussian victory in the battle of Sadowa definitely destroyed Austrian influence on Southern German States, in particular Bavaria, preparing the stage for the final leap toward the unification of Germany.

With the acquisition of the industrial, highly culturally evolute German Regions along the Rhein, Prussia, whose power was originally based on the Army, the tight structure of Public Administration and the wealth of the rich agricultural lands in the East, became extremely strong in all respects and with a clear-cut leadership on the whole of Germany. The cultural development of Germany in the XIXth century is impressive and more and more it checked the long term predominance of French Culture. In philosophy with Kant, Hegel, Schopenauer and their followers, in literature with Goethe, Heine, Schiller and others, in music with Beethoven, Brahms, Mendelssohn and a further long list of first class artists, we just see the top of a gigantic iceberg of intellectual ferment that was extremely active throughout the whole century.

German science was by no means behind and German Mathematics was by that time in a truly exploding phase. The epicentre of this explosion was the University of Göttingen where the Prince of Mathematicians, Gauss, had left a long-enduring legacy. Yet also other German Universities were coming to prominence: Berlin, first of all, but also Bonn, Leipzig, the historical Halle and, in the south, the institutions in Munich and the University of Erlangen.

In the spring of 1870 the tempest of the Franco-Prussian war, the first large scale modern war that prefigured World War One and from whose smoking guns the unification of Germany would emerge, was approaching. Two young men who had just made friendship were in Paris, attracted by their common interest for Mathematics and for the French advances in Geometry. They were Felix Klein and Sophus Lie (see Fig. 5.1).

After his first graduation in Mathematics in Christiania, the capital of Norway, Sophus Lie went through a period of uncertainty. He just knew that he wanted to make an academic career in Science but he was not sure which science to choose. For some time he was attracted by Astronomy, then he followed courses in Zoology an Botanic, finally it became clear to him that Mathematics was his mission and within the vast mathematical landscape Geometry was what mostly attracted his attention. We should now clarify what was the concept of geometry that fascinated the young

Fig. 5.2 **Julius Plucker** (1801–1868) Born in a small city near Wuppertal, Plücker was both a physicist and a geometer. Professor of Experimental Physics at Bonn University, he made important researches on the behavior of rarefied gases in magnetic fields, essentially discovering cathodic rays. He was also one of the founder of what became atomic spectroscopy. In geometry his relevant contributions arose from the studies he did in Paris in 1823 when he visited there and came in contact with the geometrical school of Gaspar Monge. Plücker's main contribution in geometry is the invention of Plücker coordinates. These are a set of homogeneous co-ordinates introduced initially to embed the set of lines in three dimensions as a quadric in five dimensions. The construction uses 2×2 minor determinants

Lie. In modern parlance we might say that it was algebraic geometry. In 1868 he fell in love with the papers on geometry written by Poncelet and Plücker. Poncelet, a French officer who was taken prisoner by the Russians in Napoleon's campaign of 1812, was the first to introduce complex numbers in projective geometry and described his results in a book published in 1822 and written while he was a prisoner on the Volga [147, 148]. In Plücker's paper, instead, geometric figures are no longer a collection of points but geometry becomes just as much a study of families of lines or of spheres and other extended surfaces [146]. Plücker's *monumental idea to create new geometries by choosing figures other than points - in fact straight lines - as elements of space* was cause of high excitement in Lie's mind. He wrote a short mathematical paper on these topics that he published at his own expenses. A year later Lie's paper was eventually accepted by the most prestigious mathematical journal of the time namely Crelle's Journal. This proved decisive for his future since this convinced the *Collegium Academicum* in Christiania to give him a research–travel grant that allowed him to go to Berlin, Göttingen and eventually to Paris.

Fig. 5.3 On the left **Gaston Darboux** (1842–1917); on the right **Marie Ennemond Camille Jordan** (1838–1922). **Gaston Darboux** taught at the College de France, at the École Normale Superiéure and eventually he was appointed on the chair of higher geometry at the *Sorbonne*. He is remembered for many results in mathematical analysis and in differential geometry. He was the biographer of Henri Poincaré. Among his students were Émile Borel and most remarkably Élie Cartan. About him Struik wrote ... *he followed the spirit of Gaspar Monge and Darboux's spirit can be detected in the work of Cartan*. **Camille Jordan**, Engineer by profession and later professor of mathematical analysis at the *École Polytechnique* has got his name associated with various important items in mathematics (Jordan normal form for matrices, Jordan's theorem in finite group theory and more). He was among the first promoters of discrete group theory and of Galois theory

In 1870, in Berlin, Sophus Lie met with Felix Klein who had studied in Bonn precisely under the supervision of Plücker, passed away two years before (see Fig. 5.2). The two young scientists had a lot of interests in common and became immediately close friends, although, as Freudenthal remarks, they had quite different characters both as humans and as mathematicians: ... *the algebraist Klein was fascinated by the peculiarities of charming problems; the analyst Lie, parting from special cases, sought to understand a problem in its appropriate generalization.* They traveled together to Paris where they met and interacted with Gaston Darboux and Camille Jordan (see Fig. 5.3). The conversations with Jordan were of the highest relevance for both Lie and Klein since the French mathematician attracted their attention to the role that group–theory could play in geometry. For Lie this was the germ of a reasoning that conducted him to the notion of *transformation groups*. Klein developed these ideas in what two years later appeared as the *Erlangen Programme*. In any case Lie and Klein discussed intensively about these issues and eventually published a common work. They lived in adjoining rooms in the same hotel and saw each other continuously.

The first stumbling of Lie on a Lie group occurred in early July 1870 when, working in his Paris hotel room, he made the discovery of what is now called Lie's line-sphere transformation. In a nut-shell it goes as follows. Arguing à la Plücker, Lie considered the space of *cycles* that can be either oriented circles (or straight lines) or points in the plane or the point at infinity (indeed the points can be thought of as circles of radius zero). It turned out that this space can be viewed as a quadric in the real five dimensional projective space $\mathbb{RP}^5$. In homogeneous coordinates x^i ($i = 1, \ldots, 6$) the equation of the quadric is the following one:

$$0 = x^i M_{ij} x^j \quad ; \quad M_{ij} = \begin{pmatrix} -1 & 0 & 0 & 0 & 0 & 0 \\ 0 & 1 & 0 & 0 & 0 & 0 \\ 0 & 0 & 1 & 0 & 0 & 0 \\ 0 & 0 & 0 & 1 & 0 & 0 \\ 0 & 0 & 0 & 0 & 0 & 1 \\ 0 & 0 & 0 & 0 & 1 & 0 \end{pmatrix} \tag{5.2.1}$$

The eigenvalues of the matrix M are $1, 1, 1, -1, -1, -1$ so that the group of linear transformations SO(3, 3) is an automorphism group of Lie's quadric. Obviously Lie did not use such a language but his transformations that can map oriented circles into points and viceversa are just the group elements of SO(3, 3). Extending his procedure from real to complex projective geometry Lie found also what in modern terms can be described as the local isomorphism of the group SL(4, $\mathbb{C}$) with the group SO(6, $\mathbb{C}$).

Klein recalls the event in this way [129][1]:

...one morning I got up early and wanted to go out right away when Lie, who still lay in bed called me to his room. He explained to me the relationship he had found during the night between the asymptotic curves of one surface and the lines of curvature of another, but in such a way that I could not understand a word. In any case he assured me that the asymptotic curves of the fourth degree Kummer surface must be algebraic curves of degree sixteen. That morning, while I was visiting the Conservatoire des Arts et Métiers, the thought came to me that these must be the same curves of degree sixteen that had appeared in my paper Theorien der Liniencomplexen ersten un zweiten grades and I quickly succeeded in showing this independently of Lie's geometric considerations. When I returned around four o'clock in the afternoon, Lie had gone out, so I left a summary of my results in a letter.

As a consequence of this, Klein and Lie wrote a paper together on the topics. The mentioned Kummer surface of fourth degree is obviously the algebraic locus K3 which has been the object of a lot of mathematical investigations and up to the present day plays an important role in string and supergravity compactifications.

Few days after these scientific events, Napoleon the third, falling into Bismarck's trap, declared war to Prussia and hostilities began (July 19th 1870). Being a citizen of Prussia, Klein had to flee immediately from France, while Lie, who was a citizen of Norway, namely of a neutral state, remained. In August, when the Prussians had already trapped part of the French Army in Metz, Lie decided to leave Paris and

---

[1] Several details of the story told in the present sections are from an article of Sigurdur Helgason entitled *Sophus Lie, the mathematician*.

hike towards Italy. When he reached Fontainebleau he was arrested as a German spy and his mathematical notes, written in German, were used as an evidence against him, regarding them as ciphered messages. He spent several weeks in prison and was finally released thanks to the intervention of Darboux who explained the case to the suspicious police. Once he was freed, Lie fled to Italy and from there he made his way back to Norway through Germany.

In 1871, back in Christiania, Lie completed his Ph.D doctoral thesis on the basis of his Paris discoveries and he was awarded his doctorate in 1872. The same year the University of Christiania created a new chair on which he was appointed.

5.2.2 The Erlangen Programme

In 1872, at the age of 23, Felix Klein was appointed Full Professor at the University of Erlangen, where he remained only three years, since in 1875 he received and accepted an offer from the *Technische Hochschule* of Münich. There he remained longer, namely five years, and accomplished important steps both in his personal and professional life. As for personal life, Münich was the city where, in August 1875, he married with Anne Hegel, the grand daughter of the philosopher Georg Wilhelm Friedrich Hegel. On the scientific side, Klein worked very much intensively in Münich and his fame as a brilliant and profound teacher spread through the world attracting there students that later became famous mathematicians and physicists: among them Max Planck, Adolf Hurwitz and Ricci Curbastro.

In these years Klein developed the ideas that he had exposed in 1872 in his inaugural address as a Professor in Erlangen. This lecture, whose German title is *Verglichende Betrachtungen über neure geometrische Forschungen* (A Comparative Review of Recent Researches in Geometry) has become known to posterity as the *Erlangen Programme* (see Fig. 5.4).

At the beginning of his lecture Klein stated the following:

Have a geometric space and some transformation group. A geometry is the study of those properties of the given geometric space that remain invariant under the transformations from this group. In other words, every geometry is the invariant theory of the given transformation group.

Up to that time geometry meant the study of geometrical figures like points, lines, triangles, circles, polyhedra. Euclidian geometry had been extended, in the course of the XIXth century, to other geometries like Lobachevsky hyperbolic geometry, or elliptic geometry. Projective geometry existed since long. In all known geometries a founding concept was, as Klein emphasized, the notion of equivalence classes. In Euclidian geometry, for instance, when you study triangles you do not distinguish this or that equilateral triangle: all equilateral triangles of the same size are equivalent and you study the properties of the class. What does it actually mean to be equivalent? It means that one triangle can be mapped into the other by means of a suitable

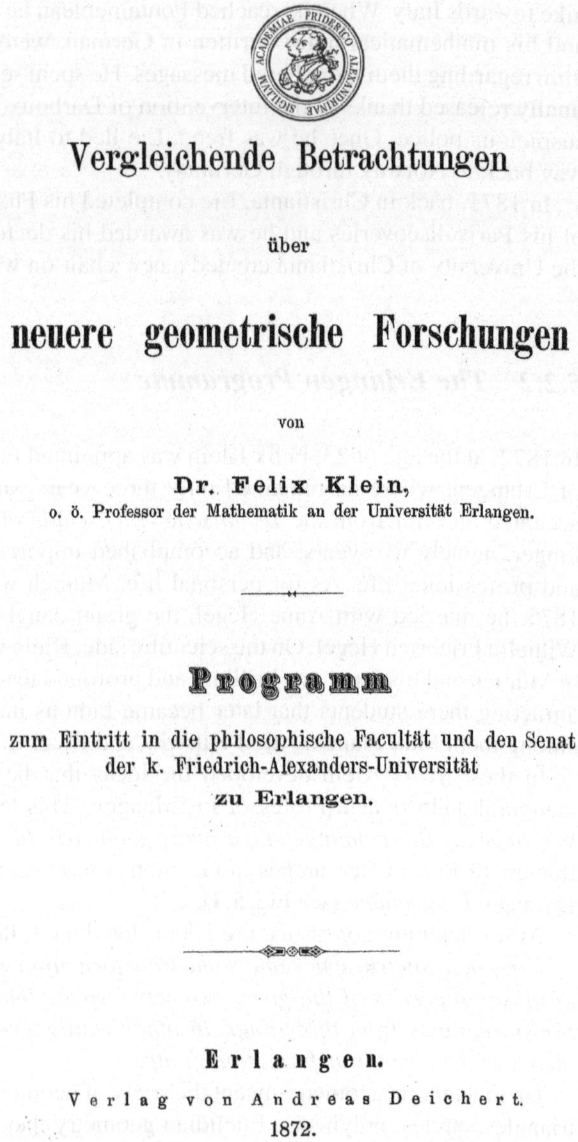

Vergleichende Betrachtungen

über

neuere geometrische Forschungen

von

Dr. Felix Klein,

o. ö. Professor der Mathematik an der Universität Erlangen.

Programm

zum Eintritt in die philosophische Facultät und den Senat
der k. Friedrich-Alexanders-Universität
zu Erlangen.

Erlangen.

Verlag von Andreas Deichert.

1872.

Fig. 5.4 The original title page of the lecture given by Felix Klein in 1872 at the University of Erlangen on the occasion of his admission as a Full Professor. The importance of the Erlangen Programme was not appreciated by the scientific community for a long time. Klein's text remained quite unknown as long it existed only as a booklet of Erlangen University. It became world-wide known later after its publication on the *Mathematische Annalen* in 1893

transformation of the chosen transformation group, namely the euclidian group $\mathbb{E}^3 = ISO(3)$, made of rotations and translations in $\mathbb{R}^3$. If the size of the two equilateral triangle is not the same, for instance they have a different height, then they are not equivalent and indeed some of their euclidian property are different, for example the *area* or the *perimeter*. Now you change the reference group enlarging for instance $\mathbb{E}^3$ with the dilatations. At this point all equilateral triangles become equivalent but the area which is not invariant against dilatation is no longer a subject of study in the new geometry.

In this way the notion of *transformation group* became the central notion in the definition of geometry and a unified viewpoint was established that could encompass all possible geometries. Not only that. Geometries could now be organized into a hierarchy. If the group $\mathscr{G}_B$ of a geometry B was a subgroup $\mathscr{G}_B \subset \mathscr{G}_A$ of the group of another geometry A, then all invariant properties that are the object of study in geometry A are invariant also in geometry B and pertain to it. Yet, since the group $\mathscr{G}_B$ is smaller, there are typically further invariant properties with respect to it that are object of study in geometry B in addition to those of geometry A.

It is obvious that the discussions hold in Paris, on the verge of the franco-prussian war, by Klein and Lie with Jordan and Darboux had a great influence in bringing the concept of *group* to the forefront in the conceptual elaborations of both the German and the Norwegian mathematician and, through them, on the subsequent evolution of Mathematics and Physics. It is historically quite significant that Bianchi and Ricci-Curbastro were in Germany in those years and participated in the seminars and lectures organized by Klein. The crucial contribution of these two Italians to the development of Differential Geometry will be reviewed later on. Einstein's theory of General Relativity needed, in order to be conceived, the geometrical language developed by Bianchi, Ricci and the student of this latter, Levi-Civita. The spring of 1870 in Paris was really a crucial moment in the history of science (Fig. 5.5).

5.2.3 *Lie Discovers Lie Algebras in Christiania*

Once appointed to professorship in Christiania in 1872, Lie started working on partial differential equations. He wrote:

... *the theory of differential equations is the most important discipline in modern mathematics.*

The influence of their *group discussions* in Paris motivated Lie in a direction different from the geometrical one pursued by Klein in Bavaria. After the interactions with Jordan he was under the strong impression of Galois theory about which he had previously heard from Sylow in his student years, without paying too much attention. He wanted to uplift to the level of differential equations what Galois had done for the algebraic ones. In a paper [136] of 1874 he wrote:

How can knowledge of a stability group for a differential equation be utilized towards its integration?

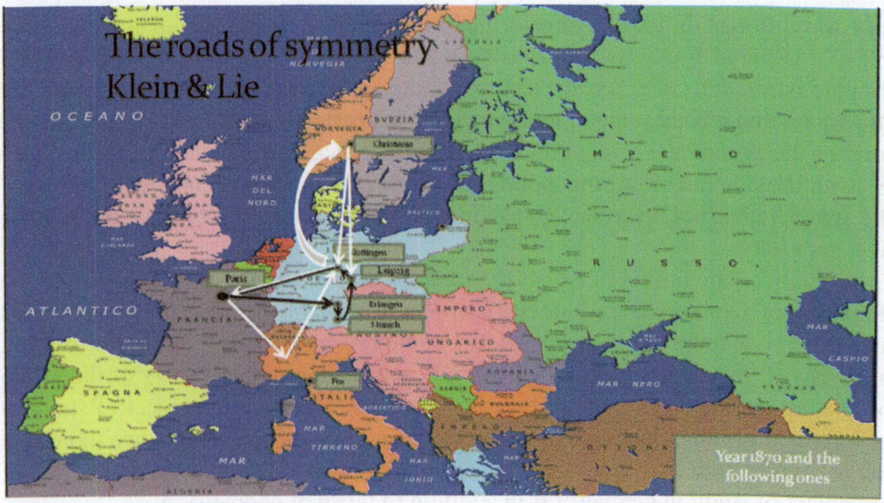

Fig. 5.5 A summary of the trips and chair shifts of Lie and Klein. First Lie went to Gottingen were he met Klein, then the two went to Paris on the eve of the Franco-Prussian war of 1870. Lie went back to Norway through Italy and Germany. Klein moved from Göttingen to Erlangen, then to Münich and after some years to Leipzig. When he was offered a chair in Göttingen and left his Leipzig chair open, this latter was offered to Lie who accepted and stayed there until his last year of life. Finally Klein moved to Berlin were he lived during World War One and died a few years after its end

By stability group of a differential equation it was meant a group of transformations whose effect was that of permuting the solutions of the equations among themselves. In the quoted paper Lie proved a famous theorem that was the beginning of Lie Group Theory. Let us describe this theorem in some detail since it illuminates the logical path that finally lead to Lie's most important discovery. We consider the real plane $\mathbb{R}^2$, whose coordinates we name x, y and the following, generically non linear, first order differential equation:

$$\frac{\mathrm{d}y}{\mathrm{d}x} = \frac{Y(x, y)}{X(x, y)} \tag{5.2.2}$$

where $X(x, y)$,$Y(x, y)$ are two generic functions. Next let us consider a one parameter group of transformations of the plane:

$$\begin{aligned} \forall t \in [0, +\infty] \quad & \phi(t) \, : \mathbb{R}^2 \to \mathbb{R}^2 \\ & \phi\,[t, \{x, y\}] = \{U(t, x, y)\,,\ W(t, x, y)\} \\ & \phi\,[0, \{x, y\}] = \{U(0, x, y)\,,\ W(0, x, y)\} = \{x, y\} \end{aligned} \tag{5.2.3}$$

We can associate a vector field to such one-parameter group, defined as follows:

$$\begin{aligned}
\boldsymbol{\Psi} &= \xi(x,y)\,\frac{\partial}{\partial x} + \eta(x,y)\,\frac{\partial}{\partial y} \\
\xi(x,y) &= \frac{\mathrm{d}U(t,x,y)}{\mathrm{d}t}\,|_{t=0} \\
\eta(x,y) &= \frac{\mathrm{d}U(t,x,y)}{\mathrm{d}t}\,|_{t=0}
\end{aligned} \tag{5.2.4}$$

Next let us introduce the vector field implicit in the differential equation (5.2.2):

$$\mathbf{Z} \equiv X(x,y)\,\frac{\partial}{\partial x} + Y(x,y)\,\frac{\partial}{\partial y} \tag{5.2.5}$$

Lie theorem states:

Theorem 5.2.1 *The transformation group $\phi(t)$ is a stability group for the differential equation (5.2.2), if and only if:*

$$[\,\boldsymbol{\Psi}\,,\,\mathbf{Z}\,] = \lambda\,\mathbf{Z} \tag{5.2.6}$$

where $\lambda = \lambda(x,y)$ is some function.

Next Lie showed how the existence of a stability group generated by the vector field $\boldsymbol{\Psi}$ allowed the construction of an integration factor and the actual integration of certain equations.

Important for us is that starting from this example and always motivated by the theory of differential equations, Lie went on to consider groups of transformation depending not on one-parameter t but on several, say r, and viewed them as acting not on $\mathbb{R}^2$ rather on $\mathbb{R}^n$ with generic n and r. He considered transformations:

$$\begin{aligned}
\phi[t] &: \mathbb{R}^n \to \mathbb{R}^n \\
x_i' &= f_i\left(x_1,\ldots,x_n|t^1,\ldots t^r\right)
\end{aligned} \tag{5.2.7}$$

where the group property was encoded in the requirements:

$$\phi[0] = I \equiv \text{Identity map}; \qquad \phi[s]\circ\phi[t] = \phi[u] \tag{5.2.8}$$

having set:

$$u = u(s,t) = \text{continuous analytic function of its arguments} \tag{5.2.9}$$

Generalizing his previous construction of the vector field $\boldsymbol{\Psi}$, Lie introduced r vector fields:

$$\boldsymbol{\Psi}_\alpha \equiv \sum_{i=1}^{n} \left.\frac{\partial f_i}{\partial t^\alpha}\right|_{t=0} \frac{\partial}{\partial x_i} \tag{5.2.10}$$

and showed that the group property (5.2.7) and (5.2.9) is satisfied if and only if:

$$[\boldsymbol{\Psi}_\alpha\,,\,\boldsymbol{\Psi}_\beta] = \sum_{\gamma=1}^{r} c_{\alpha\beta}^{\ \gamma}\,\boldsymbol{\Psi}_\gamma \tag{5.2.11}$$

where $c_{\alpha\beta}^{\ \gamma}$ are constants (hereafter we shall name them structure constants) that ought to satisfy the following constraints;

$$c_{\alpha\beta}^{\ \gamma} = -\,c_{\beta\alpha}^{\ \gamma} \tag{5.2.12}$$

$$0 = \sum_{\sigma=1}^{r} \left(c_{\alpha\beta}^{\ \sigma}\, c_{\sigma\delta}^{\ \gamma} + c_{\beta\delta}^{\ \sigma}\, c_{\sigma\alpha}^{\ \gamma} + c_{\delta\alpha}^{\ \sigma}\, c_{\sigma\beta}^{\ \gamma} \right) \tag{5.2.13}$$

What Lie had discovered was indeed Lie algebras. A set of transformations of the type (5.2.7) form a group if and only if the induced vector fields (5.2.10) span a Lie algebra.

Let us summarize in few words the notion of what a Lie Algebra is. In Sect. 3.1.13 we discussed the notion of vector spaces (see in particular Eq. (3.1.13) and Fig. 3.5). A Lie algebra $\mathbb{G}$ is a vector space (with complex or real coefficients) that is equipped with an additional internal binary operation, usually named the Lie bracket:

$$[\,,\,] \ : \ \mathbb{G} \times \mathbb{G} \longrightarrow \mathbb{G} \tag{5.2.14}$$

which satisfies the following three axiomatic properties:

1. It is antisymmetric, in the sense that for any two elements A, B of the Lie algebra we have $[A, B] = -\,[B, A]$.
2. It is linear in the sense that for any linear combination $\alpha\ A + \beta\ B$, where α, β are numbers (real or complex) and A, B are elements of $\mathbb{G}$, we have $[\alpha\ A + \beta\ B\,,\ C] = \alpha\ [A\,,\ C]\ +\ \beta\ [B\ C]$, having denoted by C an arbitrary third element of the Lie algebra.
3. For any $A, B, C \in \mathbb{G}$ the Jacobi identity is satisfied:

$$[A\,,\ [B\,,\ C]]\ +\ [B\,,\ [C\,,\ A]] +\ [C\,,\ [A\,,\ B]] = 0 \tag{5.2.15}$$

Hence Lie's problem, namely the classification of all possible transformation groups, which he viewed just as a useful tool to solve differential equations, was reduced to the problem of classifying Lie algebras. The accomplishment of such a classification is essentially the monumental work of Élie Cartan who brought to perfection previous work of Wilhelm Karl Joseph Killing.

5.2.4 Lie and Klein from 1876 to Lie's Death in 1899

The above described results of Lie are without any doubt a fundamental milestone in the history of Mathematics and also of Theoretical Physics, yet in order to appreciate this fact it required some time and a change of perspective. It was necessary to disentangle Lie theory from differential equations and enlighten both its geometrical foundations and its deep geometrical consequences, which eventually produced its entanglement with Physics.

In the first decade after the discovery, very few people took notice of Lie's results and this caused Lie's bitter disappointment. In 1884, writing to Adolf Mayer, Lie wrote:

If only I knew how to get mathematicians interested in transformation groups and their applications to differential equations. I am certain, absolutely certain, that these theories will some time in the future be recognized as fundamental. When I wish such a recognition sooner, it is partly because then I could accomplish ten times more.

By that time, Klein, whose fame had grown wider and wider, had left Münich and, since 1876 he was established in Leipzig University that offered him a prestigious Chair of Geometry. He had not forgotten his good friend Lie and, knowing about his isolation in Norway, organized to send to him his student Friedrich Engel who helped him in the course of nine years, at the beginning in Christiania, then in Leipzig to compile a three volume detailed and exhaustive exposition of the theory of transformation groups that was published in Leipzig from 1888 to 1893 [137–140].

In 1886 Felix Klein changed once again his location accepting the offer of Göttingen University, whose world leadership in Mathematics and Physics Klein strongly helped to further strengthen in particular with the appointment of David Hilbert.

The vacant Chair of Geometry in Leipzig was immediately offered to Lie who accepted and lived in Germany for twelve years up to 1898. Since 1890 he started suffering from a progressive illness, *anaemia perniciosa*, that deteriorated his health conditions steadily. In 1898, already very much physically proved he accepted a honorary professorship in Christiania and one year after his return in his home country he died.

In the last years of Lie's life his old friendship with Klein broke down and in 1892 Lie publicly attacked Klein writing:

I am no pupil of Klein, nor is the opposite the case, although this might be closer to the truth.

The cause of this sudden outbreak of enmity was attributed to Lie's mental instability caused by his illness and Klein defended his friend's behavior remarking that there is sometime a relation between being a genius and having some touches of madness. Recent biographical studies have found some possible basis for Lie's behavior in his dissatisfaction with Klein's historical reconstruction of the conceptual process that lead from Paris Spring to the Erlangen Programme. Some biographers maintain that Klein burnt all letters received from Lie up to 1877. We do not enter this controversial matter and we just remark that crediting has always been a source of bitterness for science at all times. Whatever was the case of their controversy Klein and Lie are two giants whose contributions to the development of Mathematics and eventually of Physics have few rivals in history.

5.3 The Tale of Lie Algebras Takes the Lead

In the present section the tale of Lie algebras continues.

Since the aim of this essay is to present both the logical development of the mathematical theories inherent to Symmetry and, hence, to the Modern Picture of the Physical World, together with a historical account of how these theories were born, our tale cannot proceed all the time in chronological order.

Indeed the history of ideas, just as the general political history is not a sequence of logical steps, rather it proceeds along capricious paths, notwithstanding the fact that, on the very long time scale, some rationally understandable order can be detected, but always a posteriori.

In the previous section we told the complicated tale of how the very notion of Lie groups came into being in Lie's studies of differential equations, motivated by Lie's desire to classify all possible *transformation groups*. Once the relation between *transformation groups* and Lie algebras was established, the logical task was that of classifying all Lie algebras.

Just as for finite groups the classification of all of them is a too much ambitious programme which, upon restriction to the *simple groups*, becomes a difficult yet doable task, in the same way the classification of all Lie algebras is unattainable, while that of the *simple Lie algebras* turns out to be possible and exhaustive. Historically it was accomplished by Killing and Cartan and it will be described in full in the next sections.

The philosophical question which is immediately raised in connection with this matter of fact is that the restriction to simple algebras (what simple means we will shortly explain) should have a logical basis and should avoid the criticism that we just classify what we are able to.

In the case of finite groups we advocated the viewpoint that *simple groups* are the really fundamental building blocks of group theory, while solvable groups are in some sense trivial structures that can be disassembled into smaller and smaller blocks until one reaches the abelian cyclic groups. The main point, however, is that any finite group is a semidirect product or, more generally, a splitting extension of the tensor product of simple groups with some solvable one. Is something similar true for Lie algebras as well? Indeed it is!

The basic result in connection with this issue is encoded in the *Levi decomposition theorem*: it states that the most general Lie algebra is the semidirect product of a *semisimple algebra* with a *solvable ideal.*

5.3.1 Levi's Theorem

Consider a Lie algebra $\mathbb{G}$ and define:

$$\mathscr{D}\mathbb{G} = [\mathbb{G}\,,\,\mathbb{G}] \tag{5.3.1}$$

the set of all elements $g \in \mathbb{G}$ that can be written as the Lie bracket of two other elements $g = [g_1\,,\,g_2]$. Clearly $\mathscr{D}\mathbb{G}$ is an *ideal* in $\mathbb{G}$. By definition an ideal $\mathbb{I} \subset \mathbb{G}$ of a Lie algebra is a vector subspace of $\mathbb{G}$, such that if a lies in $\mathbb{I}$ and $g \in \mathbb{G}$ is any element of the full algebra, than the Lie bracket of a with g lies in the the ideal $[a\,,\,g] \in \mathbb{I}$.

It is important to remark that the notion of ideal is the analogue for Lie algebras of the notion of normal subgroup for groups.

Consider next the sequence of ideals $\mathscr{D}^n\,\mathbb{G} = \left[\mathscr{D}^{n-1}\mathbb{G}\,,\,\mathscr{D}^{n-1}\mathbb{G}\right]$:

$$G \supset \mathscr{D}\mathbb{G} \supset \mathscr{D}^2\mathbb{G} \supset \cdots \supset \mathscr{D}^n\mathbb{G} \tag{5.3.2}$$

which is named the *derivative series* of the Lie algebra. It is the analogue for Lie algebras $\mathbb{G}$ of the subnormal series (3.3.3) for groups.

A Lie algebra $\mathbb{G}$ is named *solvable* if there exists an integer $n \in \mathbb{N}$ such that

$$\mathscr{D}^n\mathbb{G} = \{0\} \tag{5.3.3}$$

Let $\mathbb{G}$ be a Lie algebra. An ideal $\mathbb{H} \subset \mathbb{G}$ is named *maximal* if there is no other ideal $\mathbb{H}' \subset \mathbb{G}$ such that $\mathbb{H}' \supset \mathbb{H}$ except $\mathbb{H}$ itself.

The maximal solvable ideal of a Lie algebra $\mathbb{G}$ is named the radical of $\mathbb{G}$ and it is denoted Rad$\mathbb{G}$. A Lie algebra $\mathbb{G}$ is named semisimple if and only if Rad$\mathbb{G} = 0$. This is just the analogue of the definition of simple groups. We said that a group is simple if it admits no non trivial normal subgroup. Indeed we can equivalently say that a Lie algebra $\mathbb{G}$ is simple if its only ideals are $\mathbb{G}$ and $\mathbf{0}$.

It follows that any semisimple Lie algebra is the direct product of a sequence of simple Lie algebras.

Levi's fundamental theorem states that every Lie algebra $\mathbb{G}$ is the semidirect sum of a (semi)-simple Lie algebra $\mathbb{L}$ with a solvable one (the radical of $\mathbb{G}$).

Levi's theorem dates 1905 when Killing was absorbed in different deals and Cartan's doctoral dissertation was ten years old. Yet as a logical step Levi's theorem comes before and for this reason we present it here.

In a nutshell solvable Lie algebras, for which an exhaustive classification does not exist, are the trivial part, in the sense that all their linear representations are given by triangular matrices. On the contrary, for the semisimple algebras their exhaustive classification is provided by the formalism of roots and Dynkin diagrams whose story we outline in the next section.

Solvable Lie algebras are anyhow important also in the context of differential geometry. Solvable Lie algebras provide an efficient and privileged way of encoding the local geometry of non-compact homogeneous spaces.

5.3.2 Who Was Levi?

Addressing the issue of Eugenio Elia Levi's biography a lot of emotions are immediately raised. The most prominently emotional aspect of his life is just his tragic death in battle during World War One, a destiny that he shared with one of his brothers, Decio Valerio Levi. The emotions stirred by a fate that deprived the history of Mathematics of a surely talented actor, who perished definitely too young, are strongly reinforced from knowing that he participated voluntarily to the conflict having to obtain special permissions from the Rector of his University and from the Ministry to breach regulations that demanded him to remain safely home continuing his teaching and researches (Fig. 5.6).

Hence in order to give the fundamental theorem we have presented its proper historical perspective, we have to explain the environment where its discoverer grew up and we cannot avoid knitting our history of some profound mathematical ideas into the texture of general history at large.

We emphasized that the beginning of Lie group theory could be traced back to the Spring of 1870 on the verge of the franco-prussian war. The time separation between the fall of France under the strikes of Prussian Cannons in 1870 and the outbreak of World War One is only 45 years, usually considerably less than one's life span. Indeed Felix Klein was still alive during World War One and he had the time to see the German Empire defeated, the Kaiser fugitive to Holland and the advent of a new, much more democratic, course in his own country, that unfortunately lasted too shortly.

The blows received by France in 1870 were not easily forgotten and the desire of revenge continued to be nourished in several layers of French Society and Culture. The victory in World War One was such an occasion and this, quite unfortunately, contributed to create the basis of World War Two.

The Italian Kingdom, on the other hand, from 1870 to World War One continued in the process of strengthening its internal structure, the main issue being, after the creation of Italy, that of creating the Italians, as Massimo D'Azeglio said. Indeed the differences in civil development, in habits and in fundamental attitudes towards life were so gigantic among the various Regions of Italy that trying to homogenize them required an equally gigantic effort. As the *Risorgimento*, namely the movement towards italian unification, had mainly been a matter of intellectual èlites with lim-

Fig. 5.6 Eugenio Elia Levi (1883–1917) was born in Torino where he completed his secondary school studies. Then he won the national competition for selection and entered the University of Pisa and the Scuola Normale Superiore. He worked in Pisa with Bianchi and Dini and later he became Full Professor of Infinitesimal Analysis at the University of Genova. He participated as a volunteer to World War One and was killed in battle in 1917 after the Caporetto Debacle. One of his brothers was the well-known mathematician Beppo Levi

ited involvement of popular masses, also the first 45 years of united italian history continued to be marked by the same spirit who had in the liberal, anticlerical, socially advanced Piedmont a strong center.

The Kingdom of Sardinia had been the first, in Italy, and among the forerunners, in Europe, to grant to the citizens of Jewish origin complete parity, dignity and freedom of enterprise, opening also the doors of public administration to them. Furthermore the laic, anticlerical foundations of Piedmont ruling classes, that were inherited by the Kingdom of Italy, were particularly appreciated by the Piedmontese quite flourishing Jewish Community which was equally laic, well-to-do and had a strong orientation towards intellectual professions, science and mathematics in particular.

Eugenio Elia Levi was an offspring of that community.

He was born October 1883, in the *Crocetta* neighborhood of Torino, up to these days the living location of the best and richest piedmontese bourgeoisie. His father was a famous lawyer, with clearcut liberal ideas, who felt a strong commitment to the Risorgimento and to the Italian Fatherland of which he just felt a citizen, giving no relevance to religious differences. This spirit he rightly communicated to his ten children.

The University of Torino founded as a studium in 1404, under the initiative of Prince Ludovico di Savoia, has a long noble history in all fields of Natural and Human Sciences. At the time of Levi's youth, Torino University had become the center of the celebrated Italian School of Algebraic Geometry founded by Enrico D'Ovidio (also Rector of Torino University from 1880 to 1885). Professors of Torino

University were such outstanding mathematicians as Giuseppe Peano, Corrado Segre, Alessandro Terracini, Gino Fano. Due to the fame of the Torino School, other world-wide famous mathematicians like Guido Castelnuovo, Federico Amodeo, Federigo Enriques, came to Torino to accomplish their advanced education. With the exception of Giuseppe Peano, most of these mathematicians were of Jewish origin and this shows how much the environment of Eugenio Elia was intruded of mathematical thought. In his *Principia Mathematicae* Bertrand Russel openly stated that Giuseppe Peano was the greatest living mathematician.

Yet Eugenio Elia Levi did not attend Torino University. After his secondary studies in the still existing *Liceo Classico D'Azeglio*, where he was such an excellent student that he was granted his *baccalaureate cum laude* and without examination, he won the national competition to enter the *Regia Scuola Normale Superiore* of Pisa and in 1899, the year when Lie died, he enrolled in the Faculty of Mathematics of Pisa University.

In Scuola Normale he studied with Luigi Bianchi and through the strong influence of this latter he became involved with differential geometry, with analytic geometry and with theory of analytic functions. It is quite obvious that the spirit of Felix Klein and Lie, shared by Bianchi during his youth stages in Münich, came down to Levi and played some role in orienting his interest to transformation groups and Lie algebras.

Indeed in 1905, Eugenio Levi graduated from Pisa University and the same year he published the paper *Sulla struttura dei gruppi finiti e continui* (On the structure of finite and continuous groups) [134] that contains the theorem we presented in the previous section.

The subsequent career of the young Levi was fast and brilliant as its beginning promised. In the ten years between 1905 and 1915 he published 34 research papers in quite different directions of mathematics: about group theory, about automorphic functions, about differential equations, in particular the heat equation, on the calculus of variations, on multiple integrals, to mention some of them. He worked as an Assistant in Pisa until 1909 when he was included among the three winners of a competition for a Chair at the University of Messina. Consequently in the same year he was called as *Professore Straordinario* by the University of Genova of which he became *Professore Ordinario* (Full Professor) in 1912.

Having been dispensed from military service in his young age on the ground of his short stature and being a Full Professor of a Royal University, he had no obligation to serve in the Army when Italy entered World War One against Austria and Germany. Actually he could not leave his University without special permissions from the Rector and from the Ministry of Education.

Why did he ask for such permissions and insisted to be sent to the front, where in 1917 he would be killed by an Austrian bullet?

We believe that the reason must be looked for in the *risorgimental spirit* inherited from his father and in his loyalty to the Kingdom of Italy where the Jews like he was could be ordinary citizens as everyone else, also civil servants and scientists. The Austrian Empire was the historical enemy of that Kingdom which, in order to be born, had to fight battles against it and its clerical obsolete clients. Last but not least came, as a possible motivation of Levi's commitment to war the so called *Manifesto of the*

Ninety-Three a declaration signed in 1914 by 93 german intellectual and scientists, among which also Max Planck who later withdrew his signature, where German militarism was justified, the war crimes committed by the German Army in Belgium, whose neutrality had been violated, were denied and German Imperialism was praised and dubbed a *defense of european culture* in the name of Goethe, Beethoven and Kant. For a mathematician who certainly admired German mathematics, who was certainly thinking of Germany as the homeland of Gauss, Riemann, Klein, Weierstrass and the other giants of the XIXth century, this Manifesto was a disgusting treason, an unacceptable offense to truth and justice. So he wanted to fight such barbarians and in 1917 he fell when the Italian Army was routed at Caporetto.

It is bitter to remark that the main offspring of Italy's victory in World War One, for whose sake Eugenio Levi donated his own life, was the upraise of Fascism. In 1938 the Fascist Government promulgated the Racist Laws against the Jews and among several others the outstanding mathematician Beppo Levi, Eugenio's beloved brother was thrown away from his University Chair and had to emigrate to Argentina.

Having established how semisimple Lie algebras were singled out by Levi to be the hard-core of Lie Algebra theory being themselves simply the tensor product of simple Lie algebras let us turn to the story of how these latter were classified and constructed.

5.4 Killing and Cartan

In this section we step back to the XIX century and we tell the story how it happened that simple Lie algebras were classified. As it often happens in mathematics, the task was achieved by mapping the problem under consideration into another equivalent one that pertains to a different mathematical theory. In this case the equivalent problem pertains to euclidian geometry and consists of the classification of certain finite collections of vectors named **root systems**.

The classification of root systems has found several other interpretations and appears to be a deep fundamental structure of mathematics. One of these interpretations we have already met: it is the classification of finite rotation groups in three dimensions, namely Plato's problem.

The story of how the notion of root systems was found and how simple Lie algebras were classified is quite interesting and has two main actors: Wilhelm Karl Joseph Killing (see Fig. 5.7) and Élie Cartan (see Fig. 5.9).

The latter is one of the greatest mathematicians of the XXth century. His name has already occurred few times in previous chapters and will be extensively mentioned in subsequent ones. The former is instead a very remarkable figure of a school-teacher who, notwithstanding his isolation from the academic world and from the currents of frontier-research, being moreover overburden with daily tuition and administrative duties, nevertheless succeeded in making a very original mathematical discovery of

Fig. 5.7 Wilhelm Karl Joseph Killing (1847–1923). Born in a small center near Siegen, he died in Muenster. He studied first in Muenster then in Berlin and wrote his doctoral thesis under the supervision of Weierstrass. The rest of his life he was a secondary school teacher or the Principal of some school. Notwithstanding his essential isolation from the world of frontier research he made remarkable mathematical discoveries. In particular he classified simple Lie algebras and discovered also the exceptional ones prior to Cartan. He was a fervent catholic and in the last part of his life was blown by a great tragedy: the loss of his two sons during World War One

the first class, although he had neither the time nor the strength to complete all the detailed proofs and the explicit constructions required by mathematical rigour.

WILHELM KILLING

Wilhelm Killing was born in a small city, Burbach, few kilometers south of Siegen in catholic Westfalia. The education which he received both in his family and in the primary school he attended made him a fervent catholic and such he remained the whole of his life. Almost forty of age, together with his wife he even entered the laic order of franciscan tertiaries.

In Wilhelm's youth the Killing family moved around considerably in the territory of Westfalia since his father served, at different times, as mayor of different small cities of the region. The secondary education of Killing was in the Gymnasium of Brilon where he received an excellent classical education, he learnt Latin, Greek and even Hebrew, but very poor mathematics. Yet he was very much attracted by that subject and he studied it on his own, reading books of Gauss and Plücker.

After graduation from secondary school, he studied for sometime in the Royal Academy of Münster where, once again, he was dissatisfied with the level of mathematical tuition. Finally he went to Berlin where he received instead a first class scientific education by such top level mathematicians and physicists as Kummer,

Weierstrass and Helmholtz. He received his doctoral degree in 1872 (the same year as Lie did and the same year of Klein's Erlangen lecture) defending a thesis on Bundles of surfaces of the second degree (*Der Flächenbüschel zweiter Ordnung*) written under the supervision of Weierstrass.

Having become, soon after, a secondary school teacher of mathematics, but also of Greek and Latin, notwithstanding his teaching duties, he found the time to publish a few papers on curved surfaces and non-euclidian geometry. Then with a recommendation of Weierstrass, in 1882 he was appointed to a Chair of Mathematics in the *Lyceum Hosianum* located in the distant Eastern Prussia city of Braunsberg (the present name is Braniewo, a small centre of northeastern Poland almost on the border with the Russian enclave of Kaliningrad, former Königsberg).

It was there that, in complete isolation from the mathematical academic community and ignoring the work of Lie, Wilhelm Killing independently discovered Lie algebras. The idea of this mathematical structure was described in 1884 in a small booklet published by his own Lyceum with the humble name of *Programmschrift* (Notes for a Programme).

In a vein similar to that of Klein in the Erlangen Programme, Killing wanted to classify possible geometries by classifying the infinitesimal motions under which the objects of study in each geometry should be invariant. That idea of considering infinitesimal, rather than finite motions brought him directly to the notion of Lie algebra, bypassing the stage of transformation groups which was the main concern in Lie's work.

Examining his work from a modern stand point it appears that, already in the *Programmschrift*, Killing had singled out the notion of simple Lie algebra (one which has no solvable ideal) and had already formulated a strategy how to classify such algebras, yet, distracted by his original motivation of geometries, so far he had not undertaken that classification.

Right after its publication, Killing sent his *Programschrift* to Klein who immediately replied, saying that what he saw in the booklet was closely related to the structures considered by Sophus Lie, published in a series of papers that had appeared over the last ten years. Hence in August 1884 Killing forwarded his own booklet to Lie. He waited more than a year for an answer never getting it. So in October 1885 he wrote once again to Lie, requesting a copy of his papers. Lie send them to Killing under the condition that after reading he should return them to the author, which Killing did in March 1886.

Informed by Klein that his student Engel was working with Lie in Norway, Killing wrote also to this latter who, differently from his bad-tempered master, replied to Killing enthusiastically. A correspondence started between the two that was beneficial to both and encouraged Killing to push forward his researches on Lie algebras.

As we know, in 1886 Klein shifted from Leipzig to Göttingen and Lie was appointed on the Leipzig Chair of Geometry, vacant after Klein's departure. This provided an opportunity for Killing to visit both Lie and Engel in Leipzig on his way from Braunsberg to Heidelberg, whereto he was called for matters connected with his Lyceum.

The bad-tempered Lie, always very jealous of his own results and obsessed with the idea of getting insufficient recognition for his work was ill–disposed towards this humble school teacher, coming from nowhere in the far east and claiming to have independently obtained Lie algebras. The meeting was a complete failure and Killing continued his journey, remaining however on good terms with Engel.

Next year, in April 1887, Killing wrote to Engel that he had perfectioned the definition of semisimple Lie algebras and by October 18th he wrote to him again, announcing that he had found the complete list of the simple ones, any semisimple Lie algebra being a tensor sum of the latter. All of Killing's results were published between 1888 and 1889 on the prestigious journal *Mathematische Annalen* founded by Klein [125–128].

In this short series of remarkable papers, Killing invented or discovered (depending on the case) all the following items, namely:

1. The notion of Cartan subalgebra
2. The notion of root system
3. The notion of simple roots and Cartan matrix
4. The list of simple Lie algebra that might exist, including the exceptional ones, G_2, F_4, E_6, E_7, E_8

While he correctly identified the Lie algebras $A_\ell, B_\ell, C_\ell, D_\ell$ with the known classical matrix algebras, he lacked an interpretation for those associated with the exceptional root systems mentioned above. Hence, from the mathematical point of view he had only proven the possible existence, not the very existence of the exceptional Lie algebras, which required an explicit representation. Indeed as Sigurdur Helgason remarked in [113]:

The exceptional simple Lie algebras are the subject of the final paragraph 18 in Killing's paper. This is certainly his most remarkable discovery, although these algebras appeared to him at first as a kind of nuisance, which he tried to eliminate. Even Lie, who was generally critical of Killing's work, expressed in letters to Felix Klein his admiration for such a result [111].

This shows, notwithstanding his typical bad-temper the profound intellectual honesty of Sophus Lie and it is a big praise for him to have written such words to Klein (Fig. 5.8).

For the rest of his life-time, that extended until 1923, Killing was absorbed by teaching, administration and charitable work. In 1897–1898 he also temporarily served as Rector of the newly created University of Münster to which he had returned after his productive exile in the tedious foggy lands of Eastern Prussia.

It was in 1894 the turn of Cartan to continue to marvelous tale of Lie algebras.

ÉLIE CARTAN

The origins of Cartan (see Fig. 5.9) were very humble just as those of the King of Mathematicians and those of Riemann. The environment where Cartan was born was even poorer and deprived of any cultural background, since his father was a plain blacksmith in the mountain village of Dolomieu in Haute Savoie. Gauss could study thanks to the generosity of the Duke of Brunswick, Cartan obtained the very best scientific education available at the time thanks to state stipends that the French

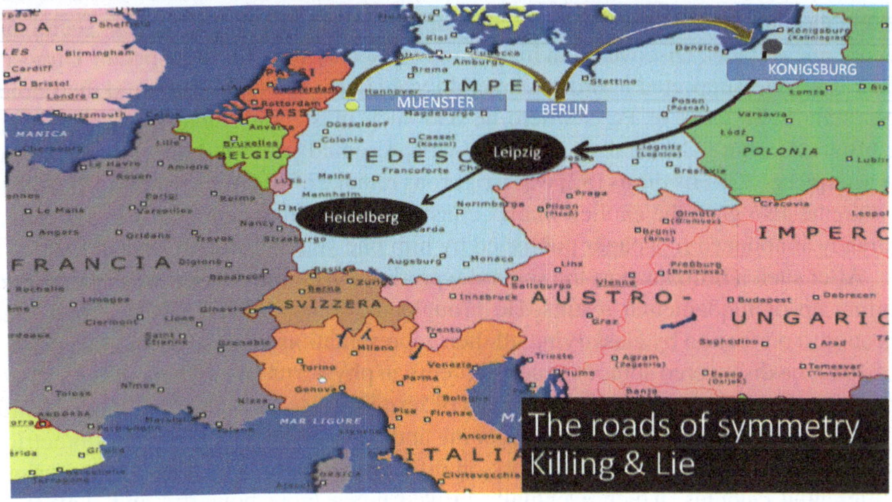

Fig. 5.8 The main trips of Wilhelm Killing. From the native Westfalia he moved to Berlin to study mathematics under the supervision of Weierstrass, then to Eastern Prussia where he became the Principal of a secondary school. He had a very strong classical culture and he was at some point also a teacher of Latin and Greek. The most momentous trip of Killing was that from Eastern Prussia to Heidelberg crossing through Leipzig where he met with Lie and Engel, remaining the rest of his life in good friendly terms with the latter. Stimulated by Engel, Killing published his extraordinary papers on the classification of simple Lie algebras based on the formalism of roots of which he is the inventor

Fig. 5.9 Élie Cartan (1869–1951). A true giant of XXth century mathematical thought. He completed the theory of Lie Algebras, invented exterior differential calculus, invented the theory of symmetric spaces, introduced the notion of mobile frames, reformulated the theory of General Relativity, discovered spinors much earlier than physicists. He gave fundamental results in the theory of differential equations and essentially invented the concept of fibre-bundles

Republic had introduced for talented people, independently from their social or economical status. Discovered in his remote village by the school inspector Dubost, Élie was state-supported in order to attend Lycée in Lyon and then entered the École Normale Superiéure of Paris where he had such masters as Picard, Darboux and Hermite. Cartan's doctoral dissertation was presented in 1894 and was already a masterpiece [32]. His thesis [32] was a rigorous remake of Killing's papers [125–128] where he also gave the explicit matrix construction of all exceptional Lie algebras, already announced in a paper published by him one year before in German [31].

After such a brilliant start the scientific production of Élie Cartan was an endless piling up of fundamental results. He brought the theory of Lie algebras and Lie groups to perfection by classifying all their representations and, so doing, in a paper of 1913, he discovered spinors, much earlier than physicists found them necessary to describe the intrinsic angular momentum of fermions. He combined Lie theory with differential geometry, founding, developing and completing the theory of symmetric spaces. Extending his very early work on differential forms, he created the exterior differential calculus, which eventually proved much more powerful and synthetic than the tensor calculus of Ricci and Levi-Civita. Right after the creation of General Relativity by Einstein, Cartan started rethinking it in terms of *mobile frames* and came to the reformulation of gravitational equations which goes under the name of Einstein–Cartan theory [34].

When Élie Cartan died in 1951, two great mathematicians, Shiing-Shen Chern and Claude Chevalley joined to write an obituary that is also an impressive summary of Cartan's mathematical work. They said: *His death came at a time when his reputation and the influence of his ideas were in full ascent. Undoubtedly one of the greatest mathematician of this century, his career was characterized by a rare harmony of genius and modesty*. They also said:*Closely interwoven with Cartan's life as a scientist and teacher has been his family life, which was filled with an atmosphere of happiness and serenity. He had four children, three sons, Henri, Jean, Louis and a daughter, Hélène*. Yet fate was quite cruel at least with two of them. Jean Cartan who studied music and was very early recognized as a prominent and talented composer was stolen to his family by premature death from an incurable illness. Louis Cartan, who was a physicist, joined the Resistance during the German occupation of France and, captured by the Nazis, was beheaded in 1943. His father suspected the terrible truth but learnt about it only in 1945: a deadly blow from which he never recovered. Henri Cartan followed his father steps and became a very prominent mathematician.

5.4.1 The General Form of a Simple Lie Algebra and the Root Systems

The final result of the historical process we have reconstructed in the previous sections is that every simple Lie algebra $\mathbb{G}$ of dimension $n = 2m + \ell$ can be described in a compact and quite inspiring way, in terms of certain systems of vectors in an ℓ dimensional Euclidian space that are named the *roots* and were invented by Killing. As anticipated above, the problem of classifying Lie algebras is mapped into another

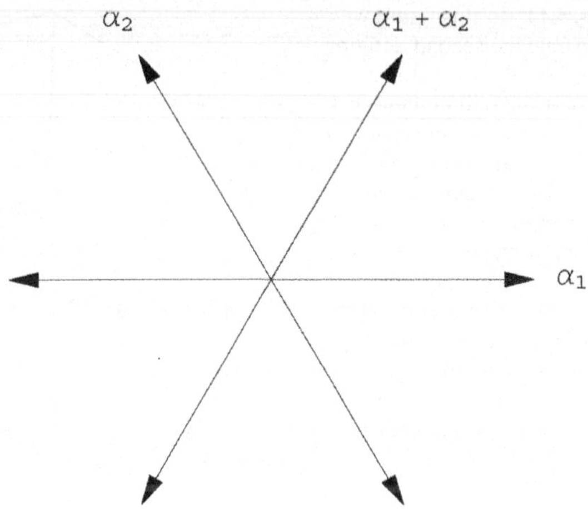

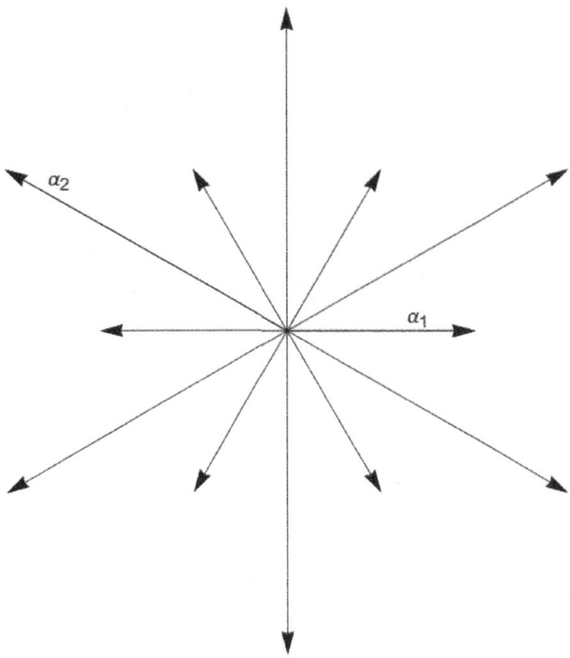

Fig. 5.10 The root systems of the $\mathfrak{a}_2$ (above) and of the $\mathfrak{g}_2$ (below) Lie algebras. In both pictures the vectors denoted $\alpha_{1,2}$ are the simple roots. That the remaining ones are linear combinations with integer coefficients (all positive or all negative) of the simple ones is something that the reader can graphically verify in an elementary way

mathematical problem which is just a problem of elementary euclidean geometry in dimension ℓ.

Let us become acquainted with these geometrical objects that, for $\ell = 2, 3$, we can even visualize in pictures. Consider for instance the plane systems of vectors shown in Fig. 5.10. They are the root systems of two of the three rank two Lie algebras the rank being the aforementioned number ℓ whose meaning we will explain below. Consider also the system of vectors displayed in Fig. 5.12 which is the root system of

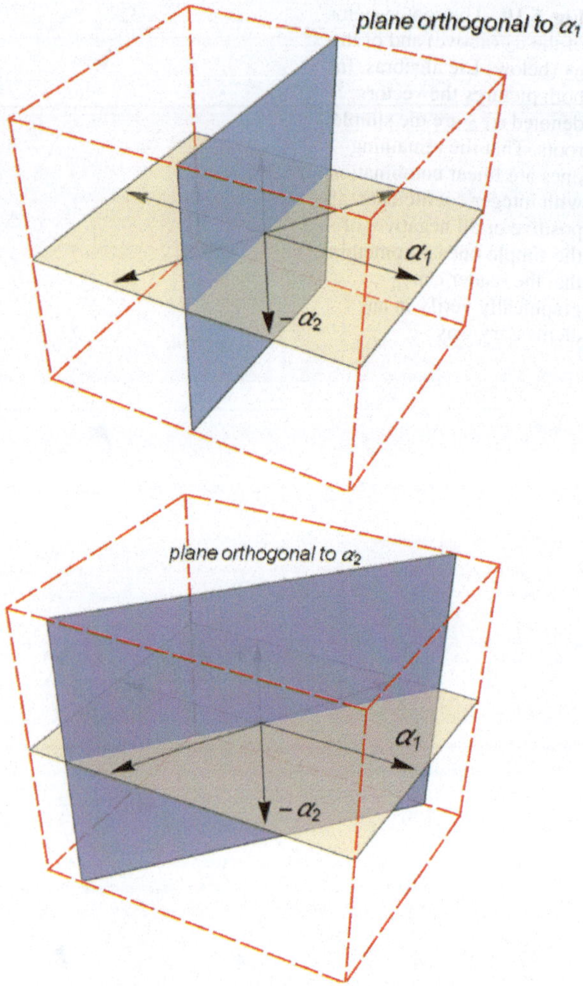

Fig. 5.11 In this figure we illustrate the second defining property of root systems, namely the requirement that the reflection of any root with respect to any other must be a root of the same system. We consider the planar case $\mathfrak{a}_2$ and we display the two planes orthogonal to the root α_1 and α_2, respectively. The mirror image of the roots that are located on one side of the planes are just the remaining roots located on the other side

the $\mathfrak{d}_3$ Lie algebra. What are the very special properties of these sets of vectors that qualify them as a *root system* Δ? We can determine such properties almost visually from the given pictures (Fig. 5.11).

(a) Given any pair of vectors α and β that belong to the root system and considering scalar products in the ordinary euclidean space $\mathbb{R}^r$, we have that the following ratio $2\,\frac{(\alpha\,,\beta)}{(\alpha\,,\alpha)}$ must be an integer number, namely:

$$2\,\frac{(\alpha\,,\,\beta)}{(\alpha\,,\,\alpha)} \in \mathbb{Z}$$

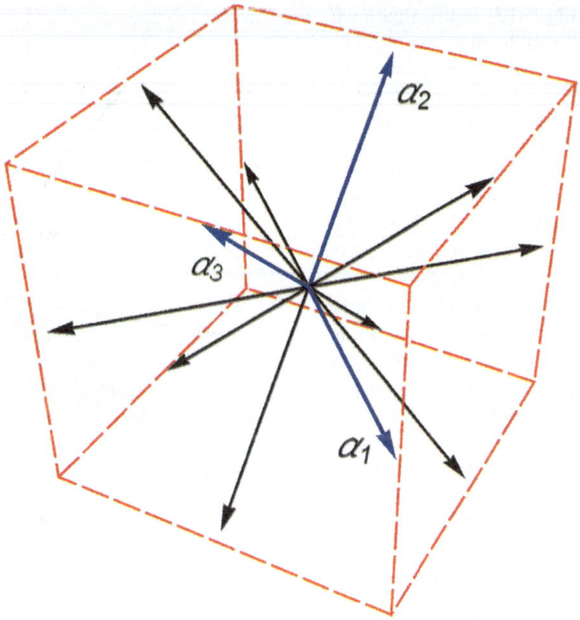

Fig. 5.12 The root systems of the $\mathfrak{d}_3$ Lie algebra. It contains 12 roots that are vectors in ordinary three-dimensional space

(b) Given any root α belonging to the root system Δ and considering the hyperplane Hyp_α orthogonal to it, the reflection of any root with respect to such an hyperplane is again a root of the same system. This property is illustrated for the planar case $\mathfrak{a}_2$ in Fig. 5.11 and it is illustrated for the three-dimensional case $\mathfrak{d}_3$ in Fig. 5.13. Mathematically the described property is formulated as follows:

$$\forall \alpha, \beta \in \Delta \quad : \quad \sigma_\alpha(\beta) \equiv \beta - 2\alpha \frac{(\alpha\,,\,\beta)}{(\alpha\,,\alpha)} \in \Delta$$

(c) In every root system Δ of rank ℓ there exists a subset $\alpha_1, \alpha_2, \ldots, \alpha_\ell$ of linearly independent roots, named the *simple roots*, that form a basis for $\mathbb{R}^\ell$ and are such that every other root $\beta = \sum_{i=1}^{\ell} n^i \alpha_i$ is a linear combination of these simple ones with integer valued coefficients n_i that are either all positive or all negative. Actually this property is not independent from the previous two, rather it can be demonstrated as a theorem starting from (a) and (b) as axioms.

What is the relation of these peculiar systems of vectors with Lie algebras? The answer is simple and it encodes the main discoveries of Killing, revisited by Cartan. Any complex simple Lie algebra $\mathbb{G}$ contains a maximal abelian subalgebra (named the Cartan subalgebra CSA) made of elements whose adjoint action on G is fully diagonalizable and whose dimension $\ell < n$ is indeed named the *rank* of $\mathbb{G}$. A basis of generators spanning the CSA is usually denoted by H_i $(i =, \ldots, \ell)$. The remaining $2m$ generators, denoted E^α are in one-to-one correspondence with a set of vectors α living in an ℓ-dimensional euclidian space and forming a root space according with the above discussion.

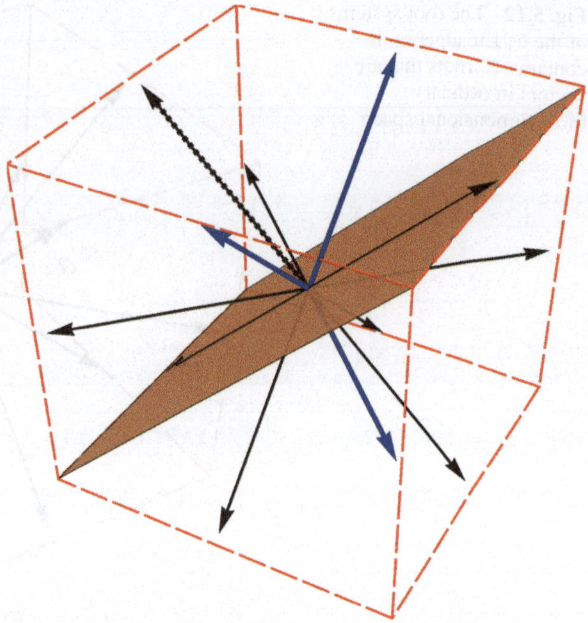

Fig. 5.13 In this figure we illustrate, in a case of rank $\ell = 3$ the second defining property of root systems, namely the requirement that the reflection of any root with respect to any other must be a root of the same system. We consider the $\mathfrak{d}_3$ root system and we draw the plane orthogonal to a particular root that is marked in a special way by dashing. The 12 roots split into three sets. Two roots lie on the plane and hence are self-mirror, one set of five roots lie on one side of the plane, the remaining five lie on the opposite side and are the mirror images of the former

Utilizing these notations and the advocated notion of root system the commutation relations of a *complex simple Lie algebra* take necessarily the following general form:

$$
\begin{aligned}
\left[H_i \,,\, H_j\right] &= 0 \\
\left[H_i \,,\, E^\alpha\right] &= \alpha_i \, E^\alpha \\
\left[E^\alpha \,,\, E^{-\alpha}\right] &= \alpha^i \, H_i \\
\left[E^\alpha \,,\, E^\beta\right] &= N(\alpha, \beta) \, E^{\alpha+\beta} \qquad \text{if} \quad \alpha + \beta \in \Delta \\
\left[E^\alpha \,,\, E^\beta\right] &= 0 \qquad \text{if} \quad \alpha + \beta \notin \Delta
\end{aligned}
\tag{5.4.1}
$$

where $N(\alpha, \beta)$ is a coefficient that has to be determined using Jacobi identities.

From now on we can associate to every complex simple Lie algebra its root system Δ. Furthermore each root system singles out a well-defined finite group, named the *Weyl group* that is obtained combining together the reflections with respect to all the roots.

5.5 Dynkin and Coxeter and the Classification of Root Systems

The main token in the classification of root systems is provided by the Cartan matrices, which we presently define. Since, as we already stated, every root system possesses a simple root basis $\alpha_1, \ldots, \alpha_\ell$ it follows that to every root system and hence to every complex Lie algebra we can associated the following $\ell \times \ell$ matrix, named the Cartan matrix:

$$C_{ij} =< \alpha_i \, , \, \alpha_j >\equiv 2 \frac{(\alpha_i \, , \, \alpha_j)}{(\alpha_j \, , \, \alpha_j)} \tag{5.5.2}$$

A simple and constructive theorem shows that from the Cartan matrix one can retrieve the entire root system and hence the simple Lie algebra.

Having established that all possible irreducible root systems Δ are uniquely determined (up to isomorphisms) by the Cartan matrix, we can classify *all the complex simple Lie algebras* by classifying all possible Cartan matrices.

This is the classification originally achieved by Killing and Cartan. Later in the XXth century the theory of Cartan matrices of root systems and of the finite reflection groups associated with them was extensively developed by three mathematicians Hermann Weyl, Harold Coxeter and Evgenij Dynkin. The next chapter is entirely devoted to Weyl and to his fundamental contributions not only to representation theory but in general to modern mathematical thought. Here we discuss both the life and the contributions of Coxeter and Dynkin who invented an extremely useful graphical representation for Cartan matrices that goes under the name of Coxeter–Dynkin diagrams.

COXETER AND DYNKIN

The two mathematicians whose name is associated with two variants of such diagrams are shown in Fig. 5.14. Both of them had a long life, beginning on the European Continent and ending in the New World, on the North American Continent. Both were very much talented and obtained a lot of recognitions during their long career, yet the character of their life was very different because of the difference of their country of origin. Born in London in 1907, the long life of Donald Coxeter, as he was usually named by all his friends, relatives and colleagues, was healthy, quiet and serene, although animated by his strong geometrical and artistic creativity. On the contrary Dynkin's life started in Sankt Peterburg, already renamed Leningrad, in the troubled times after the Bolshevik Revolution, was early marked by the ominous shadow of Stalin's purges and continued to be difficult and insecure until his final emigration to the United States in 1976.

In 1935, when Eugene was eleven of age, he and his family, of Jewish origin, were forcefully expelled from their native town and were exiled to a small city in Kazachstan. Two years later, in 1937, Eugene's father, Boris Dynkin, who in previous times had been a well-to-do lawyer, was arrested without any concrete charge and declared to be a People's Enemy. Disappeared in a Gulag, he was probably executed there, one among the several thousands that in all corners of the Soviet Union lost their life in those ominous year and in the following equally terrible ones.

In Soviet Union, being son to a People's Enemy and moreover a Jew, meant to be a priori excluded from higher education in top level universities. Notwithstanding this, Eugene Dynkin succeeded to be admitted to Moscow University in 1940, thanks to the protection of a distinguished soviet scientist. Dynkin himself wrote: *It was almost a miracle that I was admitted (at the age of sixteen) to Moscow University. Every step*

Fig. 5.14 Eugene Borisovich Dynkin (1924–2014) on the left. Harold Scott MacDonald Coxeter (1907–2003) on the right

in my professional career was difficult because the fate of my father, in combination with my Jewish origin, made me permanently undesirable for the party authorities at the university. Only special efforts by A. N. Kolmogorov, who put, more than once, his influence at stake, made it possible for me to progress through the graduate school to a teaching position at Moscow University.

Saved from military service because of his poor eyesight, Dynkin was able to continue his studies throughout World War II, graduating with a Master of Science from the Faculty of Mechanics and Mathematics in 1945. His work at this time was partly in algebra and partly in probability. Indeed he attended the seminars of Gelfand on the theory of Lie groups and those of Kolmogorov on Markov chains. These two areas of mathematics, algebra and probability, remained the focus of his interest throughout all of his life, with a strong shift toward the second in his later years. It was during his student years that Dynkin, trying to understand Weyl's writings on Lie Groups, invented the *Dynkin diagrams* to classify Cartan matrices. Similar graphs had been independently introduced by Coxeter in his study of reflection groups, presently named Coxeter groups. After graduating, Dynkin remained at Moscow University where he became a research student of Kolmogorov. For ten years he worked both on the theory of Lie algebras and on probability theory, although his main work during this period was in algebra. In 1945 he solved a problem on Markov chains suggested by Kolmogorov and his first publication in probability resulted. Dynkin received his Ph.D. in 1948 and he became an assistant professor in the Probability Department directed by Kolmogorov. Dynkin became Doctor of Physics and Mathematics in 1951 and Kolmogorov pressed for Dynkin to be awarded a chair. However there was no way that the Communist Party leaders of Moscow University would allow a person with such a background as Dynkin's to hold a chair. After Stalin's death in 1953 the grasp of Party hardliners on all sectors of Soviet Society eased a little bit. It was the time of Khrushchev's reforms. The following year, with Kolmogorov's strong support, Dynkin was appointed to a chair at the University of Moscow and he held this chair until 1968. From the time he was appointed to the chair, Dynkin's work became more and more devoted to probability theory. In 1968, the year of Prague Spring and of its suppression by the Soviet tanks sent by Brezhnev, Dynkin was removed from his chair at Moscow University because he had signed a letter in support of the two dissidents Alexander Ginzburg and Yuri Galanskov who were at that time on trial for compiling the White Book. This latter was a four-hundred page report of the infamous mock trial of the two writers Yuri Daniel and Andrei Sinyavsky condemned in 1965 to long detentions in labor camps because of *anti-soviet activities*. Both Ginzburg and Galanskov were on their turn sentenced to several years of hard labor and Galanskov died in the lager. Dynkin, removed from Moscow State University, was simply sent to the Institute of Central Economics and Mathematics of the USSR Academy of Sciences. He worked there from 1968 to 1976. At the end of 1976, Dynkin left the USSR for the United States. The decision to leave was very hard: pupils, friends, and youth were left behind. To apply for emigration was a great risk, especially for an outstanding scientist: many such applicants were denied exit visas, they lost their jobs and lived for years as outcasts of soviet society. Dynkin took the risk since life in the USSR had became more and more unbearable to him and since his only daughter had already managed to emigrate to Israel.

In the United States, Dynkin was offered a chair by Cornell University, which he accepted. He stayed in Ithaca, New York State, the rest of his life until his death in 2014.

Dynkin has been awarded many prizes for his outstanding contributions. He was elected fellow of the Institute of Mathematical Statistics in 1962 and of the American Academy of Arts and Sciences in 1978. In 1985 he was elected member of the National Academy of Sciences of the United States. He received honorary doctorates from the Pierre and Marie Curie University of Paris in 1997, the University of Warwick in 2003 and the Independent University of Moscow also in 2003.

Donald Coxeter's father, Harold, was a gas manufacturer while his mother, Lucy, was a painter. The artistic tendencies of Donald can be probably traced back to his mother's legacy. Donald was educated at the University of Cambridge, receiving his Bachelor of Science in 1929. He continued to study for a doctorate at Cambridge and this was awarded to him in 1931. He spent the next two years as a research visitor at Princeton University working with Veblen. He had a second fellowship at Princeton for the years 1934–35.

In 1936 Coxeter received an appointment from the University of Toronto in Canada that he accepted. He remained on the faculty at Toronto until his death in 2003.

Coxeter's work was mainly in Geometry. In particular he made contributions of major relevance to the theory of polytopes, non-Euclidian geometry, group theory and combinatorics. Coxeter polytopes are defined as the fundamental domains of discrete reflection groups, now called Coxeter groups, and they give rise to tessellations. In 1934 Coxeter classified all spherical and euclidian Coxeter groups. In this context he introduced Coxeter diagrams. His mathematical work was motivated by the search for beauty. Robert Moody, proposing Coxeter for an honorary degree from York University in Toronto, said: *Modern science is often driven by fads and fashions, and mathematics is no exception. Coxeter's style, I would say, is singularly unfashionable. He is guided, I think, almost completely by a profound sense of what is beautiful.*

Coxeter wrote many books not only on topics of Mathematics. He had artistic tendencies and was fascinated by the work of the Dutch painter Escher whom he met in 1954 building up with him a life-long friendship, certainly eased by the Dutch nationality of his own wife Rien. Donald had many artistic gifts, particularly in music. In fact before he became a mathematician he wanted to become a composer. However his interest in symmetry pulled him towards mathematics and into a career which he never stop loving and practicing also in his late years. Coxeter wrote: *I am extremely fortunate for being paid for what I would have done anyway.* In 2006 the Canadian journalist Siobhan Roberts wrote a biography of this outstanding mathematician with the title *King of Infinite Space: Donald Coxeter, The Man Who Saved Geometry.*

5.5.1 Dynkin Diagrams

In order to make the objects of our previous historical account concrete, let us briefly recollect the definition of Dynkin diagrams and present, in terms of them, the content of the classification theorem of simple Lie algebras. Each Cartan matrix can be given a graphical representation in the following way. To each simple root α_i we associate a circle ○ as in Fig. 5.15 having denoted θ_{ij} the angle between the two simple roots α_i and α_j. Then we link the ith circle with the jth circle by means of a line which is *simple*, *double* or *triple* depending on whether:

$$< \alpha_i \, , \, \alpha_j > < \alpha_j \, , \, \alpha_i >= 4 \cos^2 \theta_{ij} = \begin{cases} 1 \\ 2 \\ 3 \end{cases} \tag{5.5.3}$$

The corresponding graph is named a **Coxeter graph**.

If we consider the simplest case of two–dimensional Cartan matrices we have the four possible Coxeter graphs depicted in Fig. 5.16. Given a Coxeter graph if it is *simply laced*, namely if there are only simple lines, then all the simple roots appearing in such a graph have the same *length* and the corresponding Cartan matrix is completely identified. On the other hand if the Coxeter graph involves double or triple lines, then, in order to identify the corresponding Cartan matrix, we need to specify which of the two roots sitting at the end points of each multiple line is the *long* root and which is the *short* one. This can be done by associating an arrow to each multiple line. By convention we decide that this *arrow points* in the direction of the

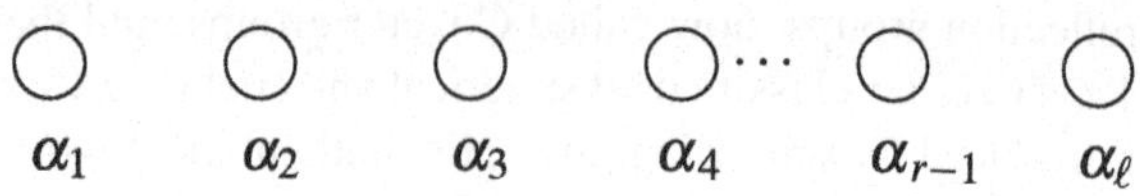

Fig. 5.15 The simple roots α_i are represented by circles

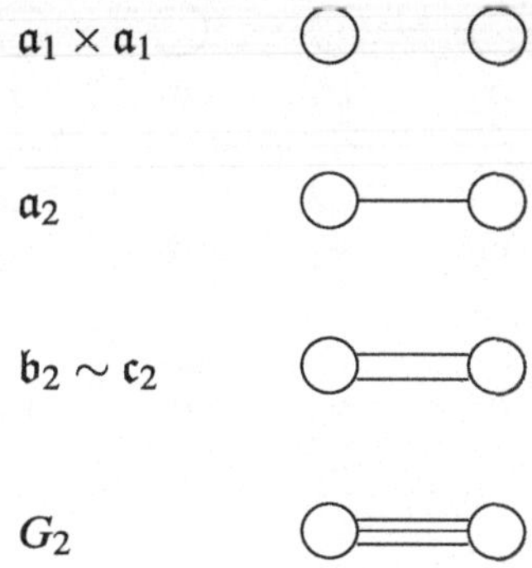

Fig. 5.16 The four possible Coxeter graphs with two vertices

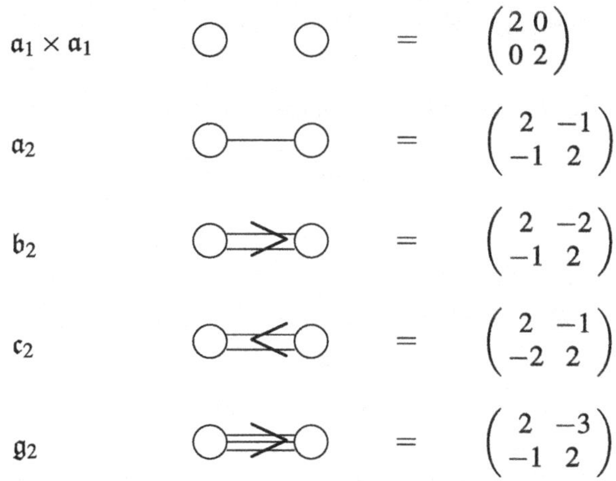

Fig. 5.17 The distinct Cartan matrices in two dimensions (and therefore the simple Algebras in rank two) correspond to the Dynkin diagrams displayed above. We have distinguished a $\mathfrak{b}_2$ and a $\mathfrak{c}_2$ matrix since they are the limiting case for $\ell = 2$ of two series of Cartan matrices the $\mathfrak{b}_\ell$ and the $\mathfrak{c}_\ell$ series that for $\ell > 2$ are truly different. However $\mathfrak{b}_2$ is the transposed of $\mathfrak{c}_2$ so that they correspond to isomorphic algebras obtained one from the other by renaming the two simple roots $\alpha_1 \leftrightarrow \alpha_2$

short root. A Coxeter graph equipped with the necessary arrows is named a *Dynkin diagram*. Applying this convention to the case of the Coxeter graphs of Fig. 5.16 we obtain the result displayed in Fig. 5.17.

5.5.2 *The Classification Theorem*

Having clarified the notation of Dynkin diagrams the basic classification theorem of *complex simple Lie algebras* is the following. If Δ is an irreducible system of roots of rank ℓ then its Dynkin diagram is either one of those shown in Fig. 5.18 or for special values of ℓ is one of those shown in Fig. 5.19. There are no other irreducible

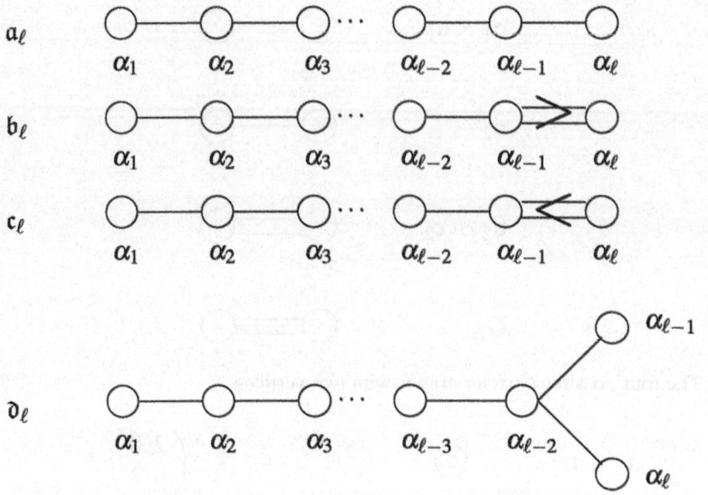

Fig. 5.18 The Dynkin diagrams of the four infinite families of classical simple algebras

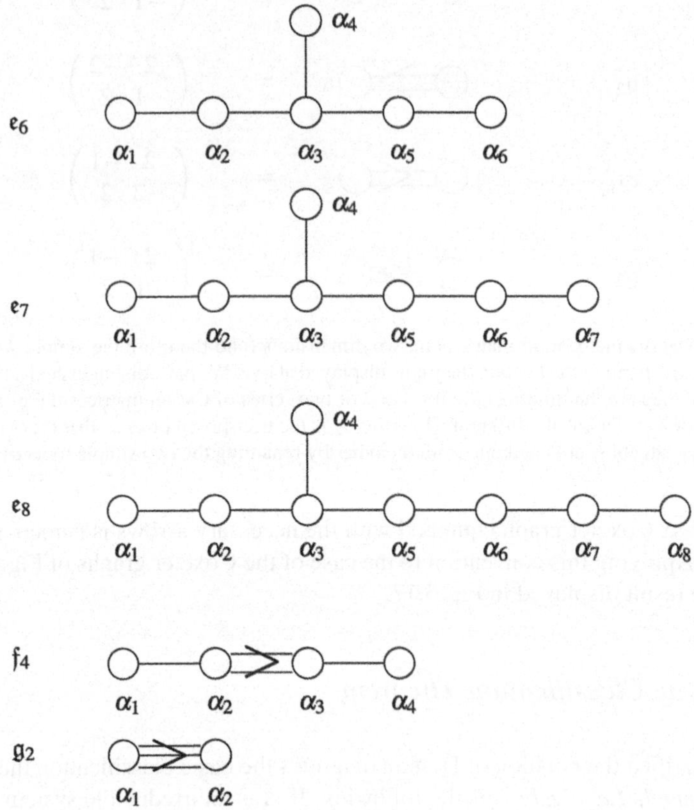

Fig. 5.19 The Dynkin diagrams of the five exceptional algebras

root systems besides these ones. The four infinite families $\mathfrak{a}_\ell, \mathfrak{b}_\ell, \mathfrak{c}_\ell, \mathfrak{d}_\ell$ correspond to the Lie algebras of the four classical matrix groups, mentioned in the same order $SL(\ell+1)$, $SO(2\ell+1)$, $Sp(2\ell)$, $SO(2\ell+1)$. By definition $SL(\ell+1)$ denotes the group of matrices in $\ell+1$ dimensions whose determinant is one, SO(n) denotes the group of $n \times n$ orthogonal matrices and $Sp(2\ell)$ denotes the group of $2\ell \times 2\ell$ symplectic matrices.

The five exceptional algebras, whose possible existence had been spotted by Killing, cannot be identified by means of algebraic conditions imposed on matrices. Indeed, as we explained, their actual existence was established by Cartan who was able to construct explicit matrix realizations for each of them.

5.6 The ADE Classification

A very interesting aspect of the root system classification relates with the subset of Dynkin diagrams, named *simply laced*, that involve only simple lines between the vertices.

At some point in the demonstration of the theorem one arrives at the stage where it is established that the only possible simply laced Dynkin diagrams have the form depicted in Fig. 5.20. A series of simple arguments in elementary euclidian geometry shows that the three integer number characterizing the diagram fulfill the following diophantine inequality:

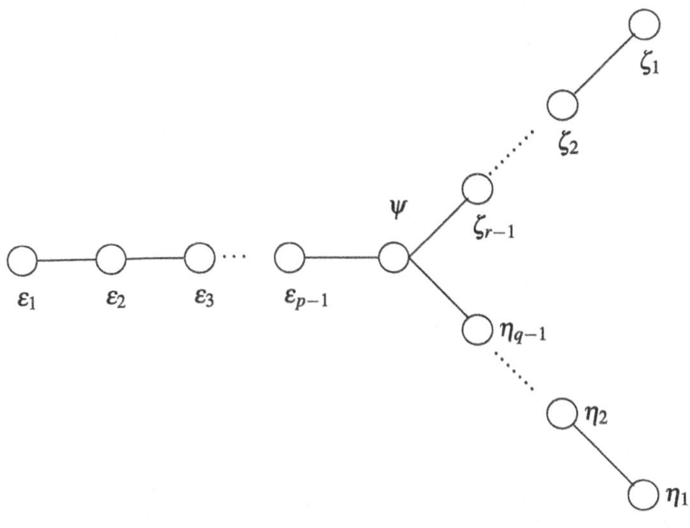

Fig. 5.20 Dynkin diagram with a node. The simple root in the node is named ψ while the roots along the three simple lines departing from the node are respectively named $\varepsilon_1, \ldots, \varepsilon_{p-1}, \eta_1, \ldots, \eta_{q-1}, \zeta_1, \ldots, \zeta_{r-1}$. The graph is characterized by the three integer numbers p, q, r that denote the lengths of the three simple lines departing from the node

$$\frac{1}{p}+\frac{1}{q}+\frac{1}{r}>1 \tag{5.6.4}$$

whose independent solutions are those displayed below:

$$(p,q,r)=\begin{cases}(\ell,1,1) & \Rightarrow \mathfrak{a}_\ell \text{ Dynkin diagrams} \quad \ell\in\mathbb{N}\\ (\ell-2,2,2) & \Rightarrow \mathfrak{d}_\ell \text{ Dynkin diagrams } 4\le\ell\in\mathbb{N}\\ (3,3,2) & \Rightarrow \mathfrak{e}_6 \text{ Dynkin diagram}\\ (4,3,2) & \Rightarrow \mathfrak{e}_7 \text{ Dynkin diagram}\\ (5,3,2) & \Rightarrow \mathfrak{e}_8 \text{ Dynkin diagram}\end{cases} \tag{5.6.5}$$

Changing the names of the variables:

$$p=k_1 \quad ; \quad q=k_2 \quad ; \quad r=k_3 \tag{5.6.6}$$

Equation (5.6.4) is the same as Eq. (4.2.20) obtained in the classification of finite subgroups of the spinor group SU(2) and of its homeomorphic image in the three-dimensional rotation group SO(3). Hence it has the same solutions. This extraordinary coincidence establishes a deep correspondence between the symmetry groups of polygons and polyhedra with simply laced Lie algebras. In particular, this correspondence associates the three symmetry groups of the five platonic solids with the exceptional Lie algebras of E-type.

The double interpretation of the same diophantine constraint is summarized in Fig. 5.21 and leads to the following correspondences between finite rotation groups and simply laced Lie algebras:

	Simple Lie Algebras	Finite subgroups of $\Gamma_b \subset \mathrm{SU}(2)$
r	number of simple chains in the Dynkin diagram	# of different types of group–element orders present in $\Gamma \equiv \omega[\Gamma_b]$
k_α	$k_\alpha - 1$ = lengths of the simple chains in the Dynkin diagram	group-element orders in $\Gamma \equiv \left(\mathrm{A},\mathrm{B} \mid (\mathrm{AB})^{k_1} = \mathrm{A}^{k_2} = \mathrm{B}^{k_3} = \mathbf{1}\right)$
$\mathscr{R}-1 \equiv \sum_{\alpha=1}^{r}(k_\alpha-1)$	$\mathscr{R}$ = rank of the Lie algebra	$\mathscr{R}+1$ = # of conjugacy classes in Γ_b

Fig. 5.21 Interpretation of the solutions of the same Diophantine equation in the case of finite subgroups of $\Gamma_b \subset \mathrm{SU}(2)$ and of simply laced Lie algebras

$$\Gamma[\ell, \ell, 1] \simeq \mathbb{Z}_\ell \Leftrightarrow \mathfrak{a}_\ell \tag{5.6.7}$$

$$\Gamma[\ell, 2, 2] \simeq \mathrm{Dih}_\ell \Leftrightarrow \mathfrak{d}_\ell \tag{5.6.8}$$

$$\Gamma[3, 3, 2] \simeq \mathrm{T}_{12} \Leftrightarrow \mathfrak{e}_6 \tag{5.6.9}$$

$$\Gamma[4, 3, 2] \simeq \mathrm{O}_{24} \Leftrightarrow \mathfrak{e}_7 \tag{5.6.10}$$

$$\Gamma[5, 3, 2] \simeq \mathrm{I}_{60} \Leftrightarrow \mathfrak{e}_8 \tag{5.6.11}$$

where $\mathfrak{a}_\ell$ is the Lie algebra associated with the Lie group $\mathrm{SL}(\ell + 1, \mathbb{C})$, $\mathfrak{d}_\ell$ is the Lie algebra associated with the Lie group $\mathrm{SO}(2\ell, \mathbb{C})$, and $\mathfrak{e}_{6,7,8}$ are the Lie algebras of three exceptional Lie groups of dimensions 78, 133 and 248, respectively. In Lie Algebra theory, as we have seen, the *rank* is the maximal number of mutually commuting and diagonalizable elements of the algebra. As we see from Fig. 5.21, the rank has a counterpart in the binary extension of the corresponding finite rotation group: it is the number of non trivial conjugacy classes of the group, except the class of the identity element. The property of Lie algebras that in Dynkin diagrams there are no nodes with more than three converging lines corresponds on the finite rotation group side to the property that in such groups there are at most three different types of group-element orders. Finite groups with more than three type of group element orders do not correspond to Dynkin diagrams of Lie algebras, although more complicated diagrams can still be associated with them that play a role in the algebraic geometry of singularity resolutions (the Mac Kay quivers, briefly discussed in Chap. 11).

5.7 Comments on the ADE Classification

The last remark in the previous section leads us to mention what follows. The fate of the diophantine inequality (5.6.4) is not accomplished with the above illustrated correspondence. A further incarnation of the ADE classification is much more recent and will be discussed in Chap. 11. It relates with HyperKähler geometry and with the so named resolution of singularities. From the point of view of singularities the ADE classification of the latter was discovered in the 1970s by Vladimir Arnold. Quotient singularities with respect to discrete group that have more than three type of group element orders provide a generalization of the ADE classification which is topical in current frontier research.

We have here a primary example of the continuous dialectics between fundamental issues in geometry, algebra and number theory that is at the heart of the mathematical conceptions of Space and Symmetry, ultimately of Physics.

A final remarkable coincidence is the following. Because of the defining property (a) the angle θ between any two roots α and β is quantized, namely:

$$\theta = \pm 30^\circ \quad \text{or} \quad \pm 60^\circ \quad \text{or} \quad 120^\circ \quad \text{or} \quad \pm 45^\circ \quad \text{or} \quad 90^\circ \quad \text{or} \quad 135^\circ \tag{5.7.12}$$

This is precisely the quantization of angles in Plato's fundamental triangles conceived by the ancient philosopher as the ultimate subconstituents of matter!

Chapter 6
Hermann Weyl and Representation Theory

My work always tried to unite the truth with the beautiful, but when I had to choose one or the other, I usually chose the beautiful....

Hermann Weyl

6.1 Conceptual Introduction

The Lie algebra tale continues in this chapter with the essential and fundamental issue of linear representations.

In previous chapters we saw how the notion of continuous groups of transformations, initiated by Lie in connection with differential equations, lead to the notion of Lie algebras, whose abstract classification was the objective pursued by Killing first and by Cartan later on. In this process exceptional Lie algebras were discovered, whose first explicit realization in terms of matrices was found by Cartan himself in his doctoral thesis. In this way exceptional Lie algebras obtained the same status as classical Lie algebras that are originally defined as vector spaces of matrices equipped with the ordinary commutator playing the role of Lie-bracket.

The realization of a Lie algebra in terms of matrices corresponds to the notion of a linear representation and it is a fundamental issue in Lie algebra and Lie group theory. The question raised by the case of exceptional Lie algebras is the same that can be raised for any other Lie algebra: *can we classify the irreducible linear representation of a Lie algebra and construct them explicitly?* It is a question absolutely analogous to the question that was already addressed in the case of finite groups and has a similar type of answer. If we restrict our attention to simple Lie algebras, just as in the case of finite groups, the set of linear representations, which now is infinite, is fully determined by the very structure of the Lie algebra and can be regarded as an integral part of its mathematical essence. From a philosophical point of view, this fact is quite significant.

Our present understanding of the fundamental interactions governing the dynamics of the fundamental constituents of matter is that they are all described by

P. G. Fré, *A Conceptual History of Space and Symmetry*,
https://doi.org/10.1007/978-3-319-98023-2_6

connections on principal fibre-bundles with specifically chosen structural Lie groups: in particular $SU(3) \times SU(2) \times U(1)$ for non-gravitational interactions, gravity being instead associated with connections on the tangent bundle determined by a pseudo-Riemannian metric. Matter fields are instead sections of associated vector bundles that are as many as the possible linear representations of the structural group. Hence the ultimate understanding of the architecture of the world relies on the discovery of some adequate superselection principle that chooses the right structural group and the right linear representations among the available ones. A definite answer to this problem is still lacking but all the steps towards a unified ultimate theory of fundamental interactions are along this line of thinking. Symmetry encodes dynamics and also the catalogue of its own possible realizations.

6.1.1 Hermann Weyl

When talking about representation theory and the accomplishment of continuous group theory the entire scene is dominated by the outstanding figure of a mathematician, philosopher and scientist who is another giant of the XXth century: Hermann Weyl (see Fig. 6.1). Weyl is beyond any doubt the great scientific personality that more than others fits into the ideal of a modern variant of the Renaissance Universal Scholar, interested with an equal degree of intensity in those aspects of Culture that are dubbed Humanities and those that are dubbed Natural Sciences. He was talented in languages and his scientific books are pieces of Art also from the literary point of view, both in German and in English. At the center of the intellectual texture in which Weyl moved at ease with a never ceasing depth of thinking there is Mathematics, strongly linked with Philosophy, and at the center of Mathematics, Weyl certainly placed *Symmetry*. Such was the title of Weyl's last book, published in 1952 and based on a lecture series given by him in Princeton in the previous several months. Coxeter reviewed this book and said: ... *The first lecture begins by showing how the idea of bilateral symmetry has influenced painting and sculpture, especially in ancient times. This leads naturally to a discussion of "the philosophy of left and right", including such questions as the following. Is the occurrence in nature of one of the two enantiomorphous forms of an optically active substance characteristic of living matter? At what stage in the development of an embryo is the plane of symmetry determined? The second lecture contains a neat exposition of the theory of groups of transformations, with special emphasis on the group of similarities and its subgroups: the groups of congruent transformations, of motions, of translations, of rotations, and finally the symmetry group of any given figure. ... the cyclic and dihedral groups are illustrated by snowflakes and flowers, by the animals called Medusae, and by the plans of symmetrical buildings. Similarly, the infinite cyclic group generated by a spiral similarity is illustrated by the Nautilus shell and by the arrangement of florets in a sunflower. The third lecture gives the essential steps in the enumeration of the seventeen space-groups of two-dimensional crystallography ... In the fourth lecture he shows how the special theory of relativity is essentially the study of the inher-*

Fig. 6.1 Hermann Klaus Hugo Weyl (1885–1955). Born in Elmshorn near Hamburg, he studied mathematics in Münich and Göttingen where he accomplished his doctorate under the supervision of David Hilbert. After a beginning as a *dozent* in Göttingen, Weyl joined the Faculty of Zürich Polytechnic ETH in 1913 where he remained until 1930 and for some years he was colleague of Einstein. In 1930, after Hilbert's retirement he was his successor in Göttingen. In 1933 when Hitler came to power, Weyl, whose wife Helene Joseph was a Jew, emigrated to the U.S.A. accepting an offer from the Institute of Advanced Study in Princeton and there he was once again Einstein's colleague. Weyl's wife was a philosopher who had been Husserl's student and this stirred Weyl's interests in philosophy. They had two sons. Weyl lived in Princeton until 1951, when he retired. After his first wife's death in 1948, Weyl married again in 1950 with the sculptress Ellen Bär and with his second wife he divided his time between Princeton and Zürich where he died in 1955 from heart attack

ent symmetry of the four-dimensional space-time continuum, where the symmetry operations are the Lorentz transformations; and how the symmetry operations of an atom, according to quantum mechanics, include the permutations of its peripheral electrons. Turning from physics to mathematics, he gives an extraordinarily concise epitome of Galois theory, leading up to the statement of his guiding principle: "Whenever you have to do with a structure-endowed entity, try to determine its group of automorphisms".

Weyl was born in 1885 in northern Germany in a small city close to Hamburg, son to a director of a bank. After secondary school he entered first the University of Münich, studying mathematics and physics and then he moved to the University of Göttingen, where he became Hilbert's student. After graduation and habilitation, obtained with two theses devoted to integral equations and to the spectral theory of Sturm–Liouville operators, already in 1913 he published a book *Die Idee der Riemannschen Fläche* (*The Idea of a Riemann Surface*) that, according to an evaluation expressed by L. Sario in 1956, *has undoubtedly had a greater influence on the development of geometric function theory than any other publication since Riemann's dissertation*. Weyl's philosophical attitude is already evident from the title, full of Platonic suggestions, yet in typical Weyl's spirit the book is not just made of beautiful words rather of the soundest mathematics providing a consistent unified approach to Riemann's function theory where geometrical, analytical and topological aspects all obtain their proper place. The philosophical inclinations of Hermann Weyl were boosted in Göttingen by the influence of Edmund Husserl that reached him both directly and through Weyl's fiancée Helene Joseph with whom he married the same year of publication of his first book. Helene had been Husserl's student and was herself a philosopher.

1913 was a very important year in Weyl's life not only because of his first book and of his marriage, but also because he was appointed professor at the Zürich Polytechnic ETH, whose Faculty he joined that year becoming Einstein's colleague. The early contact between Weyl and Einstein at a time when the latter was accomplishing the set up of the General Theory of Relativity was momentous for the former whose interest in Relativity both at the mathematical and at the philosophical level lasted all the rest of his life. Equally important for the development of Weyl's thought was his close friendship with Erwin Schrödinger who joined ETH in 1922 and with whose wife Anny, Weyl had a short-lived but intense love affair.

In Zürich where he remained until 1930, Weyl started and developed what he himself considered the major achievement of his life and found its accomplished expression in the book: *The Classical Groups*, published by him in 1939, when he was already in the United States [168]. We refer hereby to the theory of Lie Group representations that Weyl, by means of his famous character formula, brought to the same perfection attained by the theory of finite groups.

Both in the field of relativity, his interest in which was stimulated by Einstein, and in the field of quantum mechanics, to which he was attracted by Schrödinger, Weyl's contributions were pivoted around the idea of symmetry. Weyl's third book *Gruppentheorie und Quantenmechanik* (*Quantum Mechanics and Group Theory*) set up the standard universally adopted for the treatment of symmetries of the quantum hamiltonians and the treatment of their eigenfunctions.

In 1930, after Hilbert's retirement, Weyl was appointed to his chair by Göttingen University where he remained only three years. In 1933 when Hitler and the nazis came to power, Weyl accepted Princeton's offer and joined the Institute for Advanced Studies where he became once again Einstein's colleague. In Weyl's decision the Jewish origin of his wife played an important role but certainly the vulgarity of nazism was no less important given his intellectual standards and his love of the beautiful expressed in his famous sentence:

My work always tried to unite the truth with the beautiful, but when I had to choose one or the other, I usually chose the beautiful...

In Princeton Weyl spent the last active part of his life giving lecture courses that gave origin to other important books, *Elementary Theory of Invariants*, the already quoted *The Classical Groups*, *Algebraic Theory of Numbers*, *Philosophy of Mathematics and Natural Sciences* up to his *swan song*, *Symmetry*. In 1948 he suffered the loss of his wife and two years later he remarried with the swiss sculptress Ellen Bär with whom he lived his last five years dividing his time between Zürich and Princeton from which he retired in 1952. He died in Zürich from heart attack on his 70th birthday in 1955. His ashes are interred in Princeton's cemetery close to those of his son Michael Weyl passed away in 2011 at the age of 93 after a long life that saw him scholar of German literature, actor, soldier, journalist and cultural attachée in American diplomatic missions around the world.

Because of the extraordinary relevance of Weyl's views on symmetry and mathematics for the spirit of the present book we devote the next subsection to the analysis of the mathematical way of thinking according to him. This was the title of one address of his given at the University of Pennsylvanya in 1940 and published in Science the same year [169].

6.1.2 The Mathematical Way of Thinking, According to Hermann Weyl, with This Author's Comments

After some initial words that we omit Weyl says:

A movement for the reform of the teaching of mathematics, which some decades ago made a stir in Germany under the leadership of the great mathematician Felix Klein, adopted the slogan ***functional thinking***. *The important thing which the average educated man should have learned in his mathematics classes, so the reformers claimed, is thinking in terms of* ***variables and functions***. *A function describes how one variable y depends on another x; or more generally, it maps one variety, the range of a variable x, upon another (or the same) variety. This idea of function or mapping is certainly one of the most fundamental concepts, which accompanies mathematics at every step in theory and application.*

After exhibiting Galileo's law of accelerated fall in a constant gravitational field:

$$s = \tfrac{1}{2} g\, t^2 \tag{6.1.1}$$

Weyl continued:

Right from the beginning we encounter these characteristic features of the mathematical process:

1. *variables, like t and s in the above formula, whose possible values belong to a range, here the range of real numbers, which we can completely survey because it springs from our own free construction,*
2. *representation of these variables by symbols,*
3. *functions of a priori constructed mapping of the range of one variable t upon the range of another s.*

In studying a function one should let the independent variable run over its full range. A conjecture about the mutual interdependence of quantities in Nature, even before it is checked by experience, may be probed in thought by examining whether it carries through, over the whole range of independent variables. Sometimes certain simple ***limiting cases*** *at once reveal that the conjecture is untenable. Leibnitz taught us by the* ***principle of continuity*** *to consider rest not as contradictorily opposed to* *motion, but as a limiting case of motion. Arguing by continuity he was able* ***a priori*** *to refute the laws of impact proposed by Descartes.*

We clearly see in the above sentences written by Weyl himself what Sir Michael Atiyah told about him in his recent Biographical Memoir [8]:

Weyl was a strong believer in the overall unity of mathematics, not only across sub-disciplines but also across generations. For him the best of the past was not forgotten, but was subsumed and refined by the mathematics of the present. His book **The Classical Groups** *was written to bring out this historical continuity.*

Without any undue pretension to originality and with a due sense of proportions, the author of this book entirely subscribes to such an opinion as that attributed by Atiyah to Weyl. Indeed such a belief in the unity of scientific thought across generations and in the importance of a historical understanding of the development of fundamental concepts is the very motivation to write the present long essay. Actually, persuaded that Weyl would agree with it, this book is marked by a further extension of the unitarian vision advocated above for mathematics. Mathematics is linked with theoretical physics, with philosophy and with the arts and mathematical thinking just encompasses one particular very fundamental branch of the overall human thinking that cannot and never should be separated from the other branches. To confirm us in this Weyl wrote:

My own mathematical works are always quite unsystematic, without mode or connection. Expression and shape are almost more to me than knowledge itself. But I believe that, leaving aside my own peculiar nature, there is in mathematics itself, in contrast to the experimental disciplines, a character which is nearer to that of free creative art.

A little further, in the same conceptual article quoted above [169] about mathematical thinking Weyl said:

In Aristotle's logic one passes from the individual to the general by exhibiting certain abstract features in a given object and discarding the remainder, so that two objects fall under the same concept or belong to the same genus if they have those features in common. This descriptive classification, e.g., the description of plants and animals in botany and zoology, is concerned with actual existing objects. One might say that Aristotle thinks in terms of substance and accident, while the functional idea reigns over the formation of mathematical concepts. Take the notion of ellipse. Any ellipse in the $x - y$-plane is a set E of points (x, y) defined by a quadratic equation:

$$ax^2 + bxy + cy^2 = 1 \tag{6.1.2}$$

whose coefficients a, b, c satisfy the conditions:

$$a > 0, c > 0, ac - b^2 > 0 \tag{6.1.3}$$

The set E depends on the coefficients a, b, c: we have a function $E(a, b, c)$ which gives rise to an individual ellipse by assigning definite values to the variable coefficients a, b, c. In passing from the individual ellipse to the general notion one does not discard any specific difference, one rather makes certain characteristics *(here represented by the coefficients) variable over an* **a priori** *surveyable range (here described by the inequalities). The notion thus extends over all possible, rather than over all* **actually existing**, *specifications.*

In relation with the above views about the formation of mathematical concepts we can add that there is a constant historical dialectics inherent to both mathematics and theoretical physics which has, as main untangled opponents, the issue of generalization and that of specific choices. This dialectics ultimately stems from our ambition to understand Nature in purely rational terms. On one side, exactly as described by Weyl, we generalize the notion of *what exists* into a mathematically defined family of *what is possible*. Typically the *possible structures* are parameterized by variables (the quadric coefficients in Weyl's example) that have a range, namely can be thought as points in a certain space which, in contemporary mathematical physics, is customarily dubbed *the moduli space*. In this way the trend of geometrization of both physics and mathematics is generally boosted: whatever is the notion we consider, it carries, attached with it, some sort of *moduli space* and our understanding of the *virtual* is essentially encoded in our command over the *geometry of moduli spaces*. On the other hand we always would like to be able to select, among the possible structures, those that actually exist in Nature. Indeed some of Aristotle's spirit persists in us up to the present time! In mathematical terms *what actually exists* corresponds to some definite points in moduli space and our ambition is to characterize a priori such points, as special ones that we might predict. To this effect one resorts to new *functions* defined over moduli space, typically some potential or hamiltonian function, whose minima can select the special moduli points corresponding to what exists in *actuality*. The game starts at this point once again in the new rush to define the family of possible hamiltonians and their moduli spaces. In these games a fundamental issue is provided by symmetries and by their classification to which Weyl also contributed a lot. The ultimate dream of many scientists is associated with sporadic entities, for instance groups. Because of their uniqueness they have no moduli and correspond to some end point in the conceptual chain. In some sense sporadic structures are the analogue in mathematical thinking of God or better of Gods, sticking to a politheistic attitude that is historically much safer and peaceful of the monotheistic one.

In a later paragraph of his impressive article on mathematical thinking Weyl said:

Words are dangerous tools. Created for our everyday life they may have their good meanings under familiar circumstances, but Pete and the man in the street are inclined to extend them to wider spheres without bothering about whether they then still have a sure foothold in reality. We are witnesses of the disastrous effects of this witchcraft of words in the political sphere where all words have a much vaguer meaning and human passion so often drowns the voice of reason. The scientist must thrust through the fog of abstract words to reach the concrete rock of reality. It seems to me that the science of economics has a particularly hard job, and will still have to spend much effort, to live up to this principle. It is, or should be, common to all sciences, but physicists and mathematicians have been forced to apply it to the most fundamental concepts where the dogmatic resistance is strongest, and thus it has become their second nature. For instance, the first step in explaining relativity theory must always consist in shattering the dogmatic belief in the temporal terms past, present, future. You can not apply mathematics as long as words still becloud reality.

Further on, talking about relativity, Weyl continued as follows:

...the real thing emerges as soon as we replace the intuitive space in which our diagrams are drawn by its construction in terms of sheer symbols. Then the phrase that the world is a four-dimensional continuum changes from a figurative form of speech into a statement of what is literally true. At this second step the mathematician turns abstract, and there is the point where the layman's understanding most frequently breaks off: the intuitive picture must be exchanged for a symbolic construction. "By its geometric and later by its purely symbolic construction", says Andreas Speiser, "mathematics shook off the fetters of language, and one who knows the enormous work put into this process and its ever recurrent surprising successes can not help feeling that mathematics to-day is more efficient in its sphere of the intellectual world, than the modern languages in their deplorable state or even music are on their respective fronts". I will spend most of my time to-day in an attempt to give you an idea of what this magic of symbolic construction is.

Indeed, after this sentence Weyl illustrated the *constructive nature* of mathematics by explaining in some detail the combinatorial construction of topological spaces in terms of cells and infinite divisions. While doing that, in full agreement with views earlier expressed by Poincaré, he emphasized the *principle of iteration* as the most fundamental idea underlying the whole of mathematical thinking. We pause instead for a moment in order to make some comments on his previously reported sentences.

First of all it is clearly transparent from his words that mathematics and theoretical physics are in Weyl's thinking just two aspects of the very same thing. Secondly one can easily summarize Weyl's conception of the process which according to him can lead to a mathematical understanding of the laws of nature. This process occurs in three steps:

1. First one has to go to reality by dismantling the *witchcraft of words*, in particular of the abstract ones like past, present and future, higher and lower, parallel and so on.
2. Secondly one has to reformulate reality into a symbolic scheme that you treat and develop as such ignoring the interpretation of the symbolic objects.
3. Thirdly, from the developed scheme taken as a whole you can work out implications that can be compared with experimental data.

The first step in the above list is very important. It is frequently said that mathematics is the language in which Nature has written its own laws but Weyl goes beyond that. It is the very dismission of the concept of words which he brings to the forefront. Words are deceiving because they are descriptive and behind the same word everyone sees something different. In a symbolic construction the intermediate objects do not need to describe anything in particular: what is relevant are the relations and the operations one can make within the scheme. It is from these operations that one extracts the meaning of the construction and ultimately the predictions to compare with reality. In a later passage Weyl says:

We now come to the decisive step of mathematical abstraction: we forget about what the symbols stand for. The mathematician is concerned with the catalogue alone; he is like the man in the catalogue room who does not care what books or pieces of an intuitively given manifold the symbols of his catalogue denote. He need not be idle; there are many operations which he may carry out with these symbols, without ever having to look at the things they stand for.

Still further Weyl states:

The historic development of our theories proceeds by heuristic arguments over a long and devious road and in many steps from experience to construction. But systematic exposition should go the other way: first develop the theoretical scheme without attempting to define individually by appropriate measurements the symbols occurring in it as space-time coordinates, electromagnetic field strengths, etc, then describe, as it were in one breath, the contact of the whole system with observable facts.

We clearly see in this the difference between symbols and words. A word describes some supposedly real object. A symbol is an item on which you do operations or which operates on other symbols. Then partially correcting his own standpoint Weyl says:

Up to now I have emphasized the constructive character of mathematics. In our actual mathematics there lives with it the non-constructive **axiomatic method**. *Euclid's axioms of geometry are the classical prototype. Archimedes employs the method with great acumen and so do later Galileo and Huyghens in erecting the science of mechanics. One defines all concepts in terms of a few undefined basic concepts and deduces all propositions from a number of basic propositions, the axioms, concerning the basic concepts.*

…I should like to point out that since the axiomatic attitude has ceased to be the pet subject of all methodologists its influence has spread from the roots to all branches of the mathematical tree. We have seen before that topology is to be based on a full enumeration of the axioms which a topological scheme has to satisfy. One of the simplest and most basic axiomatic concepts which penetrates all fields of mathematics is that of ***group***. *Algebra with its* ***fields, rings***, *etc, is to-day from bottom to top permeated by the axiomatic spirit. …modern mathematical research often is a dexterous blending of the constructive and the axiomatic procedures.*

The last above words of Weyl are very much significant for the motivations and the entire development of the present book. When he was writing, about seventy years ago, it was absolutely clear to Hermann Weyl that the axiomatic notion of **group** is probably the most central and most relevant in the whole universe of mathematics and in physics as well. Actually group theory is just the perfect example of what was said above about the irrelevance of the objects the symbols stand for. Historically, as we have seen, continuous groups were discovered by Sophus Lie as transformation groups, yet following the work of Killing, Cartan and finally Weyl continuous groups became independent symbolic structures able to intrinsically determine their own concrete realizations as linear or non linear representations. Indeed the father of linear representation theory is Hermann Weyl who gave to continuous group theory the same status as finite group theory had already obtained few decades before through the work of Frobenius. With his book **The Classical Groups** published in 1939 Weyl set a very influential milestone for the whole subject. Groups, both continuous and finite are central and fundamental in Geometry, in Algebra and in Topology. Weyl was the first to grasp that they are the essence of physical laws as well.

How this happened is beautifully described in Weyl's Memoir recently written by Atiyah [8]. We quote again from that article:

Quantum mechanics was not Weyl's first encounter with physics. He had already learned about Einstein's general relativity, which explained gravity in geometrical terms. Weyl had the idea of extending Einstein's theory to incorporate electromagnetism, so that Maxwell's equations would also acquire geometrical significance. Weyl's idea was to introduce a scale, or gauge, that varied from point to point and whose variation round a closed path in space-time would encapsulate the electromagnetic force. Almost immediately (in fact in an appendix to Weyl's paper) Einstein criticized the idea on physical grounds. If Weyl was right, then the size of a particle would depend on its past history, whereas experiments showed that all atoms of hydrogen, say, had identical properties. One might have thought that such a telling criticism from someone of Einstein's standing would have discouraged Weyl and that he might have withdrawn his paper. It is a tribute to his mathematical insight and self-confidence that he went ahead. The idea was too beautiful to discard, and Maxwell's equations came out like magic.

As often happens, a good idea lives to fight another day and only a few years later, with the advent of quantum mechanics, a new physical interpretation was put on Weyl's calculations. Oscar Klein proposed that Weyl's gauge should be viewed as a phase and that space-time should be viewed as having a fifth dimension consisting of a very small circle. Mathematically Weyl's gauge variable gets multiplied by i (the square root of −1) and is periodic. This point of view, called the Kaluza–Klein theory (Theodor Kaluza made the first steps after Weyl) is now generally accepted. Moreover, it is just the first stage in the enlargement of ordinary spacetime. To include the other nuclear forces we need even more dimensions and current research centres on a total space-time dimension of 10 or 11.

Independently of these extra dimensions Weyl's gauge theory description of Maxwell's equations is now applied to local symmetry groups other than the circle. This leads to the non-Abelian gauge theories, which are the basis of the standard model of elementary particle physics.

This gauge theory, the infant that was nearly thrown out with the bath water, has grown up into sturdy adulthood. Not only is it the framework of modern physics but it is also one of the most novel and exciting areas in modern mathematics. One notable example is the theory of 4-dimensional manifolds due to Simon Donaldson (Donaldson and Kronheimer, 1990), which emerged from physics but has turned out to be of profound importance to geometry. More recently, an alternative interpretation uses spinors coupled non-linearly to electromagnetism, a twist that would certainly have captured the imagination of Hermann Weyl and justifies his remarks about the geometrical significance of spinors.

The past 25 years have seen the rise of gauge theories Kaluza–Klein models of high dimensions, string theories, and now M theory, as physicists grapple with the challenge of combining all the basic forces of nature into one all embracing theory. This requires sophisticated mathematics involving Lie groups, manifolds, differential operators, all of which are part of Weyls inheritance. There is no doubt that he would have been an enthusiastic supporter and admirer of this fusion of mathematics and physics. No other mathematician could claim to have initiated more of the theories that are now being exploited. His vision has stood the test of time.

Let us once again add some comments to the above beautiful telling.

The next Chap. 7 of this book is devoted to the development of differential geometry, in particular of the notions of metric and connections on fibre-bundles. As anticipated by Atiyah's words, the contemporary understanding of the physical laws envisages that all fundamental interactions, the forces binding matter together and shaping our Universe are encoded in gauge-fields, namely in connections on principal fibre-bundles. Gravity is universal since it is associated with the connection on the tangent bundle to space-time and this latter (the Levi-Civita connection) follows from a metric, namely from a structure that allows to define the lengths of curves. Although he missed an i-factor, Weyl was the first to understand that non gravitational interactions, like electromagnetism, are an yield of a symmetry group, just made local. In modern parlance this is the structural Lie group of a fibre-bundle. In addition to forces we have matter. As we illustrate at length in Chap. 7, the various types of particles are in correspondence with associated bundles, namely with linear representations of the structural Lie group. Here once again we find Weyl's legacy as stated by Atiyah. Linear representation theory of Lie groups was systematized by Weyl.

6.2 The Basic Notions in Representation Theory

According to what we discussed in Sect. 3.1.3, a linear representation of a group G, it does not matter whether discrete or continuous, is a homomorphism $D : G \to \mathrm{Hom}(V, V)$ where V denotes some vector space of dimension n; the latter number is named the dimension of the representation (see Eq. (3.1.16)). In the case of finite groups Γ, the set of *irreducible representations* D_μ, namely those where there is no non-trivial invariant subspace $W \subset V$,[1] is also finite and its cardinality $r + 1$ is equal to the number of conjugacy classes into which the group is partitioned. In the case of Lie groups the set of irreducible representations is infinite and the corresponding carrier vector spaces V_μ have to be classified. Weyl's genial idea was that of introducing a geometric description of all the existing representations in terms of a *lattice* similar to the crystallographic ones we discussed in Sect. 4.1.1.1. The basis of his idea comes from the *root system formalism* utilized by Killing and Cartan to obtain the classification of simple Lie algebras. This latter was illustrated in Sects. 5.4.1 and 5.5.2. Given the root system Δ, as defined in Sect. 5.4.1, and its basis of simple roots α_i one can easily define the *root lattice* Λ_r by setting:

[1] A subspace $W \subset V$ is named invariant if it is mapped into itself by all group elements.

$$\Lambda_{\rm root} \subset \mathbb{R}^\ell \quad / \quad \mathbf{v} \in \Lambda_{\rm root} \Leftrightarrow \mathbf{v} = n^i \alpha_i \quad , \quad n^i \in \mathbb{Z} \tag{6.2.1}$$

Given the root lattice one can construct its dual that is named the the *weight lattice* and is of the utmost relevance, in Weyl's intrinsic approach to *representation theory*:

$$\Lambda_{\rm weight} \equiv \Lambda^\star_{\rm root} \tag{6.2.2}$$

To construct the weight lattice one defines the basis of *simple weights* dual to the basis of simple roots through the following condition:

$$\langle \lambda^i \, , \alpha_j \rangle \equiv 2 \frac{(\lambda^i \, , \alpha_j)}{(\alpha_j \, , \alpha_j)} = \delta^i_j \tag{6.2.3}$$

and one sets:

$$\Lambda_{\rm weight} \subset \mathbb{R}^\ell \quad / \quad \mathbf{w} \in \Lambda_{\rm weight} \Leftrightarrow \mathbf{w} = n_i \lambda^i \quad , \quad n_i \in \mathbb{Z} \tag{6.2.4}$$

These two notions are illustrated in Fig. 6.2 with the example of the $\mathfrak{a}_2$ Lie algebra. As it is always the case the dual $\Lambda^\star$ of a lattice is larger and contains the original lattice Λ as a sublattice. Hence all points of the root lattice are also points of the weight lattice, yet there are weights that do not belong to the root lattice. What have these weight vectors got to do with the linear representations of a Lie algebra? The answer to this question is very simple and once formulated it shows the geniality of Weyl's approach. In every representation each of the Lie algebra elements is associated with a matrix. Utilizing the Cartan-Weyl basis displayed in Eq. (5.4.1) we see that we have two types of generators, the Cartan generators H_i and the step operators E^α associated with the roots. In every linear representation D the images $D(H^i)$ of the Cartan generators can be simultaneously diagonalized since they commute. Hence a basis of the representation vector space V can be provided by a set of states[2] denoted $|\mathbf{w}>$ that satisfy the condition

$$D(H^i)\,|\mathbf{w}> = w^i\,|\mathbf{w}> \tag{6.2.5}$$

where we have named $\mathbf{w}_\mu$ the vector in $\mathbb{R}^\ell$ whose components are the ℓ eigenvalues of the ℓ Cartan generators $D(H^i)$ (the number ℓ is obviously the rank of the Lie algebra).

The main result of Weyl's theory is that all the possible vectors $\mathbf{w}_\mu$, named *weights* lie in the weight lattice we have previously defined. Furthermore every irreducible representation D_μ contains a finite number s_μ of weights organized in a set $\boldsymbol{\Pi}_\mu \equiv \{\mathbf{w}_1, \ldots, \mathbf{w}_{s_\mu}\}$. In $\boldsymbol{\Pi}_\mu$ there is a unique highest weight $\mathbf{w}_{\rm highest}$ defined by

---

[2]In this context it is quite convenient to utilize the notations and the nomenclature of quantum mechanics where, as basis of the Hilbert space of physical states, one utilizes the *eigenstates* of a complete set of commuting observable operators $\mathcal{O}_{1,2,\ldots,n}$. According with Dirac, these eigenstates are denoted $|O_1, O_2, \ldots, O_n>$ naming $O_{1,2,\ldots,n}$ the eigenvalues of the considered observables. See later on in this chapter for an introduction to functional spaces and the Hilbert space.

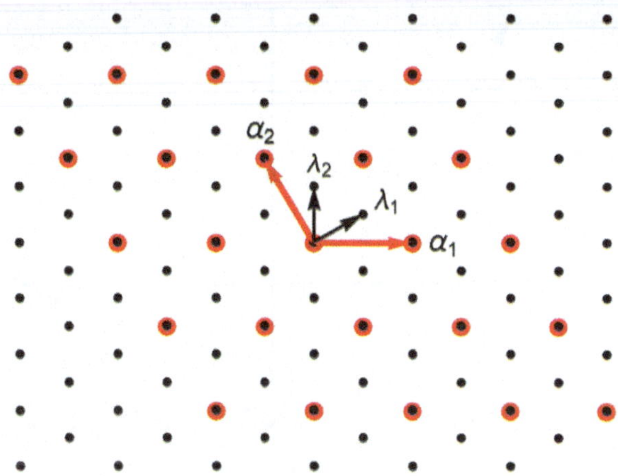

Fig. 6.2 The weight and the root lattice of the $\mathfrak{a}_2$ Lie algebra. The points of the root lattice are marked in red, while those of the weight lattice are marked in black. The simple roots $\alpha_{1,2}$ and the simple weights are also displayed $\lambda_{1,2}$

the condition that the corresponding eigenstate is annihilated by all step operators E^α associated with positive roots $\alpha > 0$:

$$E^\alpha \,|\mathbf{w}_{\text{highest}} >= 0 \tag{6.2.6}$$

The highest weight uniquely defines the irreducible representation since all the other weights can be obtained from the highest one by subtracting a suitable number of roots, until you reach a lowest weight[3] that is annihilated by all step operators E^α associated with negative roots $\alpha < 0$.

The final beauty of the formalism is the geometrical characterization of the available highest weights that are all contained in the so named *Weyl chamber*. What is this latter? It is the convex hull delimited by the *Weyl walls*, namely by the hyperplanes orthogonal to the ℓ simple roots α_i. Hence all the available representations are in one-to-one correspondence with the weight lattice points that lie in the Weyl chamber.

We exemplify the above concept utilizing once again the case of the $\mathfrak{a}_2$ Lie algebra. The Weyl chamber and the weights of one of the representations (the fundamental defining one) are displayed in Fig. 6.3. Utilizing the weight formalism, Weyl was able to bring the theory of semisimple Lie groups and of their linear representations to absolute perfection. The tale initiated many decades before in the turbolent Paris spring of 1870 reached a firm stationary point in 1939 with Weyl's masterpiece book *Classical Groups* [168]. Still the contributes of Dynkin and Coxeter to come just a few years later would provide a further improvement in notation and a powerful graphical instrument to master classical group theory and in particular all the implications of the Weyl group, yet it is fair to say that when World War II broke out the algebraic substratum of the current *episteme* encoded in points A) to E)

[3] Indeed the possible number of α root subtractions from a given weight $\mathbf{w}$ is completely determined by the scalar product of the weight with the root $< \mathbf{w}\,,\ \alpha >$.

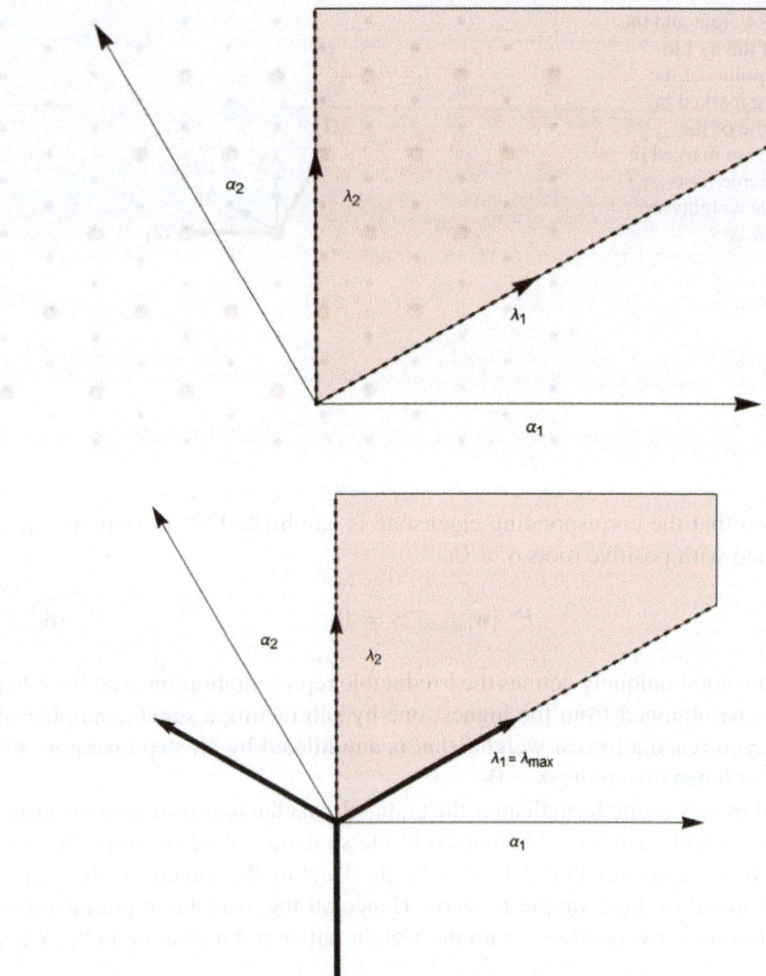

Fig. 6.3 In the picture on top the Weyl chamber of the $\mathfrak{a}_2$ Lie algebra is the shaded infinite region. In the second picture we display the three weights of the fundamental defining representation of $\mathfrak{a}_2 \sim \mathfrak{sl}(3, C)$. The weight that falls in the Weyl chamber (actually on one of its walls) is the highest weight; the other two are subdominant weights

mentioned in Chap. 1 was ready. In the next chapters we trace the historical development of the equally fundamental *geometrical pillar* of the current episteme. As we are going to advocate that pillar was also essentially ready by the very end of World War II with the establishment of the notions of fibre bundles and of the Ehresman connection on principal bundles that extends to connections on all of their associated vector bundles. Yet it still took about 40 years before these essential mathematical

Fig. 6.4 Henri Léon Lebesgue (1875–1941) was born in the city of Beauvais in Norther France in 1875. His father was a printer. He entered École Normale Supérieure in 1894 and graduated in mathematics in 1897. After some years as a school-teacher in secondary school in 1901 he wrote his most famous and most important paper Sur une généralisation de l'intégrale definie which appeared in the *Compte Rendus* the same year. This article contains the generalised definition of the definite integral based on the theory of measure. This contribution revolutionised modern integral calculus and it is the most important achievement of Henri Lebesgue. On the basis of this result he wrote his doctoral dissertation defended at the Faculty of Science of Paris in 1902. He had various academic appointments in French provincial Universities until he was appointed maître de conferérences at the Sorbonne in 1910. He published several other mathematical papers of high quality and had a distinguished academic career until his death in 1941, but for posterity his name is indissolubly entangled with the integral: the *Lebesgue integral* indeed

structures were fully integrated into the fabrics of theoretical physics. Indeed the notion of gauge–fields was parallely developed in physics and the complete identification of the mathematical leave with the physical one, according with C.N. Yang's *weltanshaung* recalled in Fig. 7.4, came later.

6.3 Infinite Dimensional Representations, Hilbert and Quantum Mechanics

Before plunging into the history of differential geometry, it is appropriate to shed some light on the parallel development, by the end of the XIXth through the beginning of the XXth century, of new ideas in the calculus of *functions*. These ideas led to *Functional Analysis*, namely to an entire subdiscipline of modern mathematics that constitutes today a major field of mathematical analysis, of high relevance in all pure and applied sciences. In the frame of our historical review, which aims at tracing the formation of the current episteme, we ought to stress that *Functional Analysis* provides the rigorous mathematical basis of Quantum Mechanics and is essential for Quantum Field Theory.

Here, once again, we intersect the intellectual path of Hermann Weyl and his major contributions to Group Theory in relation with Quantum Mechanics. Indeed functional analysis is tightly entangled both with the basic notion of *quantum states* and with the issue of *unitary representations* of a Lie group G. This happens when G is non-compact, as it is the case if G is the symmetry group of space–time. Following this way of arguing and in complete compliance with the mathematical way of thinking of Herman Weyl, one is finally led to *thrust through the fog of words and reach the solid reality of abstract symbols*, identifying the intuitive notion of a *particle* with a *unitary irreducible representation* of the Poincaré group, labeled by its two invariants, that we respectively name its *mass* and its *spin*.

This is a stimulating anticipation: let us proceed orderly.

The birth of *Functional Analysis* can be identified with some fundamental contributions of Hilbert dating 1909.

Conceptually there are several different mathematical issues that came together and crossed each other's way in this context. Let us enumerate them in some logical order.

(1) Throughout XIXth century the theory of functions underwent a spectacular development with the fundamental contributions of Fourier, Laurent, Cauchy, Riemann, Weierstrass, Fuchs and many others. At the dawn of the new century an issue came to prominence: that of integration and of integrable functions. The following idea started being conceived that we continue to cheer in the contemporary vision of mathematics. We would like to treat functions $f(x_1, \ldots, x_n)$ over a closed subset of $\mathbb{R}^n$ (an interval $[a, b] \subset \mathbb{R}$ in the one variable case) as if they were points in a suitable continuous space $\mathscr{H}$ that we are able to organize in some suitable way, introducing at the same time some notion of distance among its points. This is the proper setting in order to evaluate approximations and decide how close an approximant is to a given function. Therefore one focuses on continuous functions and looks for their integrability. Since the limit of a succession of continuous functions can be discontinuous, it came out that, in order to obtain that the limit of a succession of integrable continuous functions is also integrable one had to appropriately revise the notion of integral and this was done by Lebesgue (see Fig. 6.4).

(2) Secondly one had the issue of infinite dimensional vector spaces. Indeed, when the abstract notion of a vector space, originally envisaged by Grassman but, as we explained before, largely disregarded by the mathematical community, was finally and firmly established by Peano,[4] it appeared that it does not imply that it should be finite dimensional. If we allow for the possibility that a vector space V might admit an *arbitrary large number* of linearly indipendent vectors we arrive at the notion of an infinite dimensional vector space. Every n-dimensional real

[4]Look back at Sect. 3.2.5.

vector space $V_n(\mathbb{R})$ is isomorphic to $\mathbb{R}^n$. If we introduce a positive definite scalar product, the norm squared of a vector $\mathbf{v} \in V_n(\mathbb{R})$ is $N^2(\mathbf{v}) = \sum_{i=1}^n v_i^2$, where $v_i \in \mathbb{R}$. On the other hand every n–dimensional vector space $V_n(\mathbb{C})$ over the complex number field is isomorphic to $\mathbb{C}^n$. Introducing a hermitian scalar product the norm squared of a vector $\mathbf{v} \in V_n(\mathbb{C})$ is $N^2(\mathbf{v}) = \sum_{i=1}^n |v_i|^2$ where $v_i \in \mathbb{C}$. Then it is natural to assume that an infinite dimensional complex vector space $V_\infty(\mathbb{C})$ is isomorphic to the space of infinite successions $(v_1, v_2, \ldots, v_n, \ldots)$ of complex numbers; in other words an infinite dimensional vector space is, in a formal sense, $\mathbb{C}^\infty$. In this case, however, in order to give a meaning to such successions of complex numbers we have to assume that they define convergent series, namely:

$$\sum_{n=1}^{\infty} |v_n|^2 < \infty \tag{6.3.1}$$

(3) Issue (1) and (2) were brought together by the issue of orthogonal functions. Indeed the available functional spaces turn out to be the spaces of square integrable functions $L^2_w(\Sigma)$ that are maps from a closed domain $\Sigma \subset \mathbb{R}^n$ into $\mathbb{C}$, satisfying the following condition:

$$\begin{aligned} L^2_w(\Sigma) \ni \mathbf{f} \quad &: \quad \Sigma \longrightarrow \mathbb{C} \\ \Downarrow \quad & \\ N^2(\mathbf{f}) \quad &= \int_\Sigma |\mathbf{f}(\mathbf{x})|^2\, w(\mathbf{x})\, d^n\mathbf{x} < \infty \end{aligned} \tag{6.3.2}$$

where $w(\mathbf{x})$ is a suitable positive definite real function that is named the measure. Such spaces have a naturally defined hermitian scalar product:

$$\begin{aligned} \forall f(\mathbf{x}), g(\mathbf{x}) \in L^2_w(\Sigma) \quad ; \quad (\mathbf{f}\,,\,\mathbf{g}) &\equiv \int_\Sigma \overline{f(\mathbf{x})}\, g(\mathbf{x})\, w(\mathbf{x})\, d^n\mathbf{x} \\ (f\,,\,g) &= (g\,,\,f)^\star \end{aligned} \tag{6.3.3}$$

and the problem is raised of finding complete orthonormal basis of functions $e_m(\mathbf{x})$ such that

- $(e_n\,,\,e_m) = \delta_{n,m}$
- and $\forall f(\mathbf{x}) \in L^2_w(\Sigma)$ we can write $f(\mathbf{x}) = \sum_{n=1}^\infty a_n\, e_m(\mathbf{x})$ with $a_n \in \mathbb{C}$

When such bases are found we see that:

$$N^2(\mathbf{f}) = \sum_{n=1}^{\infty} |f_m|^2 < \infty \tag{6.3.4}$$

and we can identify the vector $\mathbf{f}$ with its succession of coefficients a_m. The issue of orthogonal functions in spaces $L^2_w(\Sigma)$, that we shall collectively name *Hilbert spaces*, is the appropriate way to reorganize a lot of XIXth century results about so named orthogonal polynomials and other classical special functions.

(4) The issue of differential equations and of self-adjoint (or hermitian) operators in a Hilbert space $\mathscr{H} = L^2_w(\Sigma)$ is the third main item in the present list. Typically an operator in a functional space is a linear differential operator[5]:

$$\forall f \in \mathscr{H} \quad : \quad \mathscr{L} f = \sum_{m=0}^{p} c_m(x) \frac{d^m f}{dx^m} \in \mathscr{L} \tag{6.3.5}$$

and one is led to define differential operators $\mathscr{L}$ that are self-adjoint with respect to the hermitian scalar product that exists on that functional space, namely such that:

$$\forall f, g \in \mathscr{H} \quad : \quad (g, \mathscr{L} f) = (\mathscr{L} g, f) \tag{6.3.6}$$

The spectrum of eigenvalues of such operators is composed of real numbers λ_n:

$$\mathscr{L} f_n(x) = \lambda_n f_n(x) \quad ; \quad \lambda_n \in R \tag{6.3.7}$$

As it happens in the case of finite dimensional vector spaces, the eigenfunctions of an operator corresponding to different eigenvalues are orthogonal among themselves and this provides a connection between this problem and the problem of orthonormal basis for functional Hilbert spaces, in particular orthogonal polynomials. This issue is strongly related with the conceptions of Quantum Mechanics. Indeed, in Quantum Mechanics the Hilbert space $\mathscr{H}$ is considered as the space of possible quantum states of some physical system and one searches for bases of $\mathscr{H}$ composed of eigenstates of a maximal set of commuting self-adjoint operators $\mathscr{O}_i$ named the *observables*. The eigenvalues of these latter constitute the labels λ_i of any given quantum state S. The representative function $\Psi_S(\mathbf{x}) \in \mathscr{H}$ is the *wave function* of the abstract quantum state S and its square modulus $|\Psi_S(\mathbf{x})|^2$ represents the probability that the quantum particle in the state S can be found at place $\mathbf{x}$ in the ambient space $\mathscr{M}$. Since the sum of probability over all possibilities must be necessarily one, the function $\Psi_S(\mathbf{x})$, properly normalized, must be square integrable:

$$\int_{\mathscr{M}} |\Psi_S(x)|^2 \, w(\mathbf{x}) \, d^n = 1 \tag{6.3.8}$$

(5) The last issue in this list which is strongly related with the issue above and with the issue (2) is that of unitary representations of non compact groups.
We introduced in Eq. (3.1.16) the notion of group representation that was resumed for the case of Lie Groups in Sect. 6.2. Let V be a vector space, any homomorphism

[5] In the definition below, for simplicity we confine ourselves to the case where the functional space is composed of functions of only one variable x.

$$D \ : \ G \longrightarrow \mathrm{Hom}(V, V)$$
$$\forall g_1 \, , \, g_2 \in G \ : \ D\left(g_1 \cdot g_2\right) = D\left(g_1\right) \cdot D\left(g_2\right) \tag{6.3.9}$$

is named a linear representation of the group G and the dimension of the vector space V is named the dimension of the representation.

Definition 6.3.1 A representation D of a group G is named **unitary** if the images $D(g)$ of all group elements are unitary operators $D(g)^\dagger = D(g)^{-1} = D(g^{-1})$.

To this effect let us recall the definition of the hermitian conjugate $\mathscr{A}^\dagger$ or adjoint of a linear operator $\mathscr{A}$ in a generic vector space V endowed with a scalar product (,):

$$\forall f, g \in V \ ; \ \overline{(g, A\, f)} = \left(A^\dagger\, g \, , \, f\right) \tag{6.3.10}$$

which applies equally well to the case of finite and infinite dimensional vector spaces.

Why is the unitarity property so much relevant? The answer is that it is essential in connection with Quantum Mechanics and, in a wider context, with Quantum Field Theory, because of what we said above on the probabilistic interpretation of the wave-functions. Suppose that a physical system has some symmetry encoded in a Lie group G. Then G will act on the space of the quantum states $\mathscr{H}$ that, by the fundamental principle of *superposition*, ought to be a vector space. Hence $\mathscr{H}$ is the carrier space of a *linear representation* $D(G) \, : \, \mathscr{H} \to \mathscr{H}$. Because of the unavoidable conservation of probability, the representation $D(G)$ has to be unitary so that all scalar products and norms of wave-functions are G-invariant.

Let us now consider the case of continuous groups, in particular of Lie groups G, that have a fundamental direct connection with the notion of Lie algebras, as extensively discussed in previous chapters. Essentially a Lie group G is:

- A group from the algebraic point of view, namely a set with an internal composition law, the product
$$\forall g_1 \, , \, g_2 \in G \quad g_1 \cdot g_2 \in G \tag{6.3.11}$$
which is associative, admits a unique neutral element e and yields an inverse for each group element.
- A smooth continuous space of finite dimension $\dim G = n < \infty$ and the two algebraic operations of taking the inverse of an element and performing the product of two elements are continuous and even analytic, namely admit a power series expansion.

The fact that a Lie Group is also a continuous space, actually a *differentiable manifold*, according with the notion discussed in next chapter, is what introduces the fundamental distinction between compact and non-compact groups. The difference between compact and non-compact spaces is best illustrated by the comparison of a sphere (prototype of a compact manifold) with a hyperboloid (prototype of a non compact smooth manifold) (see Fig. 6.5). In a compact manifold, like the sphere, all continuous curves necessarily converge to some point of the manifold, while in a

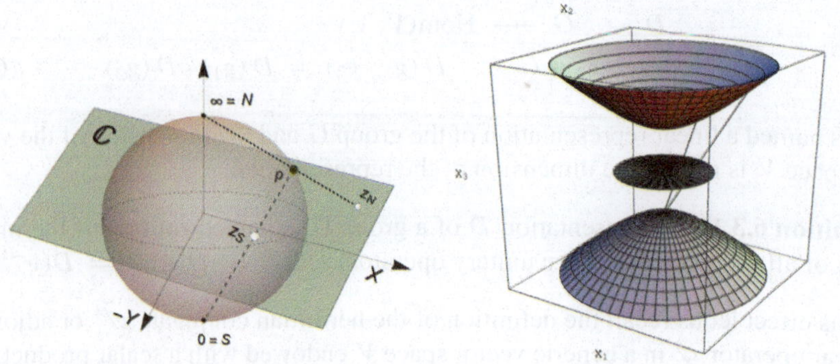

Fig. 6.5 In this figure we visually compare two paradigmatic instances of a compact manifold and of a non–compact one: the sphere and the hyperboloid

non–compact one there are continuous curves that have no limiting point and can extend indefinitely.

The connection between Lie Groups and Lie Algebras that we have repeatedly quoted, is provided by the consideration of group elements infinitesimally close to the identity element **e**. Developping in power series of a suitable parameter λ every group element $g \in G$ can be written as follows:

$$g = \exp[\lambda \mathbf{X}] \simeq e + \lambda \mathbf{X} + \mathcal{O}(\lambda^2) \quad (6.3.12)$$

where **X** is an element of a Lie algebra $\mathbb{G}$, whose dimension is equal to the dimension of the Lie group G, that is completely determined by G.

Any linear representation of the Lie group G induces a linear representation of its Lie algebra $\mathbb{G}$ and viceversa.

In the case of a unitary representation of the group G the corresponding representation of the Lie algebra $\mathbb{G}$ is provided by antihermitian operators:

$$\forall \mathbf{X} \in \mathbb{G} \quad ; \quad D(\mathbf{X})^\dagger = -D(\mathbf{X}) \quad (6.3.13)$$

which, multiplying just by an i-factor, provides a relation with self-adjoint operators.

The key point in connection with the above discussion is provided by the following general theorem of group theory that we state without proof.

Theorem 6.3.1 *Let G be a non-compact Lie group. All the unitary representations of G are necessarily infinite dimensional and for this reason they are provided by suitable* L^2_w *functional spaces. The generators* T_I *of the Lie algebra* $\mathbb{G}$ *are represented by* $i \times \mathscr{L}_I$, *where* $\mathscr{L}_I$ *are self-adjoint operators in the Hilbert space* L^2_w.

Hence most of the interesting self-adjoint differential operators in Hilbert spaces are actually generators of suitable Lie groups and the solutions of the corresponding differential equations are eigenvectors of such generators. Conversely when we consider physical space-time, its basic symmetries are provided by non compact groups

G as the Euclidian group of roto-translations in three or higher dimensions or as the Lorentz group in special relativity and its extension with translations, namely the Poincaré group.

The last sentence above provides the clue to what we anticipated concerning the rigorous formulation of the concept of a particle in Quantum Field theory. The space of quantum states of an elementary relativistic particle moving in Minkowsky space coincides with the carrier space of a *unitary, irreducible representation* of the Poincaré group and the wave-function corresponding to a specific state is a function in this functional space. So even the physical idea of a particle is just a group theoretical notion!

6.3.1 David Hilbert

The Eastern Prussian city of Königsberg is by now the fully Russian speaking town of Kaliningrad, capital of the homonimous Russian Enclave Region on the Baltic shores. It is a rather dull place where World War II seems to have ended only a few months ago and everything appears as strange as it might possibly be. Orthodox churches have an improbable gothic appearance while cottages with distinctively german spiked roofs, once upon a time clean and tidy, are still orderly lined up on dirty side roads, yet their broken glasses, their untidy courtyards, occasionally populated by chickens and uniformely filled up with all sorts of wastes, generate the impression that the strict lutheran bourgeois families that were their legal owners and inhabitants have just fled away, forced to evacuate the entire region in favor of an equally forced host of immigrants from the distant Ural Region. No one speaks German or has German ancestors, yet everywhere in the centre of the town you find german styled *Bierstuben* where, besides drinking good german ales, you can eat *bratwurst* with *sauerkraut*. Furthermore the Russian Federal University of the Baltic is pompously named after Immanuel Kant whose thumbstone stands in the churchyard of the cathedral erected on the little island at the centre of the river Pregel.

Lively capital of Eastern Prussia and one of the historic harbour cities of the Anseatic League, Königsberg was formerly a very important centre for Culture and Science and occupies a distinctively brilliant place in the history of Philosophy and Mathematics. Here, throughout most of the XVII century, lived, taught and developed his philosophy Immanuel Kant, here in the second half of the XIX century Wilhelm Killing discovered Lie Algebras independently from Sophus Lie and classified the simple ones prior to Cartan. Here were professors of mathematics for several years Ferdinand von Lindemann and Adolf Hurwitz. In Königsberg University studied both Hermann Minkowski and David Hilbert who became close friend and were later colleagues at Göttingen University.

David Hilbert (see Fig. 6.6) is an overwhelming figure in the history of modern mathematics and also of physics. His contributions have been of the highest quality and very much extensive, yet his influence goes beyond the specificity of his own results. He exerted an intellectual guiding role over the whole field. He was born in

Fig. 6.6 David Hilbert (1862–1943). The most influential mathematician of the early XXth century

Königsberg in 1862, two years earlier than his friend Minkowski, whose birth place, in the province of Kaunas within contemporary Lithuania, was at the time part of the Russian Empire, yet quite close to Königsberg. David Hilbert's father was a high ranking judge and his mother Maria, a well educated daughter of a rich merchant, with deep interests in philosophy and astronomy, cared about David's early education. He went to school at the age of eight and was not a particularly brilliant student until the time he shifted from the more classically oriented Friedrichskolleg to the Wilhelm Gymnasium, where he completed his final year of high school, prior to University, and could finally find some response to his natural inclination towards mathematics.

When Hilbert enrolled at the University of Königsberg in 1880 he started an academic career that in the course of twenty years would bring him to be professor of mathematics in Göttingen and probably the most influential mathematician of the world in his own time. Utilizing internal *ante-litteram* Erasmus programmes of the German Empire, Hilbert went for a semester to Heidelberg where he attended lectures given by Lazarus Fuchs. Back to Königsberg he could benefit from the lectures on modular forms of Adolf Hurwitz and graduated in 1884 under the supervision of Lindemann, having met among his fellow students Minkowsky who became one of his life-time closest friends.

Quite influential on Hilbert's mathematical career was the figure of Felix Klein whose lectures Hilbert attended in Leipzig before Klein shifted to Göttingen, leaving his chair of Geometry to Sophus Lie. In 1886 Klein organized a visit of Hilbert and Study to Paris where the two young and promising German mathematicians met with the *gotha* of French mathematics, namely with Henri Poincaré, Camille Jordan,

Fig. 6.7 Charles Hermite (1822–1901). Born in Dieuze, Moselle, Hermite died in Paris while being professor of the Paris Faculty of Sciences. Throughout his life Hermite suffered both physically and psychologically from a defomormity in his right foot which affected him from his birth. For instance he was dismissed from Ècole Polytechnique as not being fit to a school that was military in character. Notwithstanding such difficulties he made a brilliant career in mathematics and was eventually professor of the Ècole Polytechnique that had not accepted him as a student. Among his doctoral thesis students Henri Poincaré is the most eminent one. His major contributions to pure mathematics are in number theory in the theory of orthogonal polynomials and in that of elliptic functions. His name is associated with *Hermite Polynomials*, *Hermitian operators and metrics* and with the *Hermite normal form of matrices*

Gaston Darboux, Pierre Bonnet and in particular with Charles Hermite (see Fig. 6.7). Always in the highest esteem of Felix Klein, Hilbert benefited from the decisive promotion action of the former who twice tried to appoint him full professor of mathematics in Göttingen succeeding in his mission the second time in 1895. In the meantime, Hilbert had turned down Berlin University offer to appoint him on the vacant chair of Lazarus Fuchs. He had also proven the so named finiteness theorem which recites as follows.

Theorem 6.3.2 (Hilbert finiteness) *If* G *is a Lie group whose finite dimensional representations are completely reducible, then the ring of invariants of* G *acting on a finite dimensional vector space* V *is finitely generated.*

That above is actually part of a more abstractly formulated **Basis Theorem** proved by Hilbert in commutative algebra that turned out to be the very corner stone of modern Algebraic Geometry. The little story behind the publication of this theorem is quite revealing both historically and conceptually. Throughout the XIXth century the theory of invariants was a very hot topic studied by many scientists, Cayley and Sylvester in particular, as we already remarked in Sect. 3.2. The approach of everyone working in the subject was direct and computational and Paul Gordan, the German leading expert in this field, made no exception. Hilbert's approach was different and revolutionary, it was just abstract and purely formal. Submitted in 1888 for publication to Felix Klein as Editor of *Mathematische Annalen*, Hilbert's paper had Paul Gordan as referee. Gordan's report is traditionally reputed to include the sentence: *This is theology, not mathematics*. Upon Hilbert insistence formulated as follows: *...I am not prepared to alter or delete anything and regarding this paper, I say with all modesty, that this is my last word so long as no definite and irrefutable objection against my reasoning is raised*, Felix Klein published the paper on the Annalen in its original form. Later after a second more extended paper on the same subject was published by Hilbert, also Gordan agreed that even theology has its own merits.

Next fundamental achievements by Hilbert were in Algebraic Number Theory. The modern discipline of Class Field Theory was essentially founded by Hilbert's paper *Zahlbericht* that summarized and systematized all previous' results by Kummer, Kronecker and Dedekind, introducing into the texture a wealth of new mathematical ideas. Then he turned to Geometry that he posed into a formal axiomatic setting. His views were published in 1899 in a book *Grundlagen der Geometrie* that was reissued in many subsequent editions and exerted a major influence in promoting the axiomatic approach to mathematics throughout the XXth century.

The Second International Congress of Mathematicians, held in Paris in 1900 was the occasion for Hilbert to present his very famous list of 23 problems that challenged and still challenge mathematicians to solve fundamental questions. Several of Hilbert's problems were solved in the XXth century and each obtained solution corresponded to a major upgrading of mathematical thinking.

About the years 1909–1910 Hilbert laid the corner stones of modern *functional analysis* and introduced the concept of *Hilbert Space* which we have already discussed. In the short run, this mathematical new object that provided the appropriate conceptual frame encompassing all XIXth century results on Fourier series, Laplace transforms, orthogonal polynomials and the like, provided also the appropriate language for the formulation of Quantum Mechanics which was just about to be born.

Hilbert's interests were approaching the field of Physics closer and closer and since 1912 he got himself so much involved into it that the most influential mathematician of the world hired a private physics teacher to give him lectures so as to progress more quickly in his own self-education.

Spectacular was the result. Apart from some papers on the kinetic theory of gases, November 20th 1915, five days before the submission of Einstein's paper containing the field equations of General Relativity, Hilbert submitted an article that some one pretended to contain Einstein's field equations. Yet December 6th 1915, while correcting the proofs of his own work, Hilbert deleted Einstein's equation, if ever they were there, and added the following sentence that firmly established Einstein's priority: *The differential equations of gravitation that result are, as it seems to me, in agreement with the magnificent theory of general relativity established by Einstein in his later papers.*

The differential equations that result. So said Hilbert. Result from what? From the action of pure General Relativity that has been duely named the Hilbert-Einstein action and does not appear in Einstein's paper.

When Hilbert retired in 1930 he was at the peak of his fame and Göttingen was probably the most prestigious and active centre for mathematical physical sciences of the entire world. Three years later, when the Nazis came to power, everything changed and rapidly decayed. Jewish scientists were dismissed or fled mostly replenishing Princeton's ranks. Göttingen became an empty place and last years of Hilbert's life were somewhat gloomish in that deserted place. He died in 1943 in the middle of World War Two. On Hilbert's tombstone the following sentence of his was engraved:

Wir müssen wissen, wir werden wissen

Namely *We must know, we shall know.*

6.4 Concluding Remarks on the Idea of Functional Spaces

The first important book written by Hermann Weyl, Hilbert's most outstanding student, was entitled *The Idea of a Riemann Surface.* In Weyl's spirit I entitled the present section *Concluding Remarks on the Idea of Functional Spaces.* My goal is to outline the logic that led to Hilbert spaces, whose elements are, in a sense to be explained, *functions.* The connection of these mathematical developments with the development of Physics in the XXth century is very strong.

Classical Physics, as it evolved from the XVIIIth to the XIXth century, finally led to the notion of *phase-space.* All the possible states of a physical system are provided by the points of an even dimensional manifold $\boldsymbol{\Phi}$, whose coordinates we denote $\left\{p^i, q_i\right\}$, $(i = 1, \ldots, n)$, that is endowed with a symplectic structure encoded in a closed two form $\boldsymbol{\Omega}$ defined over it. The dynamical evolution of a physical system is described by curves in phase space, determined by $2n$ first order differential equations, the *Hamilton Equations.* With the advent of Quantum Mechanics and of Schrödinger Equation this mathematical picture changed completely. The state of a system is no longer a point in a manifold, rather it is described by a function $\Psi(q)$ of the generalized coordinates q_i or of the momenta p^i. This object, named the **wave function**, has its own support on half of the phase space $\boldsymbol{\Phi}$, while the other half of the canonical variables become a set of operators, typically differential, acting on the possible wave functions. Hence Quantum Mechanics posed the problem of characterizing the **space of states** of a physical system as a space whose points are functions. The superposition principle or, differently stated, the linearity of Schrödinger equation, implies that quantum states form a vector space. In this way the conception of infinite dimensional vector spaces whose elements (vectors) are "*functions*" emerged almost at the same time from the logical development of mathematical and physical theories.

Here we outline the mathematical logical path.

(a) The first problem to be solved arises from the observation that there are infinite successions of continuous functions defined over an interval of the real line $\mathbb{R}$ or over some closed subset of $\mathbb{R}^n$ whose limit is a discontinuous function. It follows that if we were to consider a functional space defined as the space of continuous function or of differentiable functions, it might be not complete, namely it might miss the limit of sequences of its own points. One needs some different definition.
(b) Another mathematical discomfort with spaces of functions is that the Riemann definition of an integral is not apt to cope with the case of discontinuous, yet bounded functions which, as we have seen, can be the limit of successions of Riemann integrable continuous functions.
(c) The French mathematician Lebesgue provided the solution to both the above problems introducing a generalized definition of the integral, coinciding for continuous functions with the Riemann definition, yet allowing the calculation of the integral of discontinuous functions as well. Lebesgue integration theory is based on the notion of **measure** $\mu(\mathscr{S})$ of subsets of the real line $\mathscr{S} \subset \mathbb{R}$ and introduces the notion of **measure zero sets**.
(d) Relying on the Lebesgue integral one arrives at a good definition of functional spaces that are complete, by considering the spaces $L^2_w(\Sigma)$ of those complex valued functions defined over a region $\Sigma \subset \mathbb{R}^n$ (we will mainly focus on the case $n = 1$) whose squared norm $|f(\mathbf{x})|^2$ is integrable over Σ yielding a finite result. The fundamental theorem named after Fischer Riesz states that the space $L^2_w(\Sigma)$ is complete. Since the addition of a function that is non-zero only on measure zero sets does not change the value of any Lebesgue integral, it follows that the elements of $L^2_w(\Sigma)$ are not exactly functions $f(x)$ rather they are equivalence classes of functions with respect to the following equivalence relation $f(x) \sim g(x)$ if the difference $f(x) - g(x)$ does not vanish only on measure zero sets.
(e) Having established that $L^2_w(\Sigma)$ is a complete vector space one looks for bases of orthonormal functions $\mathbf{e}_n(x)$ such that any element $f \in L^2_w(\Sigma)$ can be expanded as $f(x) = \sum_{n=0}^{\infty} a_n \, \mathbf{e}_n(x)$. Once a basis is singled out the space $L^2_w(\Sigma)$ becomes isomorphic to an abstract Hilbert space $\mathscr{H}$ whose elements are, by definition, successions of complex numbers a_i the series of whose norms is convergent $\sum_{i=0}^{\infty} |a_i|^2 < \infty$.
(f) The hint where to look for basis of functions is provided by **Weierstrass theorem** stating that a continuous bounded function can be approximated with arbitrary accuracy by a polynomial of sufficiently large degree. Since every element of $f \in L^2_w(\Sigma)$ can be seen as the limit of a succession of continuous functions, it follows that the polynomials and in particular the monomials x^n provide a basis of functions. Typically such a basis is not orthonormal.

(g) The Gram–Schmidt orthonormalization algorithm utilized in finite dimensional vector spaces is iterative and inductive. Hence it can be easily extended to the infinite dimensional case. This shows that for $L^2_w(\Sigma)$ one can find orthonormal bases composed of orthogonal polynomials.
(h) The XIXth century French mathematician Rodrigues developed an ingenious algorithm for the direct construction of families of orthogonal polynomials associated suitable weight functions $w(x)$.

The above long list of chained implications has been displayed in order to show Weyl's *mathematical way of thinking* in action. Once set on the move, mathematical ideas develop a long way, propelled by the inner force of their own logical implications. Usually the result of this process is not only a new mathematical theory, rather we also witness new additions to, or even substantial revisions of, the physical episteme. Conversely new physical ideas always produce new lines of thinking in mathematics. Pivot in these processes is always the Idea of Symmetry, which, after Galois, means Group Theory. Thus the problem of unitary representations of non–compact groups happily marries with the new visions of Physics entrained by Quantum Mechanics and leads to the Idea of Functional Spaces.

Chapter 7
A Short History of Differential Geometry

My work always tried to unite the truth with the beautiful, but when I had to choose one or the other, I usually chose the beautiful....

Hermann Weyl

7.1 Conceptual Introduction from a Contemporary Standpoint

In Chap. 5, before exposing the long and exciting history of Lie group discovery, we remarked that differential geometry is at the basis not only of General Relativity but of all those Gauge Theories by means of which XXth century Physics obtained a consistent and experimentally verified description of all Fundamental Interactions. We also noted that the central notions in differential geometry and in its application to the description of the Physical World are those which fix the geometric environment:

- Differentiable Manifolds
- Fibre-Bundles

and those which endow such environments with structures accounting for the measure of lengths and for the rules of parallel transport, namely:

- Metrics
- Connections

In this chapter we plan to outline the one century long historic process which led to establish such fundamental concepts forming nowadays integral part of the episteme as summarized in points (A)–(E).

As we already did in previous chapters, we find it convenient to begin from stating the end point result of the historical development we are concerned with by recalling, formulated in contemporary language, the definition of the mathematical structures which constitute the object of study of differential geometry, namely differentiable manifolds. This is specially important in view of the ultimate goal of the present essay

P. G. Fré, *A Conceptual History of Space and Symmetry*,
https://doi.org/10.1007/978-3-319-98023-2_7

which is to illustrate the historic evolution of the concepts of *Space* and *Symmetry*. Differentiable Manifolds is indeed the present day interpretation of the notion of Space.

7.1.1 Differentiable Manifolds

First and most fundamental in the list of geometrical concepts we need to introduce in order to address differential geometry is that of a *manifold* which corresponds, as we already explained, to our intuitive idea of a *continuous space*. In mathematical terms this is, to begin with, a *topological space*, namely a set of elements where one can define the notion of *neighborhood* and *limit*. This is the correct mathematical description of our intuitive ideas of vicinity and close-by points. Secondly the characterizing feature that distinguishes a manifold from a simple topological space is the possibility of labeling its points with a set of coordinates. Coordinates are a set of real numbers $x_1(p), \ldots, x_D(p) \in \mathbb{R}$ associated with each point $p \in \mathcal{M}$ that tell us *where* we are. Actually in General Relativity each point is an event so that coordinates specify not only its *where* but also its *when*. In other applications the coordinates of a point can be the most disparate parameters specifying the state of some complex system of the most general kind (dynamical, biological, economical or whatever).

In classical physics the laws of motion are formulated as a set of differential equations of the second order where the unknown functions are the three cartesian coordinates x, y, z of a particle and the variable t is time. Solving the dynamical problem amounts to determine the continuous functions $x(t), y(t), z(t)$, that yield a parametric description of a curve in $\mathbb{R}^3$ or better define a curve in $\mathbb{R}^4$, having included the time t in the list of coordinates of each event. Coordinates, however, are not uniquely defined. Each observer has its own way of labeling space points and the laws of motion take a different form if expressed in the coordinate frame of different observers. There is however a privileged class of observers in whose frames the laws of motion have always the same form: these are the inertial frames, that are in rectilinear relative motion with constant velocity. The existence of a privileged class of inertial frames is common to classical newtonian physics and to Special Relativity: the only difference is the form of coordinate transformations connecting them, Galileo transformations in the first case and Lorentz transformations in the second. This goes hand in hand with the fact that the space–time manifold is the *flat affine*[1] *manifold* $\mathbb{R}^4$ in both cases. By definition all points of $\mathbb{R}^N$ can be covered by one coordinate frame $\{x^i\}$ and all frames with such a property are related to each other by general linear transformations, that is by the elements of the general linear group GL(N, $\mathbb{R}$):

$$x^{i'} = A^i{}_j \, x^j \quad ; \quad A^i{}_j \in \mathrm{GL}(\mathrm{N}, \mathbb{R}) \tag{7.1.1}$$

[1]A manifold (defined in this section) is named *affine* when it is also a vector space.

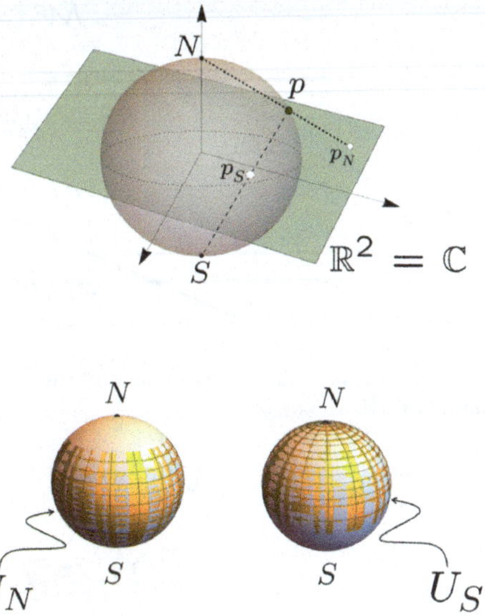

Fig. 7.1 The two open charts composing an atlas of the two-sphere: the north pole and the south pole chart described by the stereographic projection

In plain words what we have written above means that the coordinates $x^{i'}$ of the new coordinate system are linear combinations with some numerical coefficients of the coordinates in the old system:

$$x^{1'} = a^1_1 x^1 + a^1_2 x^2 + \cdots + a^1_n \,, x^n \tag{7.1.2}$$

and similarly for the others.

The restriction to the Galilei or Lorentz subgroups of GL(4, $\mathbb{R}$) is a consequence of the different *scalar product* on $\mathbb{R}^4$ vectors one wants to preserve in the two cases, but the relevant common feature is the fact that the space–time manifold has a vector–space structure (see Sect. 3.1.3 for the notion of vector space). The privileged coordinate frames are those that use the corresponding vectors as labels of each point.

A different situation arises when the space–time manifold is not flat, like, for instance, the surface of a hypersphere $\mathbb{S}^N$ (see Fig. 7.1).

As chartographers know very well there is no way of representing all points of a curved surface in a single coordinate frame, namely in a single *chart*. However we can succeed in representing all points of a curved surface by means of an *atlas*, namely by a collection of charts, each of which maps one open region of the surface and such that the union of all these regions covers the entire surface (see Fig. 7.2). Knowing the transition rule from one chart to the next one, in the regions where they overlap, we obtain a complete coordinate description of the curved surface by means of our atlas (see Fig. 7.3).

The intuitive idea of an *atlas* of *open charts*, suitably reformulated in mathematical terms, provides the very definition of a differentiable manifold, the geometrical concept that generalizes our notion of space–time, from $\mathbb{R}^N$ to more complicated non flat situations.

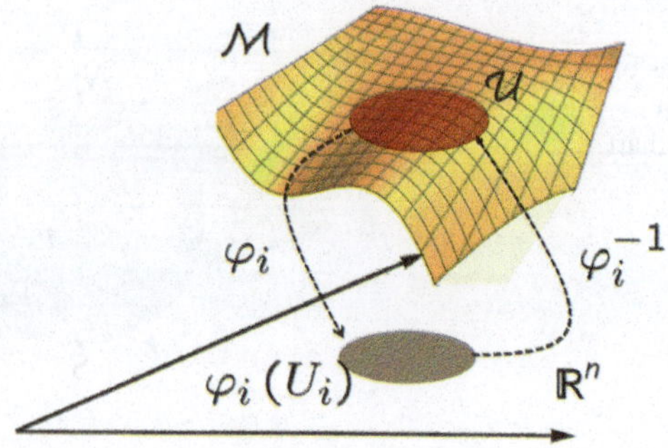

Fig. 7.2 An open chart is a homeomorphism of an open subset U_i of the manifold $\mathscr{M}$ onto an open subset of $\mathbb{R}^m$

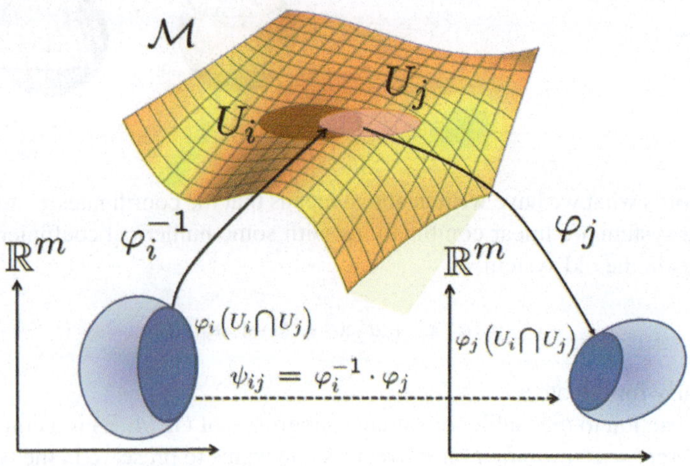

Fig. 7.3 A transition function between two open charts is a differentiable map from an open subset of $\mathbb{R}^m$ to another open subset of the same

There are many possible *atlases* that describe the same manifold $\mathscr{M}$, related to each other by more or less complicated transformations. For a generic $\mathscr{M}$ no privileged choice of the atlas is available differently from the case of $\mathbb{R}^N$: here the inertial frames are singled out by the additional *vector space* structure of the manifold, which allows to label each point with the corresponding vector. Therefore if the laws of physics have to be universal and have to accommodate non–flat space–times, then they must be formulated in such a way that they have the same form in whatsoever atlas. This is the principle of *general covariance* at the basis of General Relativity: all observers see the same laws of physics.

Similarly, in a wider perspective, the choice of a particular set of parameters to describe the state of a complex system should not be privileged with respect to any other choice. The laws that govern the dynamics of a system should be intrinsic and should not depend on the set of variables chosen to describe it.

7.2 The Second Stage in the Development of Modern Differential Geometry

Once the notions have been established, of a manifold and of a fibre-bundle, this latter to be illustrated in a later section, one considers differential calculus on these spaces and, rather than ordinary derivatives, one utilizes *covariant derivatives*. In General Relativity we mainly use the covariant derivative on the tangent bundle but it is important to realize that one can define covariant derivatives on general fibre bundles. Indeed the covariant derivative is the physicist's name for the mathematical concept of *connection* that we plan to recall at the end of this chapter. It is also important to stress that even restricting one's attention to the tangent bundle, the connection used in General Relativity is a particular one, the so called *Levi Civita connection* that arises from a more fundamental object the *metric*. A manifold endowed with a *metric structure* is a space where one can measure lengths, specifically the length of all curves. A generic connection on the tangent bundle is named *an affine connection* and the Levi Civita connection is a specific affine connection that is uniquely determined by the *metric structure*. Every connection has, associated with it, another object (actually a 2–form) that we name its curvature. The curvature of the Levi Civita connection is what we name the Riemann curvature of a manifold and it is the main concern of General Relativity. It encodes the intuitive geometrical notion of curvature of a surface or hypersurface. The field equations of Einstein's theory are statements about the Riemann curvature of space–time that is related to its energy–matter content. We should be aware that the notion of curvature applies to generic connections on generic fibre bundles, in particular on principal bundles. Physically these connections and curvatures are not less important than the Levi-Civita connection and the Riemann curvature. They constitute the main mathematical objects entering the theory of fundamental non-gravitational interactions, according to points (A)–(E) of the current *episteme* summarized in Chap. 1.

Having clarified what is the end point that was eventually reached, let us turn to illustrate the historic development of the fundamental conceptions underlying Differential Geometry.

7.3 The Development of Differential Geometry: A Historic Outline

One spring in the late seventies, C.N. Yang, the father of gauge–theories and one among the most extraordinary scientific minds of the XXth century, was invited by Scuola Normale di Pisa to give the traditional yearly series of Fermian Lectures. It was just the eve of the 1983 experimental discovery of the W and Z bosons, which finally confirmed that gauge theories are indeed the language adopted by Nature to express the fundamental forces binding matter together and driving the evolution of the physical world. Yang chose to start his recollection of gauge-theories and his

Fig. 7.4 A symbolic assessment by C.N. Yang that mathematics and physics have a common root

personal assessment of the entire subject from a symbolic picture of the kind shown in Fig. 7.4

Two leaves depart from the same stem: one is Mathematics, the other is Physics. The two leaves live parallel lives but have a common root and, as it happens in non euclidian geometry, parallels can and indeed do intersect. They frequently intersect and, equally frequently, interchange the role of guiding pivot. Examples are numberless both in recent and less recent history of science. This is a relevant but somehow trivial observation.

The most profound allusion in the typical Chinese symbolism of Yang's picture is the common stem of the two leaves. Mathematics and Physics were not that much socially and academically separated in the XIXth as they became in the XXth century and were further unified by a common denominator: the shared philosophical attitude of both the mathematician and the physicist, the latter assessing himself a natural philosopher.

The common root of Modern Mathematics and Modern Theoretical Physics is most prominently evident in the case of the concepts of *connections* and *metrics* whose history constitutes the main topics of the present chapter. From the physical side these mathematical notions encode all the *fundamental interactions* among *elementary matter constituents* as we already stressed several times. From the mathematical side they are the corner stones of Differential Geometry, Algebraic Topology and allied subjects.

The historical developments of both the notion of a *connection* and the notion of a *metric* are quite long, spreading over more than a century. They are strongly intertwined and involve both Physics and Mathematics in an alternate and entangled fashion.

Two fundamental geometrical problems are at the basis of both notions: the problem of *length* and the problem of *parallel transport*, namely how can we measure distances in a general continuous space and how do we assess the parallelism of two lines going through different points of that space. It turns out that these two apparently different problems are intimately related. Such a relation is the reason why connections and metrics, although genuinely different and independent mathematical notions can be, under certain conditions, related in a one-to-one fashion. The case where connection and metric are one-to-one related corresponds to Gravity and is the play-ground of General Relativity, while the case where the connection stands on his own feet is the case of non-gravitational interactions and encodes all the rest of Physics.

Historically the first notion of connection that was discovered is that of the *metric connection* named after its systematizer, the Italian mathematician Levi-Civita. The *Levi-Civita connection* is that appropriate to a Riemannian manifold, namely to a manifold equipped with a metric and it is analytically described by the 3-index symbols $\left\{ \begin{matrix} \mu \\ \nu\,\rho \end{matrix} \right\}$, introduced in the XIXth century by their German inventor, Elwin Bruno Christoffel. The development of such a notion is embedded in the development of Riemannian geometry, embracing the fall of the XIXth century and the dawn of the XXth, which is a mathematical tale strongly intertwined with the physical tale of Einstein's quest for General Covariance and the Theory of Gravity.

It took several decades in the first half of the XXth century and the work of several mathematicians to single out the notion of a *connection on a principal fibre bundle*, purified from association with a metric. In a completely independent way in 1954 Yang and Mills introduced the physical notion of non–abelian gauge fields which extends to all Lie groups G the structure of the electromagnetic theory based on the simplest of all such groups, namely U(1). In the course of due time the mathematical notion of a connection and the physical one of a Yang–Mills field were recognized to be identical.

Let us sketch the main outline of this crucial, century long intellectual development.

7.3.1 Gauss Introduces Intrinsic Geometry and Curvilinear Coordinates

The first appearance of a metric is in the 1828 essay of Gauss on curved surfaces (see Fig. 7.5). Written in latin, the *Disquisitiones Generales circa Superficies Curvas* [99] contains the major revolutionary step forward that was necessary to overcome the precincts of euclidian geometry and find a new *differential science* of spaces able to treat both flat and curved ones. Up to Gauss' paper, Geometry was either formulated abstractly in terms of Euclidian axioms or analytically in terms of *cartesian coordinates*. By Geometry it was meant the study of global properties of plane figures like triangles, squares and other polygons, or solids like the regular polyhedra. All such objects were conceived as immersed in an external space where it was implicitly assumed that one could always define the absolute distance $d(A, B)$ between any two given points A and B. Distance is the basic brick of the whole euclidian building and it is calculated as the length of the segment with end-points in A and B, lying on the unique straight line which goes through any such pair of distinct points.

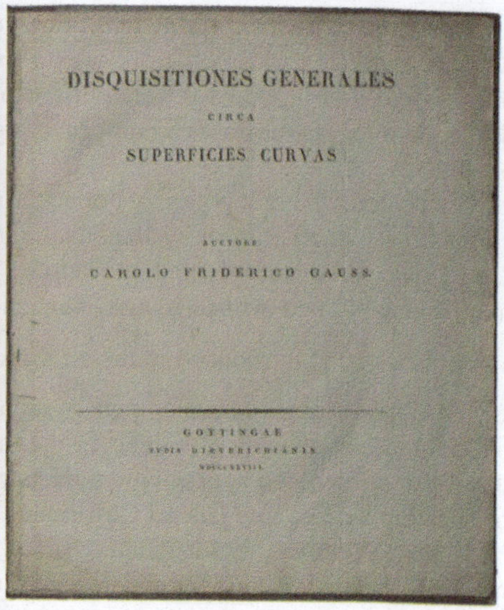

DISQUISITIONES GENERALES

CIRCA

SUPERFICIES CURVAS

CAROLO FRIDERICO GAUSS

GOTTINGAE

Fig. 7.5 Carl Friedrich Gauss (1777–1855). Gauss, *the King of Mathematicians*, was Professor at the University of Göttingen for many decades up to the very end of his long life. His contributions to all fields of mathematics were enormous and most profound

Curved surfaces were obviously known before Gauss, yet their shape and properties were conceived only through their immersion in three-dimensional space, considered unique and absolute, as pretended by Immanuel Kant who promoted euclidian geometry to an a priori truth lying at the basis of any sensorial experience. Gauss revolutionary starting point was that of reformulating the geometrical study of surfaces from an intrinsic rather than extrinsic viewpoint. He wondered how a little being, confined to live on the surface, might have perceived the geometry of his world. Rather than viewing the global shape of the surface Σ, unaccessible to his observations, the little creature would have explored its local properties in the vicinity of a point $p \in \Sigma$.

In order to study curved surfaces in these terms, Gauss understood that it was necessary to abandon cartesian coordinates as a system of point identification. In analytic geometry every point $p \in \mathbb{E} \equiv \mathbb{R}^3$ of euclidian space is singled out by a triplet of real numbers x, y, z which determine the distance of p from the three coordinate axes. As long as the points of the surface Σ are particular points of $\mathbb{E}$, they admit the labeling in terms of three cartesian coordinates, yet in such a description there is an excess of superfluous information. Why three coordinates when we are talking about a two-dimensional surface? Two should suffice. Gauss was the first to grasp the notion of curvilinear coordinates and invented *gaussian coordinates*. A very simple but revolutionary idea.

On the surface Σ let us consider a family of curves U such that each element of the family never intersects any other element of the same family, at least in the

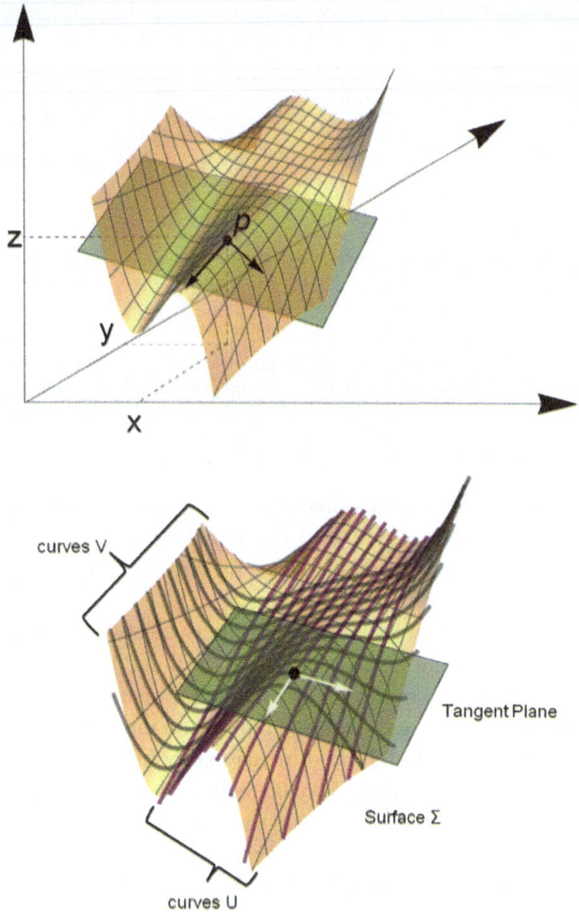

Fig. 7.6 The points p of a curved surface Σ can be labeled with the three cartesian coordinates x, y, z of the euclidian space in which M is immersed. Yet this is a redundant information. The two gaussian coordinates u and v are given through the construction of two systems of curves U and V on the surface

neighborhood $\mathcal{U}_p$ of the point p (for instance the lighter curves in Fig. 7.6) and such that the family covers the entire considered neighborhood. Let us next introduce a second family of curves V, with the same properties among themselves, yet such that each element of the family V intersects all elements of the U family at least in the neighborhood $\mathcal{U}_p$ (for instance the darker curves Fig. 7.6). Once such systems of curves have been constructed, any point $q \in \mathcal{U}_p$ in the neighborhood of p can be localized by stating on which U-curve and on which V-curve it lies. Assuming that u and v are the real parameters respectively enumerating the U and V curves, the pair of real numbers (u, v) provides the new (*gaussian*) system of coordinates to label surface points. Using these coordinates we no longer need to make any reference to the exterior space in which M is immersed that might also be non-existing!

By introducing *curvilinear gaussian coordinates* the King of Mathematicians freed the study of surfaces from their immersion in the external euclidian space $\mathbb{E}$ but he immediately had to cope with a new fundamental problem. Having abolished

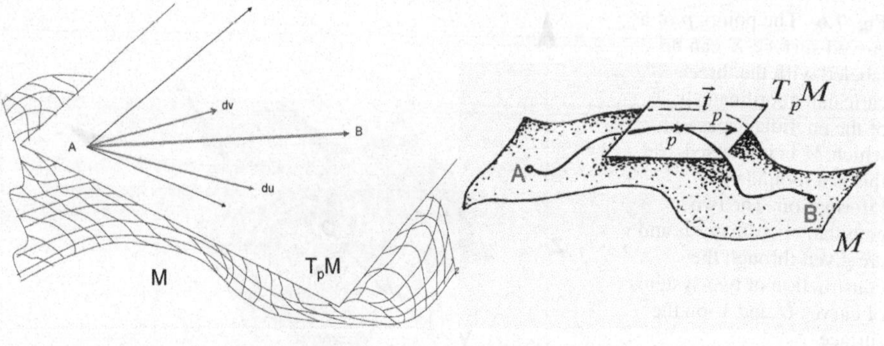

Fig. 7.7 The tangent plane to a point p of a surface Σ is a two-dimensional euclidian space $\mathbb{E}^2$ where the notion of distance is defined. In the tangent plane which approximates infinitesimal regions of the surface surrounding p, we can define the line element as the euclidian distance between two infinitesimally close points of the surface

from the list of one's mathematical instruments the straight line segments that join any two points A and B of the surface Σ, how can we calculate their distance? The great intuitions of Gauss were the *tangent plane* T_pM and the linear element ds^2, namely the metric.

Defining the absolute distance between two points A and B was no longer possible but also not interesting. In the external euclidian space, the distance between A and B is the length of the segment which joins them, but which relevance has this datum for the little creature confined to live on the surface Σ, if such a segment does not lie on it? For the little two-dimensional being the only interesting datum is the length of a road going from A to B: the length of any possible road with such a property! The small ant has exactly the same needs as any contemporary car-driver who wants to start on a journey. Both need to know the length of all possible paths from their origin to their destination (Fig. 7.7).

Hence the problem addressed by Gauss was to give an answer to the following question: *Can we define the length of any curve departing from $p \in \Sigma$ and arriving at $q \in \Sigma$ in terms of data completely intrinsic to the surface Σ?*

Gauss' answer was positive and based on the change of perspective at the basis of the new *differential geometry*.

Let us reformulate the initial question whether we might define the absolute distance between two arbitrary points $A, B \in \Sigma$ of the surface, adding the extra condition that A and B should be only infinitesimally apart from eachother. Analytically this means that if the gaussian coordinates of A are (u, v), then those of B should be $(u + du, v + dv)$ where du e dv are infinitesimal. Gauss crucial observation is that a very small portion of the surface Σ around any point $p \in M$ can be approximated by a portion of the tangent plane to the surface at the point p, namely T_pM. Smaller the considered region of Σ better the approximation. This being the case we observe that euclidian geometry makes sense in the tangent plane. Gauss remarked that the distance of two infinitesimally close points can be defined as the length of the infinitesimally short segment which joins them and which lies in the tangent plane.

Recalling Pithagora's theorem one might be tempted to say that the square length of the segments joining A and B, named ds^2, is the sum of the squared differences of the gaussian coordinates, namely: $ds^2 = du^2 + dv^2$. Yet this is not necessarily true. In order to apply Pithagora's theorem it is required that the axes $\mathbf{u}$ and $\mathbf{v}$ be orthogonal, which is generically false. Indeed a priori it is by no means guaranteed that the curve u and the curve v, meeting at point A, should intersect there at right angle. In order to calculate ds^2 one has therefore to find the components dx and dy of the infinitesimal segment AB in an orthogonal system of axes $\mathbf{x}$ and $\mathbf{y}$. Once these components are known one can apply Pithagora's theorem and write $ds^2 = dx^2 + dy^2$. The components dx and dy depend on the gaussian shifts du and dv linearly:

$$\begin{pmatrix} dx \\ dy \end{pmatrix} = \begin{pmatrix} a(u,v) & b(u,v) \\ c(u,v) & d(u,v) \end{pmatrix} \begin{pmatrix} du \\ dv \end{pmatrix} \tag{7.3.1}$$

with matrix coefficients that vary from place to place on the surface, namely are functions of the gaussian coordinate u, v. Taking this into account Gauss wrote the line element in the following way:

$$ds^2 = F(u,v)\,du^2 + G(u,v)\,dv^2 + H(u,v)dudv \tag{7.3.2}$$

where $F = a^2 + c^2$, $G = b^2 + d^2$, $H = 2ab + 2cd$.

Formula (7.3.2), written in 1828 provided the first example (a two-dimensional one) of a Riemannian metric, although Riemann was at that time only a two-year old child.

7.3.2 *Bernhard Riemann Introduces n-Dimensional Metric Manifolds*

The name of Riemann is associated in Mathematics with so many different and fundamental objects that the contemporary student is instinctively led to think about the scientific production of this giant of human thought as composed by a countless number of papers, books and contributions. Actually the entire corpus of Riemann's works is constituted only by 225 pages distributed over 11 articles published during the life-time of their author to which one has to add the 102 pages of the 4 posthumous publications. Among the latter there are the 16 pages of the *Ueber die Hypothesen, welche der Geometrie zu Grunde liegen* [152] which, in 1854, was debated by the candidate in front of the Göttingen Faculty of Philosophy as *Habilitationsschrift*. The habilitation to teach courses was the traditional first step in the academic career foreseen by most European universities all over their very long history. In XIXth century Germany the procedure to access habilitation consisted of the writing of a dissertation on a topic chosen by the Faculty from a list of three proposed by the candidate. Typical time allowed for the preparation of such a dissertation was a

Fig. 7.8 Bernhard Riemann, a genius and a giant of human thought, had a very short and not too happy life. He was Gauss' student both at the level of Diploma and of *Habilitationschrift*. The whole of his work is contained in no more than 11 papers for a total of little more than 200 pages. Yet each of his contributions was a milestone in Mathematics and set the path for century long future developments. The foundations of Riemannian geometry were laid in the 16 page long dissertation of his Habilitation. Riemann had important ties with the Scuola Normale di Pisa and died in Italy at the age of 39

couple of months and in the case of Riemann it amounted to exactly seven weeks (Fig. 7.8).

Obsessed the whole of his short life by extreme poverty and by a very poor health that eventually led him to death from pulmonary consumption at the quite young age of thirty-nine, the shy and meek Bernhard Riemann, who was nonetheless quite conscious of his own talents, had already profoundly impressed Gauss with his diploma thesis. Written in 1851 and entitled *Grundlagen für eine allgemeine Theorie der Functionen einer veränderlichen complexen Grösse* which can be translated as *Principles of a General Theory of the Functions of one complex variable*, Riemann's thesis was completely new and contained all the essentials of the theory of analytic functions as it is taught up to the present day in most universities of the world. Quite openly Gauss told his young student that for many years he had cheered the plan of writing a similar essay on that very topics yet now he would refrain from doing so since everything relevant to that province of thought had already been said by Riemann.

When three years later Riemann presented to the Göttingen Faculty his three proposals for the theme of his own *Habilitationsschrift*, two choices were in fields where the young mathematician felt quite confident, while the third, with some hesitation, was just added in order to complete the triplet and with the secret hope that it would be immediately discarded by the academic committee as something too philosophical and ill defined. The third proposed title was *Grundlagen der Geometrie*, namely the *Principles of Geometry*. Remembering the talents of the young *Herr*

Riemann, Gauss was fascinated by the idea of giving him precisely such a challenging subject as the Foundations of Geometry to see what he might come up with it. The King of Mathematicians persuaded the Faculty to make such a choice and the poor Bernhard was dismayed by the news. He wrote to his father, a poor lutheran minister, about his concerns on this matter but he also expressed him his confidence that he would not come too late and that his merits as an independent researcher would be appreciated.

Riemann had accepted the challenge and in seven weeks he produced such a masterpiece of Mathematics and Philosophy as the *Ueber die Hypothesen, welche der Geometrie zu Grunde liegen*, that is *About the Hypotheses lying at the Foundations of Geometry* (see Fig. 7.9).

With an unparalleled clarity of mind, Riemann began his essay with a profound criticism of the traditional approach to Geometry, refusing the kantian dogma that this latter is an a-priori datum and rather inclining to the idea that which geometry is the actual one of Physical Space should be determined from experience. He said: *It is known that geometry assumes, as things given, both the notion of space and the first principles of constructions in space. She gives definitions of them which are merely nominal, while the true determinations appear in the form of axioms. The relation of these assumptions remains consequently in darkness; we neither perceive whether and how far their connection is necessary, nor a priori, whether it is possible. From Euclid to Legendre (to name the most famous of modern reforming geometers) this darkness was cleared up neither by mathematicians nor by such philosophers as concerned themselves with it.*[2]

After stating this two-thousand year old stalemate, Riemann proceeded to diagnose its cause. Explicitly he said: *The reason of this is doubtless that the general notion of multiply extended magnitudes (in which space-magnitudes are included) remained entirely unworked. I have in the first place, therefore, set myself the task of constructing the notion of a multiply extended magnitude out of general notions of magnitude. It will follow from this that a multiply extended magnitude is capable of different measure-relations, and consequently that space is only a particular case of a triply extended magnitude.*

In contemporary language the *multiply extended magnitudes*[3] are simply the manifolds and the *measure relations* are just the metrics introduced for the first time by Gauss through Eq. (7.3.2).

Following the new road opened by Gauss with the *Disquisitiones*, Riemann introduced n-extended manifolds whose points are labeled by n rather than two curvilinear coordinates x^i and introduced the line element as a generic symmetric quadratic form in the differentials of these coordinates:

$$ds^2 = g_{ij}(x)\, dx^i\, dx^j \tag{7.3.3}$$

[2]The translation of Riemann's essay from German into English was done by William Clifford.

[3]In the original German text of Riemann these were named *mehrfach ausgedehnter Grossen*. In modern scientific German the notion of *manifolds* is referred to as *mannigfaltigkeiten*.

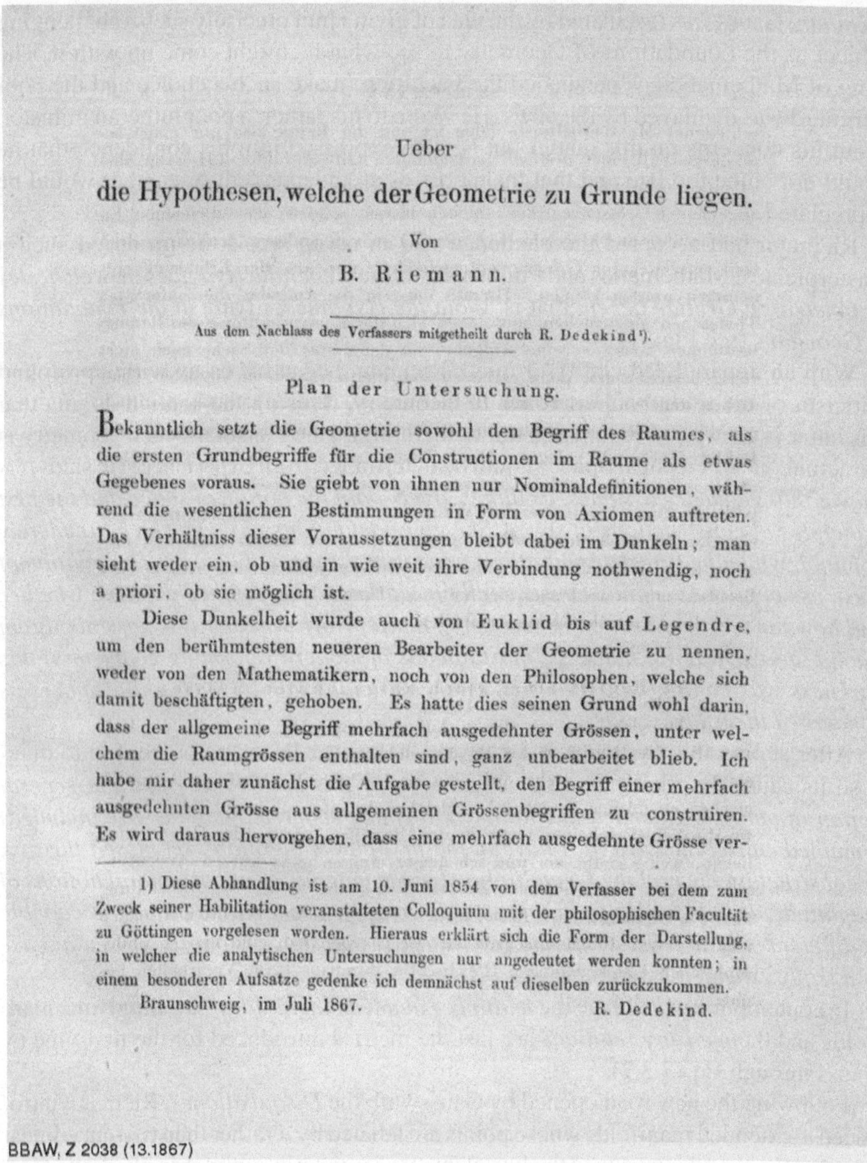

Ueber

die Hypothesen, welche der Geometrie zu Grunde liegen.

Von

B. Riemann.

Aus dem Nachlass des Verfassers mitgetheilt durch R. Dedekind[1]).

Plan der Untersuchung.

Bekanntlich setzt die Geometrie sowohl den Begriff des Raumes, als die ersten Grundbegriffe für die Constructionen im Raume als etwas Gegebenes voraus. Sie giebt von ihnen nur Nominaldefinitionen, während die wesentlichen Bestimmungen in Form von Axiomen auftreten. Das Verhältniss dieser Voraussetzungen bleibt dabei im Dunkeln; man sieht weder ein, ob und in wie weit ihre Verbindung nothwendig, noch a priori, ob sie möglich ist.

Diese Dunkelheit wurde auch von Euklid bis auf Legendre, um den berühmtesten neueren Bearbeiter der Geometrie zu nennen, weder von den Mathematikern, noch von den Philosophen, welche sich damit beschäftigten, gehoben. Es hatte dies seinen Grund wohl darin, dass der allgemeine Begriff mehrfach ausgedehnter Grössen, unter welchem die Raumgrössen enthalten sind, ganz unbearbeitet blieb. Ich habe mir daher zunächst die Aufgabe gestellt, den Begriff einer mehrfach ausgedehnten Grösse aus allgemeinen Grössenbegriffen zu construiren. Es wird daraus hervorgehen, dass eine mehrfach ausgedehnte Grösse ver-

1) Diese Abhandlung ist am 10. Juni 1854 von dem Verfasser bei dem zum Zweck seiner Habilitation veranstalteten Colloquium mit der philosophischen Facultät zu Göttingen vorgelesen worden. Hieraus erklärt sich die Form der Darstellung, in welcher die analytischen Untersuchungen nur angedeutet werden konnten; in einem besonderen Aufsatze gedenke ich demnächst auf dieselben zurückzukommen.

Braunschweig, im Juli 1867. R. Dedekind.

BBAW, Z 2038 (13.1867)

Fig. 7.9 Riemann's *Habilitationschrift* entitled *Ueber die Hypothesen, welche der Geometrie zu Grunde liegen*, that is *about the hypotheses lying at the foundations of geometry*

The coefficients of this quadratic form $g_{ij}(x)$ were later known as the Riemannian *metric tensor*.

Riemann grasped the main point, namely that the geometry of manifolds is encoded in the possible metric tensors or measure relations, as he called them, and made the following bold statement: *Hence flows as a necessary consequence that the propositions of geometry cannot be derived from general notions of magnitude, but that the properties which distinguish space from other conceivable triply extended magnitudes are only to be deduced from experience. Thus arises the problem, to discover the simplest matters of fact from which the measure-relations of space may be determined; a problem which from the nature of the case is not completely determinate, since there may be several systems of matters of fact which suffice to determine the measure-relations of space.*

In other words, the young genius was aware that the same manifold could support quite different metrics and thought that this applied in particular to *Space*, i.e. to the 3-dimensional physical world of our sensorial experience. He posed himself the question which should be the metric of Space and came to the conclusion that such a question could only be answered through experiment. This amounted to say that the geometry of the world is a matter of Physics and not of a priori Philosophy or Mathematics. Such a sentence of Riemann must have influenced Einstein quite deeply. Indeed the final outcome of Einstein Theory of Relativity is that the geometry of space-time is dynamically determined by its matter content through Einstein field equations.

In considering such a question as what is the preferred metric to be selected for a given manifold, Riemann formulated the basic problem of *invariants*. The *matter of facts*[4] to which he alluded are the intrinsic properties encoded in a given metric tensor namely its invariants and he formulated the problem of determining, for instance, the minimal complete number of invariants able to select Euclidian geometry. Riemann's views anticipate from a different wider angle Klein's Erlangen Programme (see Sect. 5.2.2) that linked the specification of a geometry with an invariance group. In Riemann's approach Klein's group would be the isometry group (see Sect. 7.6). Yet as we discuss later there are metrics with a small or no non-trivial isometry: hence the question as posed by Riemann is more general and more profound. What are the invariants (or covariants) that completely characterize a metric and hence a geometry?

In this quest for invariants Riemann came to the notion of the *Riemann curvature tensor* that he outlined in his very dissertation. The curvature form of connections will be the subject of a subsequent section.

As we already recalled, Riemann died young and had no time to develop the new theory of differential geometry that he had founded. Yet he had the time to come to Italy and, through his contact with the Scuola Normale di Pisa and the research group of Enrico Betti, of whom he was a close friend since the time they met in Göttingen, to plant the seeds of the *absolute differential calculus* in the Italian Peninsula where, later, they were strongly developed by Gregorio Ricci Curbastro, Luigi Bianchi and Tullio Levi Civita.

---

[4] *Einfachsten Thatsachen* in the original German text.

7.3.3 Parallel Transport and Connections

The idea of connections developed in the XIXth century along two different routes which merged only in the XXth century. One route comes directly from Riemann and through Christoffel, Ricci and Levi Civita went to Einstein. We can name this the *metric route*. It resulted in the notion of the Levi Civita connection, which is the parallel transport defined by the existence of a metric structure. The other route was independently started in France by the work of Frenet and Serret and went down to Élie Cartan: it is the route of the *mobile frames* (or *répères mobiles* as they are named in French). A metric structure is not required in the second approach and the connection deals with the parallel transport of a basis of vectors from one point of a manifold to another one, along a curve. Connections of this type are named *Cartan connections*. There is obviously a relation between Cartan connections and metric connections. Such a relation is established through a so named *soldering* of the *frame bundle* with the *tangent bundle*, which also provides the conversion vocabulary between two different formulations of General Relativity: the original metric one of Einstein and the *vielbein* reformulation by Cartan. From a modern perspective the vielbein formulation appears to be the most fundamental since it includes the metric formulation, yet, differently from this latter, allows also for the coupling to gravity of the fermionic fields, the spinors. A modern revisitation of Cartan's viewpoint is encoded in the notion of *group-structure* on manifolds, Riemannian metrics corresponding essentially to $SO(n)$-structures on n-dimensional manifolds.

The further generalization of Cartan connections to principal connections on generic fibre bundles was finally elaborated by Ehresmann in the post-war years of the middle XXth century and this mathematical development came almost in parallel with the introduction of non-abelian gauge theories in physics: the two leaves of C.N. Yang's drawing protruding from the same stem!

7.3.4 The Metric Connection and Tensor Calculus from Christoffel to Einstein, via Ricci and Levi Civita

As we emphasized in the previous section, the primary concern of the new differential geometry, founded by Riemann as a generalization of Gauss work on surfaces, was that of defining the length of curves on arbitrary *manifolds*. This leads to the notion of the metric $g_{ik}(x)$, introduced in Eq. (7.3.3). Once the metric is established, a natural way arises of transporting vectors along any given curve $x^\mu(\lambda)$ (see Fig. 7.10).

We can say that a vector is parallel-transported along an arc of curve $\mathscr{C}(t) = \lambda(t)$ if the angle between the transported vector and the tangent vector to the curve remains constant throughout the entire transport (see Fig. 7.11).

This notion is meaningful if we can define and measure angles. This is precisely what the existence of a metric allows. Indeed, in presence of $g_{\mu\nu}(x)$, we can generalize the relations of Euclidean geometry by defining the norm of any vector as:

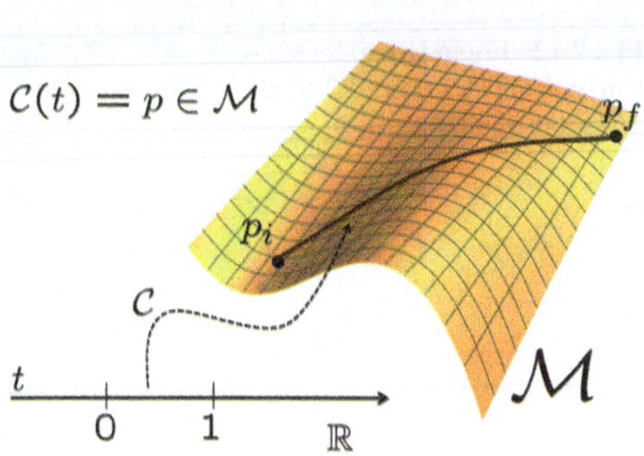

Fig. 7.10 A curve $\mathscr{C}(t)$ in a manifold $\mathscr{M}$ is just a continous map of the interval [0, 1] of the real line R into $\mathscr{M}$. The image of 0 through this map is the initial point $p_i \in \mathscr{M}$, the image of 1 is the final point p_f

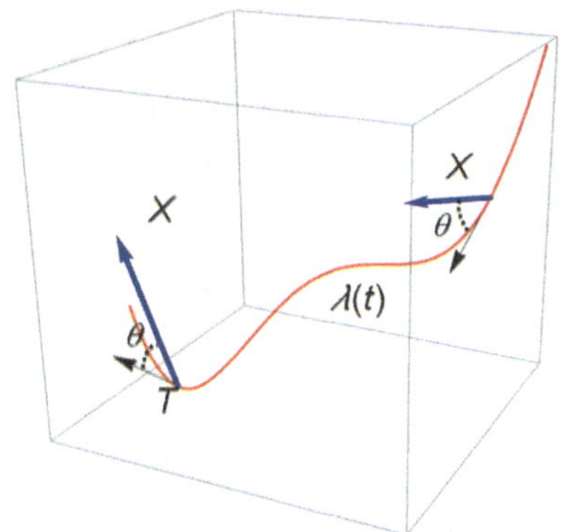

Fig. 7.11 The parallel transport of a vector X along a curve is defined through the preservation of its angle θ with the tangent vector T to the curve $\lambda(t)$ at each point of the curve. This definition is possible if we have a metric and hence the notion of scalar product of local vectors

$$||v|| \equiv \sqrt{v^\mu(x)\, v^\nu(x)\, g_{\mu\nu}(x)} \tag{7.3.4}$$

and the angle between any two vectors $v_1^\mu(x)$ and $v_2^\nu(x)$ as

$$\cos\theta \equiv \frac{< v_1 \, , \, v_2 >}{||v_1||\, ||v_2||} \tag{7.3.5}$$

where the scalar product $< v_1 \, , \, v_2 >$ is:

$$< v_1 \, , \, v_2 > \equiv v_1^\mu(x)\, v_2^\nu(x)\, g_{\mu\nu}(x) \tag{7.3.6}$$

The metric connection is that infinitesimal displacement of a vector X along the direction singled out by another one Y which is so defined as to fulfil the property of preserving angles. It was first conceived by Christoffel.

Elwin Bruno Christoffel was born in 1829 in Montjoie, near Aachen, that was renamed Monschau in 1918. After attending secondary schools in Cologne, he enrolled at the University of Berlin, where he had such teachers as Eisenstein and Dirichlet. Particularly the latter is duely considered his master. Christoffel's doctor dissertation, dealing with the motion of electricity in homogeneous media was defended in 1856, just two years after Riemann's presentation of the *Ueber die*

Fig. 7.12 Elwin Bruno Christoffel (Montjoie 1829, Strasbourg 1918)

Hypothesen. Having spent a few years out of the academic world, Christoffel returned to Mathematics in 1859, obtaining his *habilitation* from Berlin University. In the following years he was professor at the Polytechnic of Zurich, at the newly founded Technical University of Berlin and finally at the University of Strasbourg which had become German after the defeat of Napoleon III in the 1870 war. According to opinions reported by contemporaries, *Elwin Bruno Christoffel was one of the most polished teachers ever to occupy a chair. His lectures were meticulously prepared and his delivery was lucid and of the greatest aesthetic perfection...* The core of his course was the theory of complex functions that he developed and presented according to Riemann's approach (Fig. 7.12).

Although he wrote papers on several different topics like potential theory, differential equations, conformal mappings, orthogonal polynomials and still more, the most relevant and influential of Christoffel's contributions with the furthest reaching consequences was his invention of the three-index symbols that bear his name. Defined in terms of a metric $g_{\mu\nu}(x)$ and of its inverse $g^{\rho\sigma}(x)$, the symbols:

$$\left\{ \begin{array}{c} \lambda \\ \mu\,\nu \end{array} \right\} \equiv \frac{1}{2}\, g^{\lambda\sigma} \left(\partial_\mu\, g_{\nu\sigma} + \partial_\nu\, g_{\mu\sigma} - \partial_\sigma\, g_{\mu\nu} \right) \tag{7.3.7}$$

are the first example of *connection coefficients*, actually those of the Levi-Civita connection, that preserves angles along the parallel transport it defines [43].

Christoffel symbols are the key ingredients in the definition of the *covariant derivative* of a tensor (in particular of a vector):

$$\nabla_\mu t_{\lambda_1\ldots\lambda_n} \equiv \partial_\mu t_{\lambda_1\ldots\lambda_n} - \left\{ \begin{array}{c} \rho \\ \mu\,\lambda_1 \end{array} \right\} t_{\rho\,\lambda_2\ldots\lambda_n} - \left\{ \begin{array}{c} \rho \\ \mu\,\lambda_2 \end{array} \right\} t_{\lambda_1\rho\ldots\lambda_n} \cdots - \left\{ \begin{array}{c} \rho \\ \mu\,\lambda_n \end{array} \right\} t_{\lambda_1\ldots\lambda_{n-1}\rho} \tag{7.3.8}$$

The word tensor was introduced for the first time by Hamilton in 1846, but tensor calculus was developed around 1890 by Gregorio Ricci Curbastro under the title of *absolute differential calculus* and was made accessible to mathematicians by the publication of Tullio Levi Civita's 1900 classic text of the same name, originally written in Italian, later republished in French with Ricci [135] (see Fig. 7.13).

We report the words that open Levi Civita and Ricci's paper of 1899. They are particularly inspiring in view of what was to follow: *M. Poincaré a écrit que dans les Sciences mathématiques une bonne notation a la même importance philosophique qu'une bonne classification dans les Sciences naturelles. Évidemment, et même avec plus de raison, on peut en dire autant des méthodes, car c'est bien de leur choix que dépend la possibilité de forcer (pour nous servir encore des paroles de l'illustre géomètre français) une multitude de faits sans aucun lien apparent à se grouper suivant leurs affinités naturelles.*

Indeed it happens, as stressed by Weyl (see Chap. 6), that *the scientist must thrust through the fog of abstract words to reach the concrete rock of reality* and *the real thing emerges as soon as we replace the intuitive space in which our diagrams are drawn by its construction in terms of sheer symbols.*

The intuitive notion of curvature and of curved surfaces is as old as the dawn of human thought, yet it was only through the work of Gauss and Riemann that *curvature* found the beginning of its intrinsic mathematical definition. In order to develop in a full–fledged fashion the vision pioneered by Riemann, who first introduced the curvature coefficients and counted them, the sheer symbols were needed and a new mathematical language was required that would accommodate them. The *sheer symbols* were *la bonne notation* introduced by Ricci and Levi Civita and the mathematical language in which they fit was the *tensor calculus* or *absolute differential calculus* as they named it.

Ricci Curbastro

Gregorio Ricci Curbastro (see Fig. 7.14) was son in an aristocratic family of Lugo di Romagna. On the house where he was born in 1853 there stands a plate with the following words *Diede alla scienza il calcolo differenziale assoluto, strumento indispensabile per la teoria della relativitá generale, visione nuova dell'universo.*[5]

He began his studies at Rome University but he continued them at *Scuola Normale di Pisa* and finally graduated from the University of Padova in 1875. As his younger friend Luigi Bianchi, born in Parma in 1865 and also student of Scuola Normale, in the Pisa years he was deeply influenced by the teaching of Ulisse Dini

[5]He gave to science Absolute Differential Calculus, essential instrument of the Theory of General Relativity, a new vision of the Universe.

Méthodes de calcul différentiel absolu et leurs applications.

Par

M. M. G. Ricci et T. Levi-Civita à Padoue.

Table des matières.

Chapitre I.

Algorithme du calcul différentiel absolu.

Chapitre II.

La géométrie intrinsèque comme instrument de calcul.

Chapitre III.

Applications analytiques.

Fig. 7.13 The first page of Ricci and Civita's 1899 paper on tensor calculus

and Enrico Betti, the founder of modern topology. Through Betti, both Ricci and Bianchi captured the seeds of differential geometry planted by Riemann few years before. After graduation, Ricci obtained a fellowship that allowed him to spend some years in Munich, in Germany. There he came in touch with the new conception and classification of geometries, based on symmetry groups, developed by Felix Klein and magisterially summarized by him in the celebrated *Erlangen Programme*. These ideas had an analogous strong impact on Luigi Bianchi.

Fig. 7.14 Gregorio Ricci Curbastro (1853–1925). Born in Lugo di Romagna, Ricci Curbastro studied first in Rome, than in Pisa and finally graduated from Padua of which university he later became full-professor. There he had as student Tullio Levi-Civita. Together they wrote the fundamental papers establishing the general form of tensor calculus

Promoted to the position of full-professor at the University of Padova in 1880, Ricci had there an exceptionally talented graduate student: Tullio Levi Civita who was born in that city in 1873. Ricci, Bianchi and Levi-Civita constructed the mathematical language used by Einstein to formulate General Relativity, which is also the most common language for classical differential geometry. The key ingredients of that language are just the tensors $t^{\mu_1\dots\mu_m}_{\lambda_1\dots\lambda_n}$, whose defining property is that of transforming from one coordinate patch x^μ to another one $\tilde{x}^\sigma$, according to:

$$\tilde{t}^{\mu_1\dots\mu_m}_{\lambda_1\dots\lambda_n}(\tilde{x}) = \frac{\partial\tilde{x}^{\mu_1}}{\partial x^{\rho_1}}\cdots\frac{\partial\tilde{x}^{\mu_m}}{\partial x^{\rho_m}}\frac{\partial x^{\sigma_1}}{\partial\tilde{x}^{\sigma_1}}\cdots\frac{\partial x^{\sigma_n}}{\partial\tilde{x}^{\sigma_n}}\,t^{\rho_1\dots\rho_m}_{\sigma_1\dots\sigma_n}(x) \tag{7.3.9}$$

In contemporary mathematical language, a tensor with m upper indices and n lower indices is just a section of the m-th power of the tangent bundle and at the same time a section of the n-th power of the cotangent bundle.[6] Hence the absolute differential calculus of Ricci, Levi Civita and Bianchi is just the differential calculus for sections of those fibre bundles whose transition functions are completely determined by the very manifold structure of their base-manifold $\mathcal{M}$, namely $T\mathcal{M}$ and $T^*\mathcal{M}$. The concept of covariant differentiation was formally developed by Ricci and Levi

[6]The fundamental notion of fibre bundle is described in simple words in Sect. 7.4

Système de Riemann. — Relations entre les éléments du deuxième système dérivé d'un système covariant quelconque.

Soit

$$\varphi = \sum_{1}^{n}{}_{rs}\, a_{rs}\, dx_r\, dx_s$$

la quadrique fondamentale et posons

$$2a_{rs,t} = \frac{\partial a_{rt}}{\partial x_s} + \frac{\partial a_{st}}{\partial x_r} - \frac{\partial a_{rs}}{\partial x_t},$$

$$a_{rs,tu} = \frac{\partial a_{rt,s}}{\partial x_u} - \frac{\partial a_{ru,s}}{\partial x_t} + \sum_{1}^{n}{}_{pq}\, a^{(pq)} (a_{ru,p}\, a_{st,q} - a_{rt,p}\, a_{su,q}).$$

Les symboles $a_{rs,tu}$ sont les éléments d'un système quadruple covariant, qui a une grande importance dans la théorie des quadriques de différentielles. On les trouve dans la *Commentatio mathematica* de Riemann *) (à un facteur numérique près) et c'est à cause de cela que nous désignerons ce système par le nom de *système covariant de Riemann.* — Les expressions $a_{rs,tu}$ furent rencontrées avant la publication du Mémoire cité du grand géométre par M. Christoffel **), qui en mit en évidence les propriétés fondamentales. Il suffira ici de rappeler que le nombre de ces expressions, qui ne sont liées entre elles par aucune relation linéaire, est $N = n^2(n^2-1):12$.

Fig. 7.15 The formula for the Riemann tensor in Ricci's and Levi Civita's paper of 1899

Civita and, as already stressed above, by using the Christoffel symbols, it realizes the idea of parallel transport preserving the angles defined by a metric structure.

THE RIEMANN CURVATURE TENSOR

Once the covariant differentiation ∇_μ is given, one can consider its antisymmetric square and this leads to the Riemann–Christoffel curvature tensor (see Figs. 7.15 and 7.20):

$$R_{\lambda\sigma\nu}{}^{\mu} \equiv \partial_\lambda \left\{ \begin{matrix} \mu \\ \sigma\,\nu \end{matrix} \right\} - \partial_\sigma \left\{ \begin{matrix} \mu \\ \lambda\,\nu \end{matrix} \right\} + \left\{ \begin{matrix} \mu \\ \lambda\,\theta \end{matrix} \right\} \left\{ \begin{matrix} \theta \\ \sigma\,\nu \end{matrix} \right\} - \left\{ \begin{matrix} \mu \\ \sigma\,\theta \end{matrix} \right\} \left\{ \begin{matrix} \theta \\ \lambda\,\nu \end{matrix} \right\} \quad (7.3.10)$$

which, sketched by Riemann in the *Ueber die Hypothesen* and analytically defined by Christoffel [43], realizes for an arbitrary manifold the idea of intrinsic curvature devised by Gauss in the 1828 *Disquisitiones*.

In Ricci and Levi Civita's paper of 1899, the Riemann tensor was referred as *le Systeme de Riemann* and was denoted $a_{rs,tu}$ (see Fig. 7.15). The authors said that these coefficients formed *les elements d'un systeme quadruple covariant*. The *Commentatio Mathematica* quoted by Ricci and Levi Civita in their paper (see Fig. 7.15) is an essay in Latin written by Riemann in 1861 and submitted to the Paris Academy, in order to compete for a prize relating to the conduction of heat. The second part of this essay contains some of the mathematical analysis underlying Riemann's *Habilitationssrhift* and in particular his way of writing what became the Riemann tensor.

Here are some of the crucial passages of the *Commentatio*:

De transformatione expressionis $\sum \beta_{\iota,\iota'}\, ds_\iota\, ds_{\iota'}$ *in formam datum* $\sum \alpha_{\iota,\iota'}\, dx_\iota\, dx_{\iota'}$.

......

Quantitatibus $p_{\iota,\iota',\iota''}$ iterum differentiatis obtinetur

$$\frac{\partial p_{\iota,\iota',\iota''}}{\partial s_{\iota'''}} - \frac{\partial p_{\iota,\iota',\iota'''}}{\partial s_{\iota''}} = 2\sum_\nu \frac{\partial^2 x_\nu}{\partial s_{\iota'}\,\partial s_{\iota''}} \frac{\partial^2 x_\nu}{\partial s_\iota\,\partial s_{\iota'''}} - 2\sum_\nu \frac{\partial^2 x_\nu}{\partial s_{\iota'}\,\partial s_{\iota'''}} \frac{\partial^2 x_\nu}{\partial s_\iota\,\partial s_{\iota''}},$$

unde tandem prodit, substitutis valoribus ......

......

partes laevas harum aequationum designabimus per

$$(\iota\iota', \iota''\iota''').$$

......

Quo pacto haec expressio invenietur

$$\Delta_2 = (\iota\iota', \iota''\iota''')(ds_\iota\,\delta s_{\iota'} - ds_{\iota'}\,\delta s_\iota)(ds_{\iota''}\,\delta s_{\iota'''} - ds_{\iota'''}\,\delta s_{\iota''}). \tag{7.3.11}$$

Although the notation is quite involute, for our contemporary eyes it is wrapped in deep darkness, yet the coefficients

$$(\iota\iota', \iota''\iota''')$$

are just the entries of the Riemann tensor. The Riemann tensor appears also in the first edition of Bianchi's *Lezioni di Geometria differenziale*) (see Fig. 7.20) and it is named there th 4–index symbol.

Considering a vector V^ρ we find:

$$\left[\nabla_\mu\,,\,\nabla_\nu\right] V^\rho \;=\; V^\sigma\, R_{\mu\nu\sigma}{}^{\rho} \tag{7.3.12}$$

The geometrical meaning of this relation is exemplified in Fig. 7.16.

Consider an infinitesimally small rectangle whose two sides are given by the two vectors X^μ and Y^μ (also of infinitesimally short length), departing from a given point p. Consider next the parallel transport of a third vector V^ρ to the opposite site of the rectangle. This parallel transport can be performed along two routes, both arriving at the same destination. The first route follows first X and then Y. The second route does the opposite. The image vectors of these two transports are based at the same point, so they can be compared, in particular they can be subtracted. So doing we find:

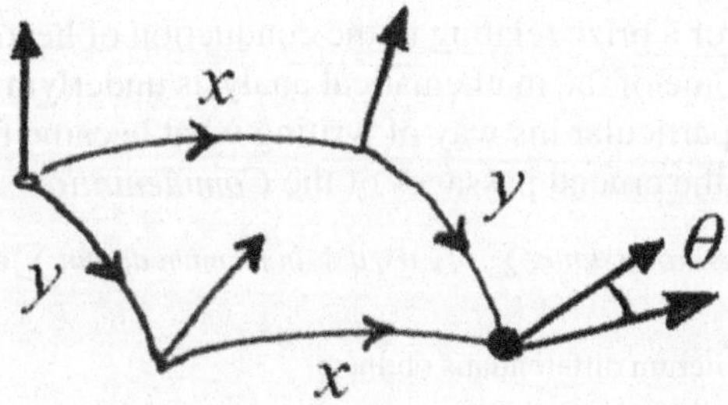

Fig. 7.16 The images of the same vector parallel transported along two different paths that converge at the same point of a manifold, generically differ by a rotation angle. That angle is a measure of the intrinsic curvature of the manifold and it is the information codified in the Riemann–Christoffel tensor

$$\begin{aligned}\nabla_X \nabla_Y V^\rho - \nabla_Y \nabla_X V^\rho &= X^\mu Y^\nu \left[\nabla_\mu \nabla_\nu\right] V^\rho + (\nabla_X Y^\sigma - \nabla_Y X^\sigma)\, \nabla_\sigma V^\rho \\ &= X^\mu Y^\nu \left(V^\sigma R_{\mu\nu\sigma}{}^\rho + T^\sigma{}_{\mu\nu} \nabla_\sigma V^\rho\right) + [X\,,\,Y]^\sigma\, \nabla_\sigma V^\rho\end{aligned} \tag{7.3.13}$$

where the Riemann–Christoffel tensor was defined above and where:

$$T^\rho_{\mu\lambda} \equiv \left\{\begin{matrix}\rho \\ \mu\,\lambda\end{matrix}\right\} - \left\{\begin{matrix}\rho \\ \lambda\,\mu\end{matrix}\right\} \tag{7.3.14}$$

is another tensor named the torsion. In the case of the Christoffel symbols the torsion is identically zero, yet for more general connections it can be different from zero and Levi-Civita correctly singled out the vanishing of the torsion as one of the two axioms from which the metric connection can be derived. In any case comparison of Eq. (7.3.13) with Fig. 7.16 enlightens the geometrical meaning of both torsion and curvature. If torsion is zero the parallel transport along the two different paths produces two vectors that differ from one another by a rotation angle and the Riemann tensor $R_{\mu\nu\sigma}{}^\rho$ encodes all these possible angles. If torsion is not zero the two images of parallel transport differ also by a displacement and the torsion tensor $T^\rho_{\mu\lambda}$ encodes all such possible displacements. In a flat space, just as a plane, parallel transport produces no rotation angle and no displacement. Hence the Riemann–Christoffel tensor $R_{\mu\nu\sigma}{}^\rho$ measures the intrinsic curvature of a metric manifold (Fig. 7.17).

The Ricci tensor

In a paper of 1903 [151], Gregorio Ricci introduced a new tensor, later named after him, which is obtained from the Riemann–Christoffel tensor through a contraction of indices. The Ricci tensor is defined as follows:

$$\text{Ric}_{\mu\nu} \equiv \sum_{\rho=1}^{N} R_{\mu\nu\rho}{}^\rho \tag{7.3.15}$$

Fig. 7.17 Tullio Levi Civita (1873–1941) Born in Padova, died in Rome. He graduated in 1892 from the University of Padua, faculty of mathematics, writing his laurea thesis under the supervision of Ricci Curbastro. In 1898 he was appointed to the Padua Chair of rational mechanics. He remained in his position at Padua until 1918, when he was appointed to the chair of higher analysis at the University of Rome. In 1936, receiving an invitation from Einstein, Levi-Civita traveled to Princeton, United States and lived there with him for a year. Then he returned to Italy. The 1938 Fascist race laws deprived Levi-Civita of his professorship and of his membership of all scientific societies. Einstein used to say that the best things in Italy were spaghettis and Levi Civita

and, on a metric manifold, measures the first deviation of its volume form from the euclidian value. Just for this reason it was originally considered by its inventor. Yet such tensor was doomed to play a major role in the development of XXth century scientific thought and in the birth of General Relativity.[7]

Bianchi and the Bianchi identities

Preparatory to this great future of the Ricci tensor were the algebraic and differential identifies it satisfies. They were derived by Luigi Bianchi (see Fig. 7.18) in 1902. Actually, according to Levi Civita, the same identities had already been discovered

[7]The very first embryonal idea of the Ricci tensor actually appeared as early as 1892 in another publication of its inventor [150].

Fig. 7.18 Luigi Bianchi (1865–1928)

by Ricci as early as 1880 but they had been discarded by their author as not relevant. The first of Bianchi identities states that the Ricci tensor is symmetric:

$$\mathrm{Ric}_{\mu\nu} = \mathrm{Ric}_{\nu\mu} \tag{7.3.16}$$

the second, differential identity, states that its divergence is equal to one half of the gradient of its trace:

$$\nabla^{\mu}\, \mathrm{Ric}_{\mu\nu} = \tfrac{1}{2}\nabla_{\nu}\mathrm{R} \tag{7.3.17}$$

where, by definition, we have posed:

$$\mathrm{R} = g^{\mu\nu}\, \mathrm{Ric}_{\mu\nu} \tag{7.3.18}$$

which is named the curvature scalar and we have set $\nabla^{\mu} = g^{\mu\sigma}\, \nabla_{\sigma}$.

The Bianchi identities were precisely the clue that lead Einstein, with the help of Marcel Grossman, to single out the form of the field equations of General Relativity. Combined in a proper way, they suggest the form of a covariantly conserved tensor, the Einstein tensor, which plays the role of left hand side in the propagation equations, the right hand side being already decided on physical grounds, namely the conserved stress energy-tensor.

Actually the identities discovered by Bianchi on the Ricci tensor are the particular form taken, in the case of the Levi-Civita connection, by very general and fundamental identities satisfied by the curvature 2-form of any connection on any principal fibre bundle. In the present chapter we will emphasize this concept, which has the highest significance both in Physics and in Mathematics.

After his *laurea* in Mathematics from the University of Pisa, which he obtained in 1877, Bianchi remained in that city for other two years as student of the *Corso di Perfezionamento* of the *Scuola Normale Superiore*. He graduated in 1879, defending a thesis on *helicoidal surfaces*. Then, just following the steps of Ricci, he was in Germany, first in Munich and then in Göttingen, were he attended courses and seminars given by Felix Klein. As already stated, he was deeply influenced by Klein's group-theoretical view of geometry and one of his major achievements is precisely along that line. In a paper of 1898 [18], Bianchi classified all tridimensional spaces that admit a continuous group of motions. Actually, so doing, he classified all Lie algebras of dimension three. This classification, which is organized into nine types, turned out to be quite relevant for Cosmology in the framework of General Relativity, since it amounts to a classification of all possible space-times that are spatially homogeneous (see Chapter 5 of Volume Two of [89]). Since 1882, Bianchi was *internal professor* at the Scuola Normale and in 1886 he won the competition for the chair of Projective Geometry at University of Pisa, where he was full-professor for the rest of his life. In 1894 he published the first edition of his *Lezioni di Geometria Differenziale* [17], which is the very first comprehensive treaty on the new discipline pioneered by Riemann and also the first place where the name *Differential Geometry* appeared (see Figs. 7.19, 7.20 and 7.21).

Bianchi died in 1928 and he is buried in the *Cimitero Monumentale, Piazza dei Miracoli* of Pisa. Since the later 1880s up to the end of his life he was an extremely prominent and influential mathematician of the then flourishing Italian School of Geometry. In 1904 Bianchi was member of the committee appointed by the *Accademia Nazionale dei Lincei* to select the winning paper for the 1901 Royal Prize of Mathematics. Ricci's ambitions on that Prize had already been manifested some years before, when he presented his works to the committee then headed by Eugenio Beltrami. Notwithstanding Beltrami's very favorable impressions, the final verdict of the jury on the relevance of *tensor analysis* had been hesitating and the Prize had not been attributed. Similar conclusion obtained the competition of 1904. Luigi Bianchi showed a great appreciation for the mathematical soundness and vastity of Ricci's methods but concluded that tensor analysis had not yet demonstrated its relevance and essentiality. He utilized Kronecker's words to say that he preferred new results found with old methods rather than old results retrieved with new, although very powerful, techniques. These events are quite surprising in view of the fact that two years before, in 1902, Bianchi had published his paper [19] containing those identities on the Ricci tensor for which his name is mostly remembered.

The Royal Prize for Mathematics, denied to Ricci Curbastro, was attributed few years later, in the 1907 edition, to Ricci's former student Tullio Levi Civita, by a committee that once again included Luigi Bianchi, together with other distinguished mathematicians such as Vito Volterra and Corrado Segre. This time the usefulness of the tensor methods had been made absolutely undoubtable by the vastity of Levi-Civita's results.

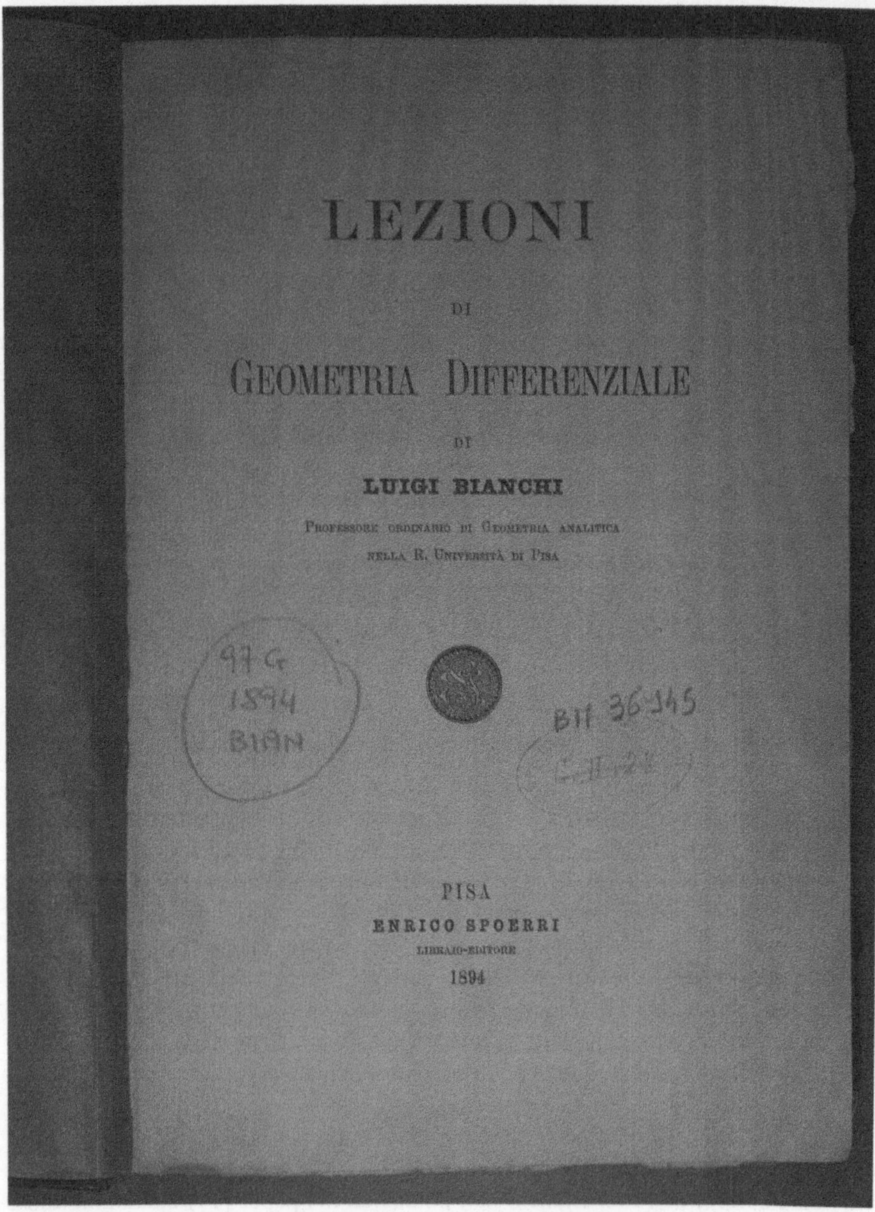

Fig. 7.19 The first edition (1894) of Luigi Bianchi's lectures on differential geometry. Frontespice. (Courtesy of the *Bliblioteca Peano - Dipartimento di Matematica - University of Torino*)

Although a little bit dismayed by the failure to get the Royal Prize, Ricci Curbastro ended his life in 1925 surrounded by the appreciation of his colleagues and of his fellow citizens both as a scientist and as a politician. Indeed he was nominated

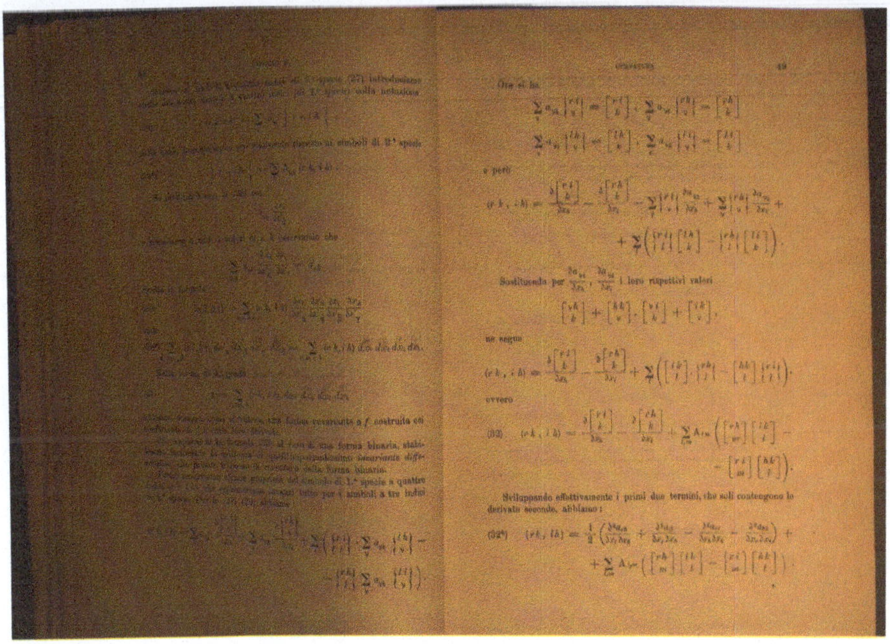

Fig. 7.20 The first edition (1894) of Luigi Bianchi's lectures on differential geometry. Pages containing the formula of the Riemann tensor, named 4-index symbol. (Courtesy of the *Bliblioteca Peano - Dipartimento di Matematica - University of Torino*)

member of several academies, including the most prestigious one, that of *Lincei* and also occupied positions in the local administration of his native city, Lugo di Romagna. On the contrary his genial student Levi-Civita, who was professor at the University of Rome *La Sapienza*, notwithstanding the Royal Prize and other honours, suffered, under the fascist racial laws of 1938, the removal from his chair because of his Jewish origin. Depressed and completely isolated from the scientific world he died from sorrow in 1941. It is a luminous shot in a dark and barbarous time that when he was removed from his Chair at la Sapienza, Levi-Civita was offered a chair by the Academia Pontificia.

7.3.5 Mobiles Frames from Frenet and Serret to Cartan

As everyone knows Einstein's paper on Special Relativity is dated 1905, while that on General Relativity was published at the end of 1915. Einstein's theory of gravitation, which is the main theme of a previous book of the present author [89], was formulated in the language of differential geometry as developed by Ricci and Levi-Civita and centered around the notion of metric, firstly introduced by Gauss for two-dimensional surfaces and then extended to all dimensions by Riemann.

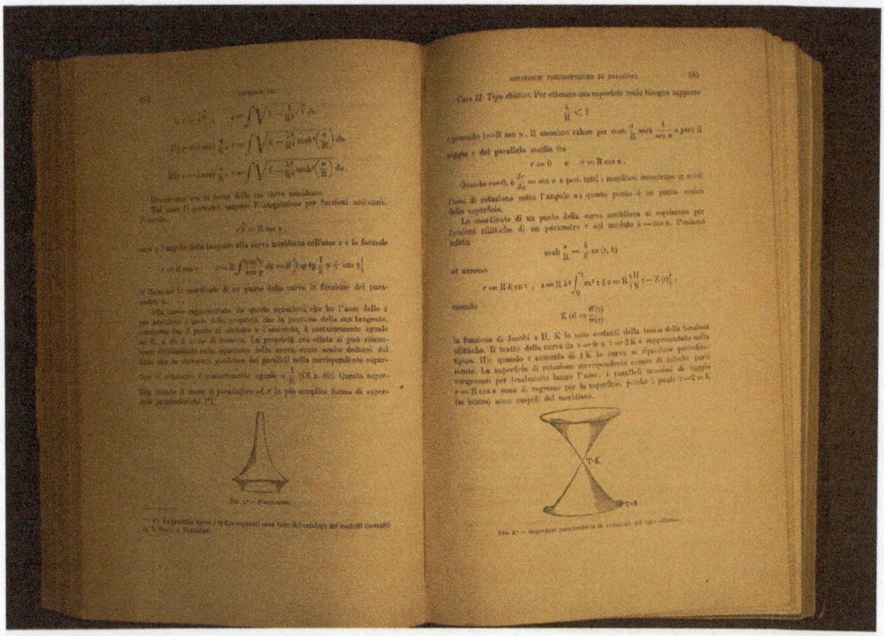

Fig. 7.21 The first edition (1894) lk,of Luigi Bianchi's lectures on differential geometry. Pages containing the discussion of Beltrami pseudosphere. (Courtesy of the *Bliblioteca Peano - Dipartimento di Matematica - University of Torino*)

Yet as early as 1901, a relatively young scholar, who was to be a giant of XXth century mathematical thought, had already introduced a different approach to differential geometry, whose greater depth and power started to be appreciated only later on.

The genius we are referring to is Élie Cartan about whom we already said a lot in Sect. 5.4 and the above mentioned 1901 paper actually deals with the theory of first order partial differential equations. Just as Sophus Lie, dealing with the same classical problem of analysis, developed the notion of Lie groups, in the same way Cartan, who had already given unparalleled contributions to the completion of Lie's theory, reconsidering differential systems from a new view-point, introduced the notion of exterior differential forms and laid the basis of modern differential geometry. In a subsequent paper of 1904 [33], Cartan introduced a particular set of 1-forms that, in modern scientific literature, bear his name together with that of the German mathematician Ludwig Maurer.[8] Maurer–Cartan one-forms are associated with Lie groups and play an essential role in the general theory of connections on fibre-

[8]Ludwig Maurer (1859–1927) obtained his Doctorate in 1887 from the University of Strassburg (at the time under German rule after the defeat of France in the 1870 war) and became professor of Mathematics at the University of Tübingen. His doctoral dissertation *Zur Theorie der linearen Substitutionen* [142] happens to contain a germ of the idea of Maurer–Cartan forms developed by Cartan in 1904.

bundles, namely in establishing the notion of gauge-fields, presently identified with the mediators of all fundamental forces of Nature. Cartan's 1904 paper, that preceded both Special and General Relativity, is fundamental not only for the actual result it contained about covariantly closed one-forms on Lie group manifolds, but also for the general view-point on geometry that such structures advocated. Using Cartan's phrasing, this view-point is that of *repères mobiles*, mobile frames in English. As all great and simples ideas, that of mobile frames developed slowly, through the contribution of more than one mind and, in this case, all the minds were French.

Jean Frédéric Frenet was born in Périgueux in 1816 and died in the same city in 1900. He entered the École Normale Supérieure in 1840 but then he quitted it and studied at the University of Toulouse. Frenet's doctoral thesis, submitted in 1847 had the intriguing title *Sur le fonctiones qui servent à determiner l'attraction des sphéroides quelconques. Programme d'une thèse sur quelque propriétés des courbes a double courbure.*[9] This thesis presented the idea of attaching a frame to each point of an arbitrary curve that develops in three-dimensional space. As this frame moves along the curve, we can look at its rate of change to determine how the curve turns and twists, two evolutions which completely determine the geometry of the considered curve.

At the time of his writing, matrix notation and matrix calculus were not yet in general use and Frenet wrote six formulae which just correspond to six entries of a 3×3 matrix. The latter has obviously nine entries and it was Serret's historical mission to write all of them independently from Frenet.

Joseph Alfred Serret was three year younger than Frenet. Born in Paris in 1819 he died in Versailles in 1885. While Frenet had been a missed issue of École Normale Superiéure, Serret was an achieved product of École Polytechnique, from which he graduated in 1848. Appointed professor of Celestial Mechanics at the Collège de France in 1861, Serret was later offered the chair of differential and integral calculus at the Sorbonne. Frenet instead, after his start in Toulouse, was appointed professor of mathematics at the University of Lyon, where he was also director of the astronomical observatory.

Frenet–Serret formulae, as history decided they should be named, were published (nine and six) in two independent papers with a very similar title, at one year distance one from the other. Serret's paper [157] appeared in 1851 and was named *Sur quelques formules relatives à la théorie des courbes à double courbure*. Frenet published in 1852 the results of his 1847 doctoral thesis under the title: *Sur quelques proprétés des courbes à double courbure* [91].

Consider Fig. 7.22. Given a curve $\mathscr{C}(s)$ in three-dimensional Euclidean space, we can easily fix the following rule which uniquely defines a basis of three orthonormal vectors attached to each point of the curve. Let $\mathbf{r}$ be the standard cartesian coordinates of $\mathbb{R}^3$. Any curve can be described by giving $\mathbf{r} = \mathbf{r}(s)$ where s is the length parameter:

[9] *About the functions which help determining the attraction of general spheroids. Programme for a thesis about some properties of curves with a double curvature.*

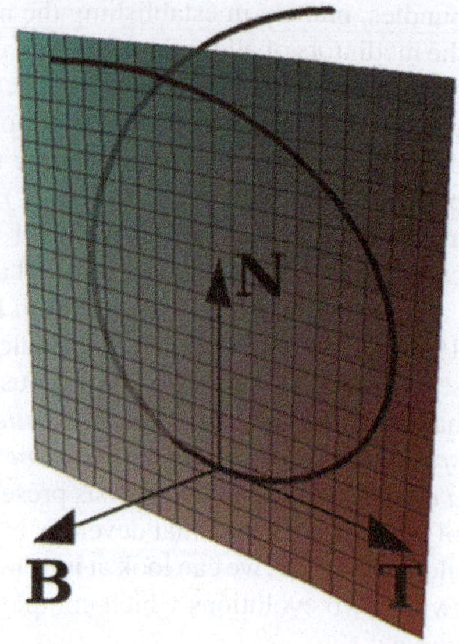

Fig. 7.22 The moving frames of Frenet and Serret

$$s = \int_0^s \left| \frac{d\mathbf{r}}{ds'} \right| ds' \tag{7.3.19}$$

By definition the tangent vector to the curve is:

$$\mathbf{T}(s) = \frac{d}{ds}\mathbf{r}(s) \tag{7.3.20}$$

and we can define the normal vector by means of the normalized second derivative:

$$\mathbf{N}(s) = \frac{1}{|\mathbf{t}(s)|}\frac{d}{ds}\mathbf{t}(s) \tag{7.3.21}$$

Finally we can complete the orthonormal system by means of the exterior product of $\mathbf{N}(s)$ and $\mathbf{T}(s)$ which is historically named the binormal vector:

$$\mathbf{B}(s) = \mathbf{T}(s) \wedge \mathbf{N}(s) \quad \Leftrightarrow \quad B^i(s) = \varepsilon^{ijk}\, T^j(s)\, N^k(s) \tag{7.3.22}$$

Since $\mathbb{E}(s) = \{\mathbf{T}(s), \mathbf{B}(s), \mathbf{N}(s)\}$ provides a basis for three-vectors, it is clear that also the derivative of this basis can be reexpressed in terms of itself, namely we can write:

$$\frac{d}{ds}\mathbb{E}(s) = \Omega(s)\,\mathbb{E}(s) \tag{7.3.23}$$

where $\Omega(s)$ is a 3×3 matrix which is necessarily antisymmetric since $\mathbb{E}^T\,\mathbb{E} = \mathbf{1}$. Indeed from such a relation it follows $\frac{d}{ds}\mathbb{E}\,\mathbb{E}^T = \Omega = -\mathbb{E}\,\frac{d}{ds}\mathbb{E}^T = -\,\Omega^T$.

In the component language of the middle XIXth century, Frenet and Serret proved that this matrix is not only antisymmetric but it has also a special *normal form*, in terms of two-parameters that they identified with the curvature κ and the torsion τ of the curve:

$$\Omega(s) = \begin{pmatrix} 0 & \kappa(s) & 0 \\ -\kappa(s) & 0 & \tau(s) \\ 0 & -\tau(s) & 0 \end{pmatrix} \tag{7.3.24}$$

The reason of these names becomes obvious looking at Fig. 7.22. The span of the vectors $\mathbf{T}, \mathbf{N}$ defines the *osculating plane* to the curve. If the torsion parameter τ vanishes the curve lies at all times in such a plane and the tangent and normal vectors undergo, from one point to the next one, an infinitesimal rotation of an angle $\delta\theta = \kappa$. When the torsion τ is different from zero, the tangent and normal vectors undergo not only an infinitesimal rotation but also a displacement in the direction of the third vector $\mathbf{B}$ and this displacement is precisely proportional to τ. The geometry of the curve is fully determined by the knowledge of the parameters κ and τ. For instance all planar curves are characterized by $\tau = 0$ and among them the circles are those with constant curvature κ. Similarly when both κ and τ are non vanishing but constant we have a spiral and so on.

In these very elementary geometrical facts the astonishing mathematical insight of Élie Cartan perceived a far reaching perspective of incredible richness. Refining the idea of the moving frames he started a conceptual revolution of the same amplitude as that done by Gauss with the *Disquisitiones* of 1828.

Right after the creation of General Relativity by Einstein, he started rethinking it in terms of *mobile frames* and came to the reformulation of gravitational equations which goes under the name of Einstein–Cartan theory [34]. While the original metric formulation is inadequate to incorporate fermionic fields, the new one can do that and is therefore more fundamental. Moreover it is much simpler from the algorithmic point of view and leads to extremely elegant and compact formulae. Cartan's viewpoint is centered around the idea of mobile frames which comes down from Frenet and Serret. It also leads to a simple formulation of the notion of connection that Charles Ehresman, one of the most brilliant students of Cartan, finally brought to mathematical perfection in the fifties of the XXth century.

Élie Cartan grasped from Frenet and Serret formulae the message that the geometry of an m-dimensional manifold $\mathcal{M}$ could be described by attaching an orthonormal frame to each of its points. The evolution of the orthonormal frame from one point to the next one, which can occur in m directions, encodes all information about the intrinsic curvature and torsion of the manifold. Relying on the exterior differential calculus he had created, Cartan introduced a system of m one-forms:

$$E^a = E^a_\mu(x)\, dx^\mu \tag{7.3.25}$$

which at each point constitute an orthonormal reference frame for controvariant vectors. In modern words $\{E^a\}$ are a basis of sections of the cotangent bundle

$T^*\mathscr{M} \stackrel{\pi}{\Longrightarrow} \mathscr{M}$. Calculating the exterior derivative of the forms E^a we can write the following formula, just analogous to Frenet and Serret's formulae:

$$dE^a = -\,\omega^{ab} \wedge E^b + \mathfrak{T}^a \tag{7.3.26}$$

where $\omega^{ab} = -\omega^{ba}$ is an antisymmetric-valued one-form, named the *spin connection* and

$$\mathfrak{T}^a = T^a_{bc}(x)\, E^b \wedge E^c \tag{7.3.27}$$

is a 2-form which will be named the torsion. If we have no information on the torsion, Eq. (7.3.26) is undetermined, namely there are many solutions for the one-form ω^{ab}. This latter falls into the mathematical category of connections. According to the definition we shall provide in next section, ω^{ab} is a connection on a principal bundle $P(\mathscr{M}, \mathrm{SO(m)})$ where the base manifold is the considered manifold $\mathscr{M}$ and where the structural group is SO(m), namely the Lie group which rotates orthonormal frames among themselves. If we have geometrical or physical reasons to prescribe the value of the torsion, for instance zero ($T^a = 0$), but also any other externally given 2-form, then Eq. (7.3.26) uniquely determines the SO(m)-connection ω^{ab}. In this case we say that the SO(m)-bundle is soldered to the tangent bundle and the SO(m)-connection ω^{ab} becomes a *Cartan-connection*. Just as in the Frenet-Serret case, the geometry of the manifold, in particular its curvature, is revealed through the calculation of the second derivative, which Cartan immediately understood should be another exterior derivative. Indeed, according to the general theory of principal connections, one can calculate the natural 2-form associated with any connection, which in this case reads as follows:

$$\mathfrak{R}^{ab} = d\omega^{ab} + \omega^{ac} \wedge \omega^{cb} \tag{7.3.28}$$

and the expansion of $\mathfrak{R}^{ab}$ along the orthonormal frame provides a tensor $R^{ab}{}_{cd}$ which represents the curvature of the manifold:

$$\mathfrak{R}^{ab} = R^{ab}{}_{cd}\, E^c \wedge E^d \tag{7.3.29}$$

Indeed $R^{ab}{}_{cd}$ has a unique invertible one-to-one relation with the Riemann tensor introduced in Eq. (7.3.10).

In this way Cartan's was able to reformulate the entire setup of Riemannian geometry, and with it General Relativity, into the language of exterior differential forms, avoiding the metric tensor of Gauss and Riemann, replaced by the notion of mobile-frames. From the physical viewpoint, Cartan's approach is also very natural since it can be rephrased as Einstein's beloved equivalence principle, asserting that we can always find locally inertial reference frames.

There is another important aspect of Cartan's formulation of gravitational theory: it suppresses its diversity from the other gauge-theories describing non gravitational interactions. Adopting Cartan's viewpoint, the fundamental fields describing gravity are also encoded into a *connection on a principal bundle* as it happens for all other forces. What is special about gravity is the soldering phenomenon, namely the possibility of solving for one part of the connection (ω^{ab}) in terms of the other (E^a), by imposing an external condition on the torsion T^a. In Einstein–Cartan formulation the condition on the torsion becomes a field equation streaming from the same variational principle which yields Einstein equations on the Riemann tensor and so everything is unified in a consistent, powerful scheme.

7.4 Fibre-Bundles and Principal Connections

The concept of *fibre bundle* emerged slowly in the mathematical literature of the XXth century and came to full ripeness at the time of World War II when the theory of characteristic classes was developed by Chern and Weil. Telling the story of this development will be my concern in Chap. 9, as part of the general fresco depicting the evolution of geometry in the first half of the XXth century.

Here we sketch and discuss the mathematical definition of fibre-bundles in order to clarify what the mathematical conceptions whose evolution we discuss actually are.

As a matter of fact the concept of fibre bundle is absolutely central both in contemporary mathematics and physics as it provides the appropriate mathematical framework to formulate modern field theory. All the fields one can consider are either *sections* of *associated bundles* or *connections* on *principal bundles*. There are indeed two kinds of fibre bundles:

1. principal bundles
2. associated bundles

The notion of a principal fibre bundle is the appropriate mathematical concept underlying the formulation of *gauge theories* that provide the general framework to describe the dynamics of all non–gravitational interactions. The concept of a connection on such principal bundles codifies the physical notion of the bosonic particles mediating the interaction, namely the gauge bosons, like the photon, the gluon or the graviton. Indeed, gravity itself is a gauge theory although of a very special type. On the other hand the notion of associated fibre bundles is the appropriate mathematical framework to describe *matter fields* that interact through the exchange of the *gauge bosons*. The reader recognizes here points (B) and (C) of the episteme as we formulated it in Chap. 1.

Also from a more general viewpoint and in relation with all sort of applications the notion of fibre bundles is absolutely fundamental. As we already emphasized, the points of a manifold can be identified with the possible states of a complex system specified by an m–tuplet of parameters $x_1, \ldots x_m$. Real or complex functions of such parameters are the natural objects one expects to deal with in any scientific theory that explains the phenomena observed in such a system. Yet, as we already anticipated, calculus on manifolds that are not trivial as the flat $\mathbb{R}^m$ cannot be confined to functions, which correspond to a too restrictive notion. The appropriate generalization of functions is provided by the *sections* of fibre bundles. Locally, namely in each coordinate patch, functions and sections are just the same thing. Globally, however, there are essential differences. A section is obtained by gluing together many local functions by means of non trivial transition functions that reflect the geometric structure of the fibre bundle.

7.4.1 *Definition of Fibre-Bundles*

To introduce the mathematical definition of a fibre bundle we need the notion of Lie group whose history was thoroughly discussed in Chap. 5. A Lie group G is:

- A group from the algebraic point of view, namely a set with an internal composition law, the product
$$\forall g_1 , g_2 \in G \quad g_1 \cdot g_2 \in G \tag{7.4.1}$$
which is associative, admits a unique neutral element e and yields an inverse for each group element.
- A smooth manifold of finite dimension $\dim G = n < \infty$ whose transition function are not only infinitely differentiable but also real analytic, namely they admit an expansion in power series.
- In the topology defined by the manifold structure the two algebraic operations of taking the inverse of an element and performing the product of two elements are real analytic (admit a power series expansion).

Coming now to fibre bundles let us begin by recalling that an artistic illustration of such spaces is provided by the celebrated picture by Escher of an ant crawling on a Mobius strip (see Fig. 7.23).

The basic idea is that if we consider a piece of the bundle this cannot be distinguished from a trivial direct product of two spaces, an open subset of the base manifold and the fibre. In Fig. 7.24 the base manifold is a circle and the fibre is a

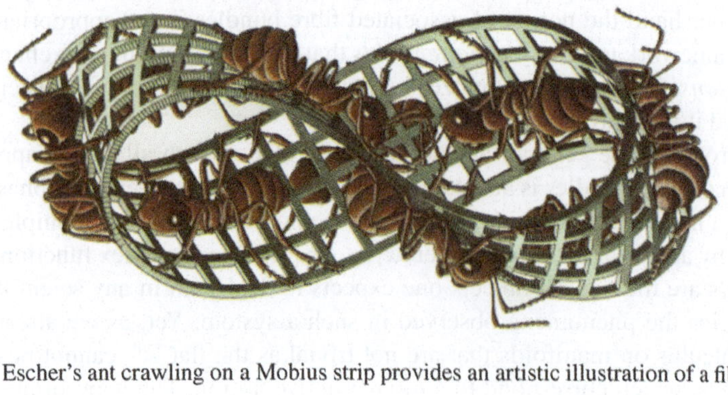

Fig. 7.23 Escher's ant crawling on a Mobius strip provides an artistic illustration of a fibre bundle

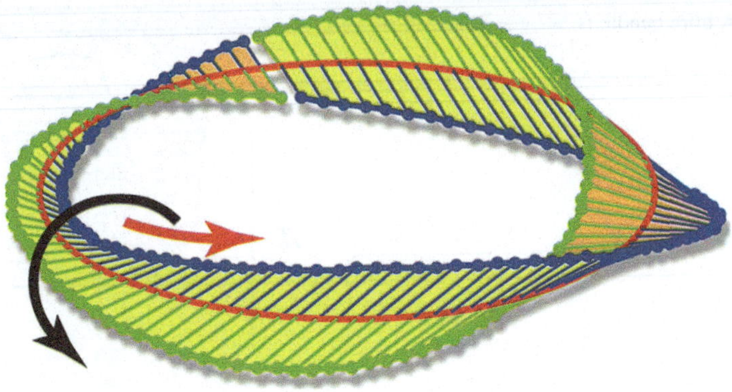

Fig. 7.24 Mobius strip provides a pedagogical example of a fibre bundle

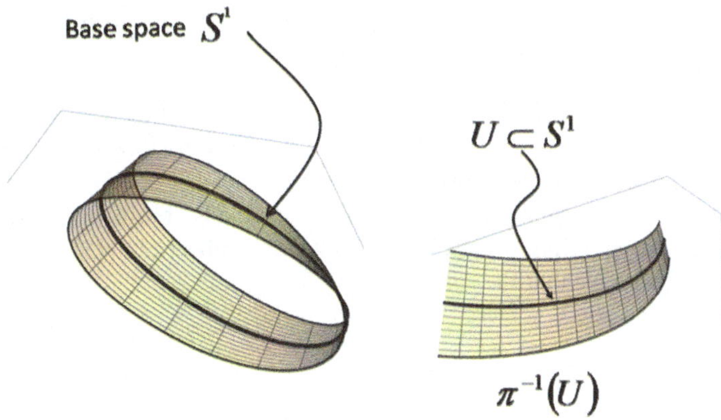

Fig. 7.25 Local triviality of an open piece of the Mobius strip

segment I $\equiv [-1, 1]$. Locally the space is the direct product of an open interval of $U =]a, b[\subset \mathbb{R}$ with the standard fibre I, as it is evident from Fig. 7.25.

However, the relevant point is that, *globally*, the bundle *is not a direct product of spaces*.

Hence the notion of fibre bundle corresponds to that of a differentiable manifold P with dimension $\dim P = m + n$ that locally *looks like* the direct product $U \times F$ of an open manifold U of dimension $\dim U = m$ with another manifold F (the standard fibre) of dimension $\dim F = n$. Essential in the definition is the existence of a map:

$$\pi \quad : \quad P \rightarrow \mathcal{M} \tag{7.4.2}$$

named *the projection* from the *total manifold* P of dimension $m + n$ to a manifold $\mathcal{M}$ of dimension m, named the *base manifold*. Such a map is required to be continuous. Due to the difference in dimensions the projection cannot be invertible. Indeed to

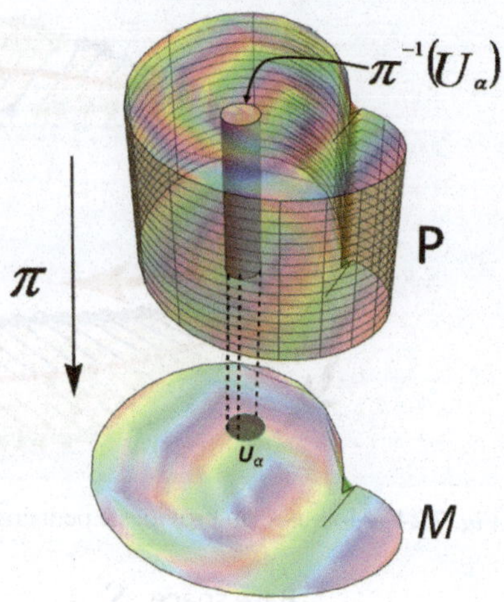

Fig. 7.26 A fibre bundle is locally trivial

every point $\forall\, p \in \mathcal{M}$ of the base manifold the projection associates a submanifold $\pi^{-1}(p) \subset P$ of dimension $\dim \pi^{-1}(p) = n$ composed by those points of $x \in P$ whose projection on $\mathcal{M}$ is the chosen point p: $\pi(x) = p$. The submanifold $\pi^{-1}(p)$ is named the *fibre over* p and the basic idea is that each fibre is homeomorphic to the *standard fibre* F. More precisely for each open subset $U_\alpha \subset \mathcal{M}$ of the base manifold we must have that the submanifold

$$\pi^{-1}(U_\alpha)$$

is homeomorphic to the direct product

$$U_\alpha \times F$$

This is the precise meaning of the statement that, locally, the bundle looks like a direct product (see Fig. 7.26). Explicitly what we require is the following: there should be a family of pairs (U_α, ϕ_α) where U_α are open charts covering the base manifold $\bigcup_\alpha U_\alpha = \mathcal{M}$ and ϕ_α are maps:

$$\phi_\alpha \quad : \quad \pi^{-1}(U_\alpha) \subset P \to U_\alpha \otimes F \tag{7.4.3}$$

that are required to be one–to–one, bicontinuous (=continuous, together with its inverse) and to satisfy the property that:

$$\pi \circ \phi_\alpha^{-1}(p, f) = p \tag{7.4.4}$$

Namely the projection of the image in P of a base manifold point p *times* some fibre point f is p itself.

Each pair (U_α, ϕ_α) is named a *local trivialization*. As for the case of manifolds, the interesting question is what happens in the intersection of two different local trivializations. Indeed if $U_\alpha \bigcap U_\beta \neq \emptyset$, then we also have $\pi^{-1}(U_\alpha) \bigcap \pi^{-1}(U_\beta) \neq \emptyset$. Hence each point $x \in \pi^{-1}(U_\alpha \bigcap U_\beta)$ is mapped by ϕ_α and ϕ_β in two different pairs $(p, f_\alpha) \in U_\alpha \otimes F$ and $(p, f_\beta) \in U_\alpha \otimes F$ with the property, however, that the first entry p is the same in both pairs. This follows from property (7.4.4). It implies that there must exist a map:

$$t_{\alpha\beta} \equiv \phi_\beta^{-1} \circ \phi_\alpha \quad : \quad \left(U_\alpha \bigcap U_\beta\right) \otimes F \to \left(U_\alpha \bigcap U_\beta\right) \otimes F \tag{7.4.5}$$

named the *transition function*, which acts exclusively on the fibre points in the sense that:

$$\forall p \in U_\alpha \bigcap U_\beta, \quad \forall f \in F \quad t_{\alpha\beta}(p, f) = \left(p, t_{\alpha\beta}(p).f)\right) \tag{7.4.6}$$

where for each choice of the point $p \in U_\alpha \bigcap U_\beta$,

$$t_{\alpha\beta}(p) \quad : \quad F \mapsto F \tag{7.4.7}$$

is a continuous and invertible map of the standard fibre F into itself (see Fig. 7.27).

The last bit of information contained in the notion of fibre bundle is related with the *structural group*. This has to do with answering the following question: where are the transition functions chosen from? Indeed the set of all possible continuous invertible maps of the standard fibre F into itself constitute a group, so that it is no restriction to say that the transition functions $t_{\alpha\beta}(p)$ are group elements. Yet the group of all homeomorphisms Hom(F, F) is very large and it makes sense to include into the definition of fibre bundle the request that the transition functions should be chosen within a smaller hunting ground, namely inside some finite dimensional Lie group G that has a well defined action on the standard fibre F.

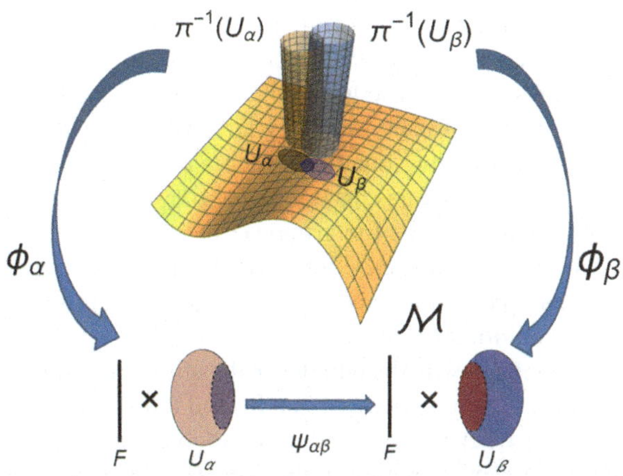

Fig. 7.27 Transition function between two local trivializations of a fibre bundle

Just as manifolds can be constructed by gluing together open charts, fibre bundles can be obtained by gluing together local trivializations. Explicitly one proceeds as follows.

1. First choose a base manifold $\mathcal{M}$, a typical fibre F and a structural Lie Group G whose action on F must be well–defined.
2. Then choose an atlas of open neighborhoods $U_\alpha \subset \mathcal{M}$ covering the base manifold $\mathcal{M}$.
3. Next to each non–vanishing intersection $U_\alpha \bigcap U_\beta \neq \emptyset$ assign a transition function, namely a smooth map:

$$\psi_{\alpha\beta} \quad : \quad U_\alpha \bigcap U_\beta \mapsto G \tag{7.4.8}$$

from the open subset $U_\alpha \bigcap U_\beta \subset \mathcal{M}$ of the base manifold to the structural Lie group. For consistency the transition functions must satisfy the two conditions:

$$\begin{array}{lcl} \forall\, U_\alpha, U_\beta \;/\; U_\alpha \bigcap U_\beta \neq \emptyset & : & \psi_{\beta\alpha} = \psi_{\alpha\beta}^{-1} \\ \forall\, U_\alpha, U_\beta, U_\gamma \;/\; U_\alpha \bigcap U_\beta \bigcap U_\gamma \neq \emptyset & : & \psi_{\alpha\beta} \cdot \psi_{\beta\gamma} \cdot \psi_{\gamma\alpha} = \mathbf{1}_G \end{array} \tag{7.4.9}$$

Whenever a set of local trivializations with consistent transition functions satisfying Eq. (7.4.9) has been given a fibre bundle is defined. A different and much more difficult question to answer is to decide whether two sets of local trivializations define the same fibre bundle or not. We do not address such a problem whose proper treatment is beyond the scope of this essay. We just point out that the classification of inequivalent fibre bundles one can construct on a given base manifold $\mathcal{M}$ is a problem of global geometry which can also be addressed with the techniques of algebraic topology and algebraic geometry.

Typically inequivalent bundles are characterized by topological invariants that receive the name of *characteristic classes* (see Sect. 8.2).

In physical language the transition functions (7.4.8) from one local trivialization to another one are the *gauge transformations*, namely group transformations depending on the position in space–time (*i.e.* the point on the base manifold).

A principal bundle $P(\mathcal{M}, G)$ is a fibre bundle where the standard fibre coincides with the structural Lie group $F = G$ and the action of G on the fibre is the action of the group on itself either by left or by right multiplication. The name principal is given to the fibre bundle in such a definition since it is a "*father*" bundle which, once given, generates an infinity of *associated vector bundles*, one for each linear representation of the Lie group G.

An *associated vector bundle* is a fibre bundle where the standard fibre $F = V$ is a vector space and the action of the structural group on the standard fibre is a linear representation of G on V.

The reason why the bundles in the above definition are named associated is almost obvious. Given a principal bundle and a linear representation of G we can immediately construct a corresponding vector bundle. It suffices to use as transition functions the linear representation of the transition functions of the principal bundle:

$$\psi_{\alpha\beta}^{(V)} \equiv D\left(\psi_{\alpha\beta}^{(G)}\right) \in \mathrm{Hom}\,(V, V) \tag{7.4.10}$$

For any vector bundle the dimension of the standard fibre is named the *rank* of the bundle.

7.4.2 Ehresmann and the Connections on a Principal Fibre Bundle

The mathematician who brought the definition of a connection on principal fibre-bundle to perfection, providing in this way the rigorous basis of physical gauge–theories is Charles Ehresman (see Fig. 7.28).

He was born in German speaking Alsace in 1905 from a poor family. His first education was in German, but after Alsace was returned to France in 1918 as a result of Germany's defeat in World War I, Ehresman attended only French schools and his University education was entirely French. Indeed in 1924 he entered the École Normale Superieure from which he graduated in 1927.

After that, he served as a teacher of Mathematics in the French colony of Morocco and then he went to Göttingen that in the late twenties and beginning of the thirties was the major scientific center of the world, at least for Mathematics and Physics.

The raising of Nazi power in Germany dismantled the scientific leadership of the country, caused the decay of Göttingen and obliged all the Jewish scientists, who so greatly contributed to German culture, to emigrate to the United States.

Fig. 7.28 Charles Ehresman (1905–1979)

Ehresman also fled from Göttingen to Princeton where he studied for few years until 1934. In that year he returned to France to obtain his doctorate under the supervision of Élie Cartan. Charles Ehresman was professor at the Universities of Strasbourg and Clermont Ferrand.

In 1955 a special chair of *Topology* was created for him at the University of Paris which he occupied up to his retirement in 1975. He died in 1979 in Amiens where his second wife, also a mathematician held a chair.

Charles Ehresman was one of the creators of differential topology and, as we already stressed, he greatly contributed to the development of the notion of fibre-bundles [69] and of the connections defined over them [70]. He founded the mathematical theory of categories.

7.4.2.1 The Notion of a Connection

Let $P(\mathcal{M}, G)$ be a principal fibre-bundle with base-manifold $\mathcal{M}$ and structural group G. Let us moreover denote π the projection:

$$\pi : P \to \mathcal{M} \tag{7.4.11}$$

Consider the action of the Lie group G on the total space P. By definition this action is vertical in the sense that

$$\forall u \in P \,,\, \forall g \in G \quad : \quad \pi(g(u)) = \pi(u) \tag{7.4.12}$$

namely it moves points only along the fibres. Given any element $X \in \mathbb{G}$ where we have denoted by $\mathbb{G}$ the Lie algebra of the structural group, we can consider the one-dimensional subgroup generated by it $g_X(t) = \exp[t\,X]$, where t is a real parameter. There is a curve $\mathscr{C}_X(t,u)$ in the manifold P obtained by acting with $g_X(t)$ on some point $u \in P$, namely: $\mathscr{C}_X(t,u) \equiv g_X(t)(u)$. Because of the vertical action of the structural group, this curve develops along the fibres and every point of the curve has the same projection on the base manifold:

$$\pi\left(\mathscr{C}_X(t,u)\right) = p \in \mathcal{M} \quad \text{if} \quad \pi(u) = p \tag{7.4.13}$$

This feature is illustrated in Fig. 7.29.

By means of this construction, to every Lie algebra element $\mathbf{X} \in \mathbb{G}$ we associate a vector field $\mathbf{X}^{\#}$ defined at each point u of the total space P which is just the tangent vector to the vertical curve $\mathscr{C}_X(t,u)$. In short what we have achieved is a map from the structural group Lie algebra $\mathbb{G}$ to the tangent space to the bundle at each of its points u:

$$\#_u \quad : \quad \mathbb{G} \to T_u P \tag{7.4.14}$$

Clearly the map $\#_u$ is not surjective: its image is what we can name the vertical subspace $V_u P$ since it spans the fibre–directions.

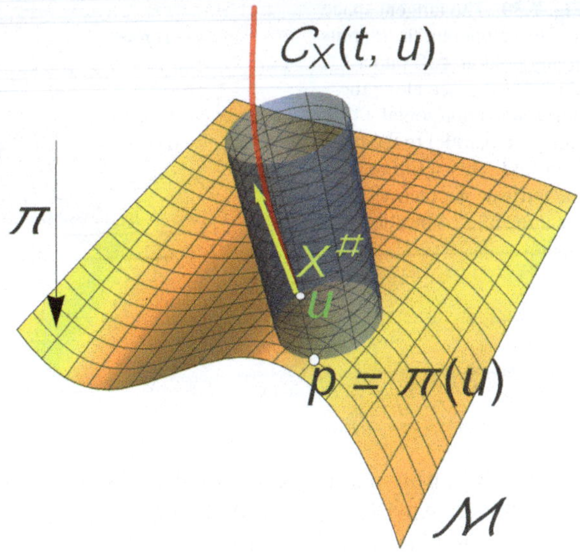

Fig. 7.29 Every element $\mathbf{X} \in \mathbb{G}$ of the structural group Lie algebra generates a curve $\mathscr{C}_{\mathbf{X}}(t, u)$ in the total space P, that starting from a point u is completely vertical, namely all points of the curve have the same projection p on the base manifold $\mathscr{M}$

Having defined the vertical subspace of the tangent space one would be interested in giving a definition also of the horizontal subspace $H_u P$, which, intuitively, must be somehow parallel to the tangent space $T_p\mathscr{M}$ to the base manifold at the projection point $p = \pi(u)$. The dimension of the vertical space is the same as the dimension of the fibre, namely $n = \dim G$. The dimension of the horizontal space $H_u P$ must be the same as the dimension of the base manifold $m = \dim\mathscr{M}$. Indeed $H_u P$ should be the orthogonal complement of $V_u P$. Easy to say, but orthogonal with respect to what? This is precisely the point. Is there an a priori intrinsically defined way of defining the orthogonal complement to the vertical subspace $V_u \subset T_u P$? The answer is that there is not. Given a basis $\{\mathbf{v}^{\mu}\}$ of n vectors for the subspace $V_u P$, there are infinitely many ways of finding m extra vectors $\{\mathbf{h}^i\}$ which complete this basis to a basis of $T_u P$. The span of any such collection of m vectors $\{\mathbf{h}^i\}$ is a possible legitimate definition of the orthogonal complement $H_u P$. This arbitrariness is the root of the mathematical notion of a *connection*. Providing a fibre bundle with a connection precisely means introducing a rule that uniquely defines the orthogonal complement $H_u P$.

Let $P(M, G)$ be a principal fibre-bundle. A *connection* on P is a rule which at any point $u \in P$ defines a unique splitting of the tangent space $T_u P$ into the vertical subspace $V_u P$ and into a horizontal complement $H_u P$ satisfying the following properties:

(i) $T_u P = H_u P \oplus V_u P$
(ii) Any smooth vector field $\mathbf{X}$ separates into the sum of two smooth vector fields $\mathbf{X} = \mathbf{X}^H + \mathbf{X}^V$ such that at any point $u \in P$ we have $\mathbf{X}^H_u \in H_u P$ and $\mathbf{X}^V_u \in V_u P$
(iii) the horizontal spaces along the same fibre are related to each other by the action of the structural group G on the fibre bundle.

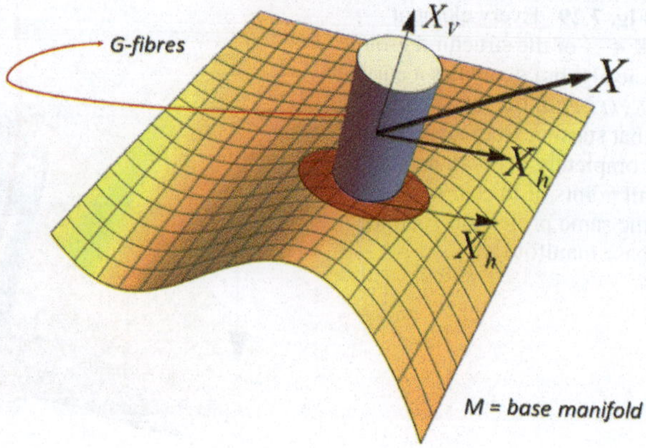

Fig. 7.30 The tangent space to a principal bundle P splits at any point $u \in P$ into a vertical subspace along the fibres and a horizontal subspace parallel to the base manifold. This splitting is the intrinsic geometric meaning of a connection

This beautiful purely geometrical definition of the connection illustrated in Fig. 7.30 is due to Ehresmann [70]. It emphasizes that it is just an intrinsic attribute of the principal fibre-bundle. In a trivial bundle, which is a direct product of manifolds, the splitting between vertical and horizontal spaces is done once for ever. The vertical space is the tangent space to the fibre, the horizontal space is the tangent space to the base. In a non trivial bundle the splitting between vertical and horizontal directions has to be reconsidered at every next point and fixing this ambiguity is the task of the connection.

7.4.2.2 The Connection One-Form

The algorithmic way to implement the splitting rule advocated by the Ehresmann definition is provided by introducing a connection one-form $\mathbf{A}$ which is just a Lie algebra valued differential one-form on the bundle P satisfying two precise requirements:

(i) $\forall X \in \mathbb{G} \quad : \quad \mathbf{A}\left(\mathbf{X}^{\#}\right) = X$
(ii) $\forall \mathrm{g} \in G \quad : \quad \mathrm{g}^{*}\mathbf{A} = \mathrm{g}^{-1}\,\mathbf{A}\,\mathrm{g}$

Given the connection one-form $\mathbf{A}$ the splitting between vertical and horizontal subspaces is performed in the following way. At any $u \in P$, the horizontal subspace $H_u P$ of the tangent space to the bundle is the kernel of $\mathbf{A}$, namely

$$H_u P \equiv \{\mathbf{X} \in T_u P \,|\, \mathbf{A}(\mathbf{X}) = 0\} \tag{7.4.15}$$

Elaborating the consequences of the above definitions one arrives at the conclusion that, in that in any local trivialization $u = (x, \mathrm{g})$ the connection one-form has the following structure:

$$\mathbf{A} = \mathrm{g} \cdot \mathscr{A} \cdot \mathrm{g}^{-1} + d\mathrm{g} \cdot \mathrm{g}^{-1} \tag{7.4.16}$$

where

$$\mathscr{A} = \mathscr{A}^I_\mu(x)\, dx^\mu\, T_I \qquad (7.4.17)$$

is a $\mathbb{G}$ Lie algebra valued one-form on the base manifold $\mathscr{M}$, having denoted by T_I a basis of generators of the Lie algebra $\mathbb{G}$.

In contemporary theoretical physics the objects like that displayed in Eq. (7.4.17) are named *gauge fields* and encode the various mediators of fundamental interactions.

7.5 Conclusive Considerations on Gauge Fields

Reconsidering what we have learnt from our historical survey we can say the following. Gauge fields and connections on fibre-bundles are just the very same thing. The first is the name utilized in the early physical literature, the second that introduced by the early mathematical literature. Nowadays the identification of the two conceptions is fully perceived both by physicists and by mathematicians. Hence the two denominations are utilized in an interchangeable way in both communities. We see in this the deep significance of the two leaves departing from the same stem presented in Fig. 7.4.

Strictly speaking the very first to introduce a connection was Christoffel, whose paper on the coefficients named after him dates 1869. The Christoffel symbols encode the components of an affine connection derived from a metric, the Levi-Civita connection which we interpret as encoding gravitational interactions. Almost immediately after him, in 1873, Maxwell was the second to introduce a connection, this time on a U(1) principal bundle. In his famous *Treatise on Electricity and Magnetism*, published in 1873, Maxwell utilized a vector potential **A** which is indeed a connection one form–on a U(1) bundle and describes magnetic interactions.[10]

In 1923 Cartan formalized the notion of *affine connections*, while *Ehresman connection* on a principal bundle was introduced in 1950. The famous paper by Yang and Mills introducing non-abelian gauge fields dates 1954 and it is completely independent from Ehresmann's paper. As early as 1929, Hermann Weyl had introduced his peculiar gauge theory based on scale transformations rather than phase transformations, as it is appropriate for electromagnetism.

From this short summary not only we can fully appreciate the meaning of Yang's picture but we also learn another important lesson.

Observing the history of science on a longer time-scale we see that the Galilean Method consisting of the three phases:

(a) Interrogation of Nature
(b) Formulation of a Theory to explain Observed Phenomena
(c) Verification or Falsification of the further predictions of the Theory

is very important and valuable but it is not the end of the story.

---

[10]Obviously Maxwell was not aware of the mathematical significance of the vector potential **A**, yet that **A** is a U(1)–connection, is a fact beyond any doubt.

Indeed there is not only Nature that has to be interrogated, but also Abstract Human Thought which finds its most efficient way of expression in the language of Mathematics. There exists, historically, an independent logical development of mathematical notions and constructions, whose point of origin is of philosophical nature, rooted in a System of Thought which is civilization dependent. Fundamental steps forward in physics occur quite often through a process of agnition: an existing mathematical structure is recognized to be the category encompassing fundamental concepts elaborated in physics. At that moment all the conceptual implications for physical thought of that mathematical structure are activated and a new vision emerges which not only contributes to a modification of the *episteme* but it is also capable of reorienting item (a) of the Galilean method, namely the Interrogation of Nature. Posing a question, even to Nature, always requires a language, and languages do not exist elsewhere than in the human mind.

The history of Gauge Fields is a paradigmatic example of what we said above.

The notion of connection, just as that of fibre-bundle to which it refers, has a long independent history in mathematics and pertains to the mathematical development of the *philosophical* concept of *Space*, namely it pertains to *Geometry* in its most comprehensive definition. Starting from the rigidity of Kant's conceptions it took about one and half century to enlarge the notion of *Space* by including non-euclidian geometries, to recognize that the propositions about geometrical figures are propositions about invariants with respect to some symmetry group, to smooth rigid spaces into differentiable manifolds endowed with the features of curvature, finally to glue together differentiable manifolds into smoothly twisted fibre-bundles. These developments were motivated by the urge to answer two philosophical-mathematical questions:

1. How do we define distances?
2. How do we define parallelism and parallel transport?

During the same period of time, physicists were interrogating Nature in a Galilean way about electric and magnetic phenomena. All the laws experimentally determined in this way were summarized by Maxwell in his four differential equations whose further consequence is the existence of electromagnetic waves, leading to the understanding of light, to Marconi and to the radio.

All that was Galilean.

However, when Weyl tried to interpret Electromagnetism as a gauge theory, the detachment from the Galilean method was complete. Similarly it was complete in the identification of the gravitational field with a riemannian metric pursued by Einstein. Yang and Mills succeeded in generalizing the U(1)-gauge-structure of electromagnetism to non abelian groups G and this was also a non galilean operation which, nonetheless, opened the door to the contemporary Standard Model of non gravitational interactions. It was a conceptual operation inside a mathematical category extracted from the empirically established Maxwell Theory. Yet the deeper mathematical significance of the gauge transformations was still missing in the physical thought, while it was already established in mathematics through Ehresman's work. The agnition came later and produced a quite significant upgrading of physical thought.

7.6 Isometries: Back from Geometry to Groups

The word isometry comes from the Greek word ἡ ἰσομετρία which means the equality of measures.

The origin of the modern concept of isometry is rooted in that of *congruence* of geometrical figures that Euclid never introduced explicitly, yet implicitly assumed when he proceeded to identify those triangles that can be superimposed one onto the other.

As we already explained, it was indeed the question about what are the transformations that define such congruences what led Felix Klein to the Erlangen Programme. Klein understood that Euclidian congruences are based on the transformations of the Euclidian Group and he came to the idea that other geometries are based on different groups of transformations with respect to which we consider congruences.

Such a concept, however, would have been essentially empty without an additional element, the *metric*. The area and the volume of geometrical figures, the length of sides and the relative angles have to be measured in order to compare them. These measurements can be performed if and only if we have a metric g, in other words if the *substratum* of the considered geometry is a *Riemannian* or a *pseudo Riemannian* manifold $(\mathcal{M}, g)$.

Therefore the group of transformations which, according to the vision of the Erlangen Programme, defines a geometry, is the *group of isometries* $\mathrm{G}_{\mathrm{iso}}$ of a given Riemannian space $(\mathcal{M}, g)$, the elements of this group being diffeomorphisms:

$$\phi \ : \ \mathcal{M} \to \mathcal{M} \tag{7.6.1}$$

such that their pull-back on the metric form leaves it invariant:

$$\forall \phi \in \mathrm{G}_{\mathrm{iso}} \quad : \quad \phi^{\star}\left[g_{\mu\nu}(x)\, dx^{\mu}\, dx^{\nu}\right] = g_{\mu\nu}(x)\, dx^{\mu}\, dx^{\nu} \tag{7.6.2}$$

Quite intuitively it becomes clear that the structure of $\mathrm{G}_{\mathrm{iso}}$ is determined by the manifold $\mathcal{M}$ and by its metric g, so that the concept of geometries is now identified with that of Riemannian spaces $(\mathcal{M}, g)$.

A generic metric g has no isometries and hence there are no congruences to study. (Pseudo)-Riemannian manifolds with no isometry, or with few isometries, are relevant to several different problems pertaining to physics and also to other sciences, yet they are not in the vein of the Erlangen Programme, aiming at the classification of geometries in terms of groups. Hence we can legitimately ask ourselves the question whether such a programme can be ultimately saved, notwithstanding our discovery that a geometry is necessarily based on a (pseudo)-Riemannian manifold $(\mathcal{M}, g)$. The answer is obviously yes if we can invert the relation between the metric g and its isometry group $\mathrm{G}_{\mathrm{iso}}$. Given a Lie group G can we construct the Riemannian manifold $(\mathcal{M}, g)$ which admits G as its own isometry group $\mathrm{G}_{\mathrm{iso}}$? Indeed we can; the answers are also exhaustive if we add an additional request, that of transitivity. A group G acting on a manifold $\mathcal{M}$ by means of diffeomorphisms:

$$\forall \gamma \in \mathrm{G} \ \ \gamma : \mathcal{M} \to \mathcal{M} \tag{7.6.3}$$

has a transitive action if and only if any two points p, q of the manifold are mapped one into the other by some element of the group G, namely if

$$\forall p, q \in \mathcal{M} \ , \quad \exists \gamma \in \mathrm{G} / \gamma(q) = p \tag{7.6.4}$$

If the Riemannian manifold $(\mathcal{M}, g)$ admits a transitive group of isometries it means that any point of $\mathcal{M}$ can be mapped into any other by means of a transformation that is an isometry. In this case the very manifold $\mathcal{M}$ and its metric g are completely determined by group theory: $\mathcal{M}$ is necessarily a *coset manifold* G/H, namely the space of equivalence classes of elements of G with respect to multiplication (either on the right or on the left) by elements of a subgroup $\mathrm{H} \subset \mathrm{G}$. The metric g is induced on the equivalence classes by the Killing metric of the Lie algebra, defined on $\mathbb{G}$.

In the mathematical-philosophical perspective of the present tale the geometry of coset manifolds is extensively discussed in my parallel book [90]. Among coset manifolds particular attention is there given to the so named *symmetric spaces* characterized by an additional reflection symmetry whose nature will become clear to the reader of [90].

7.6.1 Symmetric Spaces and Élie Cartan

The full-fledged classification of all symmetric spaces was the gigantic achievement of Élie Cartan. The classification of symmetric spaces is at the same time a classification of the real forms of the complex Lie algebras and it is the conclusive step in the path initiated by Killing in his papers of 1888, 1889. At the same time the geometries of non-compact symmetric spaces can be formulated in terms of other quite interesting algebraic structures, the *normed solvable Lie algebras*. The class of these latter is wider than that of symmetric spaces and this provides a generalization path leading to a wider class of geometries, all of them under firm algebraic control. This topic is also thoroughly discussed in a dedicated chapter of [90] which is propaedeutical to the developments of the subsequent chapters of that book.

7.6.2 Where and How do Coset Manifolds Come into Play?

By now it should be clear to the reader that, just as we have the whole spectrum of linear representations of a Lie algebra $\mathbb{G}$ and of its corresponding Lie group G, in the same way we have the set of *non-linear representations* of the same Lie algebra $\mathbb{G}$ and of the same Lie group G. These are encoded in all possible coset manifolds G/H with their associated G–invariant metrics.

Where and how do these geometries pop up?

The answer is that they appear at several different levels of analysis and in connection with different aspects of physical theories. Let us enumerate them and discover a conceptual hierarchy.

(A) A first context of utilization of coset manifolds G/H is in the quest for solutions of Einstein Equations in $d = 4$ or in higher dimensions. One is typically interested in space-times with a prescribed *isometry* and one tries to fit into the equations G/H metrics whose parameters depend on some residual coordinate like the time t in cosmology or the radius r in black-hole physics. The field equations of the theory reduce to few parameter differential equations in the residual space.

(B) Another instance of utilization of coset manifolds is in the context of σ-models. In physical theories that include scalar fields $\phi^I(x)$ the kinetic term is necessarily of the following form:

$$\mathscr{L}_{kin} = \frac{1}{2} \gamma_{IJ}(\phi)\, \partial_\mu \phi^I(x)\, \partial_\nu \phi^J(x)\, g^{\mu\nu}(x) \tag{7.6.5}$$

where $g^{\mu\nu}(x)$ is the metric of space-time, while $\gamma_{IJ}(\phi)$ can be interpreted as the metric of some manifold $\mathcal{M}_{\text{target}}$ of which the fields ϕ^I are the coordinates and whose dimension is just equal to the number of scalar fields present in the theory. If we require the field theory to have some Lie Group symmetry G, either we have linear representations or non linear ones. In the first case the metric γ_{IJ} is constant and invariant under the linear transformations of G acting on the $\phi^I(x)$. In the second case the manifold $\mathcal{M}_{\text{target}} = \text{G/H}$ is some coset of the considered group and $\gamma_{IJ}(\phi)$ is the corresponding G-invariant metric.

(C) In mathematics and sometimes in physics you can consider structures that depend on a continuous set of parameters, for instance the solutions of certain differential equations, like the self-duality constraint for gauge-field strengths or the Ricci-flat metrics on certain manifolds, or the algebraic surfaces of a certain degree in some projective spaces. The parameters corresponding to all the possible deformations of the considered structure constitute themselves a manifold $\mathcal{M}$ which typically has some symmetries and in many cases is actually a coset manifold. A typical example is provided by the so named Kummer surface K3 whose Ricci flat metric no one has so far constructed, yet we know a priori that it depends on 3×19 parameters that span the homogeneous space $\frac{\text{SO}(3,19)}{\text{SO}(3)\times\text{SO}(19)}$.

(D) In many instances of field theories that include scalar fields there is a scalar potential term $V(\phi)$ which has a certain group of symmetries G. The vacua of the theory, namely the set of extrema of the potential usually fill up a coset manifold G/H where $\text{H} \subset \text{G}$ is the residual symmetry of the vacuum configuration $\phi = \phi_0$.

7.6.3 The Deep Insight of Supersymmetry

In supersymmetric field theories, in particular in supergravities that are supersymmetric extensions of Einstein Gravity coupled to matter multiplets, all the uses listed above of coset manifolds do occur, but there is an additional ingredient whose consequences are very deep and far reaching for geometry: supersymmetry itself. Consistency with supersymmetry introduces further restrictions on the geometry of target manifolds $\mathcal{M}_{\text{target}}$ that are required to fall in specialized categories like *Kähler manifolds*, *special Kähler manifolds*, *quaternionic Kähler manifolds* and so on. These geometries, that we collectively dub *Special Geometries*, require the existence of complex structures and encompass both manifolds that do not have transitive groups of isometries and homogeneous manifolds G/H. In the second case, which is one of the main focuses of interest for the companion book [90], the combination of the special structures with the theory of Lie algebras produces new insights in homogenous geometries that would have been inconceivable outside the framework of supergravity. This is what we call the deep geometrical insight of supersymmetry. In this history oriented book and in his mathematically constructive companion [90] we neither discuss the construction of supergravity theories, nor we derive the constraints imposed by supersymmetry on geometry. Our commitment is simply to present the vast wealth of geometrical lore that supergravity Occam's razor has introduced, or systematically reorganized, in the field of mathematics.

Chapter 8
Geometry Becomes Complex

Mathematics, however, is, as it were, its own explanation; this, although it may seem hard to accept, is nevertheless true, for the recognition that a fact is so is the cause upon which we base the proof

Girolamo Cardano

8.1 History of Algebra and Complex Numbers

In our conceptual journey from the algebraic notions to the geometrical ones, journey that may be deemed to be the very heart of the present history essay, we have vastly emphasized the idea that groups exist abstractly as algebraic structures, yet are concretely realized as symmetries of geometrical structures, in particular of smooth *manifolds*. Conversely, possible geometries can be characterized in terms of the symmetries they admit, i.e. of the groups of transformations that preserve some of their fundamental properties.

Along a different line, we have illustrated the historical path that, starting from Gauss' new conception of *curvilinear coordinates* led to the idea of gluing together different open charts, in this way giving birth to the very notions of *differentiable manifold* and *fibre bundle*.

At this point an attentive and unbiased reader should note that from geometry we are back to algebra, since the following question arises. *Which mathematical beasts are the coordinates*, by means of which, in every local chart, we label the points of a geometrical space?

To begin with, the answer is that they are real numbers and this choice was the basis of the definition of manifolds recalled in Chap. 7. Yet they might be other numbers, for instance complex numbers. A different choice of the *numbers*, leads to different geometries, the deep philosophical question therefore being *what are the numbers?* No doubt, this is an algebraic question.

The history of the concept of real numbers is a long tale, hallmarked by the quest to master the notion of the *infinitesimals* and that of the *limits*. It probably started in the third century B.C. with *Archimedes*, came down the ages to Newton and Leibnitz

P. G. Fré, *A Conceptual History of Space and Symmetry*,
https://doi.org/10.1007/978-3-319-98023-2_8

Fig. 8.1 Abu Jafar Muhammad ibn Musa al-Khwarizmi

and mixed up with the invention of differential and integral calculus. It found its final *ubi consistam* in the XIXth century through the axiomatic construction of real numbers in the related works of *Cantor*, *Cauchy*, *Dedekind* and *Weierstrass*.

The history of complex numbers is of a completely different type and it has a clear-cut algebraic profile.

Algebra is a word which came into being through the latin transliteration of an arabic verb that means *to complete*. This transliteration appears for the first time in the *Liber algebrae et almucabala* providing the translation from arabic, performed by Robert of Chester, of the main work of the IXth century persian mathematician al-Khwarizmi (see Fig. 8.1).

Abu Jafar Muhammad ibn Musa al-Khwarizmi lived in Bagdad in the first half of the IXth century. Estimated dates of his birth and death are 780 A.D. and 850 A.D. respectively. Of persian origin he worked at the court of the Abbasid Caliph al-Mamun, who appointed him director of his rich library. Astronomer, geographer, mathematician, al-Khwarizmi shares with Diophantus the title of Father of Algebra. Benefiting from the cultural aliveness of the contemporary Abbasid capital which allowed him to meet with Indian scientists and with the heirs of hellenistic science, he both made original researches and compiled precious systematic summaries of known results in mathematics. His works were all written in arabic. His most famous book *al-Kitab al-mukhtasar fi hisab al-jabr wa al-muqabala* which deals with the solution of first and second degree equations was translated in the XIIth century from arabic into latin by western scholars who used to come to Spain to study in those centers that were at the time the most scientifically advanced of the world. The first translation was done by the English arabist Robert from Chester who worked in Segovia in the years around 1140.

This is *Liber algebrae et almucabala* mentioned above. The latin title does not reveal the actual meaning of the original title, literally *The Compendious Book on Calculation by Completion and Balancing*, yet, by transliterating into latin the arabic word *al-jabr*, which means *to complete* it is responsible for the introduction in western mathematics of the very notion of *algebra*.

A second translation of al-Khwarizmi's book, which obtained larger popularity in the latin-germanic world, was performed by the italian arabist Gerard of Cremona who came to Toledo and lived there in the years from A.D. 1134 to A.D. 1178. In that city recently reconquered to Christianity by Alfonso VI of Castilla, the Hebrew and Arabic scholars were allowed to continue their work and to meet with scholars from the West who purposely came there to meet them, learn the arabic language and approach by this token the sources of the lost hellenistic and oriental science. Gerard learnt arabic and not only translated al-Khwarizmi's book but also the *Almagesto*, namely the *Opera Omnia* of Tolomeus on Astronomy. The original Greek treatise composed in the IInd century by Claudius Tolomeus was known as $M\varepsilon\gamma\iota\sigma\tau\eta$ (*the large one*) and in the arabic translation it became the *Almagesti*.

8.1.1 The Middle Age Conception of Algebra

The main issue of what we came to know as algebra was the solution of equations, viewed in Antiquity and in the Middle Ages as *puzzles* to be solved with special sagacity, the frequent main obstacle being the lack of the appropriate entity which corresponds to the solution. The methods utilized by the IXth century Arabs, who probably learnt them from the Indians and from lost treatises of the hellenistic period, came to be known to the XIIIth century italian scholars via the latin translations of al-Khwarizmi's book, especially that of Gerard of Cremona.

So it happened that about 1225, when Leonardo Fibonacci (see Fig. 8.2) was received by the Emperor Frederick II at his court in Sicily, a local mathematician, probably of arabic culture, posed several problems to him one of which was a cubic equation:

$$x^3 + 2x^2 + 10x = 0 \tag{8.1.1}$$

At the time a general formula for the solution of the cubic equation was not known, yet particular solutions were occasionally constructed.

Fig. 8.2 Leonardo Pisano named *il Fibonacci* since his father's last name was *Bonacci* was born in Pisa about september 1175 and died an unknown date in the period between 1240 and 1250. Thanks to the mercantile activity of the father, who traded with North African partners, he had the opportunity to visit some cities on the shore of Algeria and come in touch with arabic mathematicians of the time learning from them the Hindu-Arabic numeral system for whose introduction in Europe he is responsible especially by means of his *Liber Abbaci*. Famous for the sequence of integer numbers that is named after him he gave several contributions to the early development of algebra and algebraic equations

8.1.2 The Cubic and Quartic Equation

It seems that the first who arrived at the solution formula of the cubic equation was *Scipione del Ferro*, professor in the University of Bologna who kept it as a secret. When he died in 1526 he confided it to his pupil Antonio Maria Fiore who, relying on such a secret weapon, challenged Tartaglia (see Fig. 8.3) to a mathematical contest, refereed by a public notary.

Unfortunately for Fiore, Tartaglia had separately discovered the solution formula of the equation $x^3 + px + q = 0$ and solved all the problems posed by Fiore, while the latter solved none of those posed by his opponent.

The news of this contest spread around and Tartaglia, who had become quite famous, was invited to visit Milano by Cardano. This latter learned from his guest

Fig. 8.3 Niccoló Fontana (Brescia 1499 – Venice 1557), nicknamed *Tartaglia* since he was stammering because of severe face blows received in his infancy during the brutal French sack of Brescia. Tartaglia, born in a very power family and orphan of his father since his youth had a much quieter life than Cardano. He was a teacher in Verona where he wrote his famous book *General trattato di numeri et misure* containing his even more famous *triangle* (see Fig. 8.5)

about the cubic equation formula, under the condition that he would not reveal it to anyone. Cardano worked further on it with his own student Ludovico Ferrari who arrived at the solution formula also for the quartic equation. Since Tartaglia had not yet published his results and Cardano came to know about the previous discovery of the same formula by Scipione del Ferro, he felt free from his promise to Tartaglia and published the solution formulae of both the cubic and the quartic equations in his book *Artis Magnae* (see Fig. 8.4).

As it might be expected, this publication was the origin of a ten year long and very harsh *querelle* between Tartaglia on one side and Cardano and Ferrari on the other. Notwithstanding their enmity the solution formula for the cubic equation was named by posterity the *Cardano–Tartaglia formula.*

According to the author of [167], *Cardano was the first to introduce complex numbers* since he wrote solutions of some cubic equation in the form $5 + \sqrt{-5}$, yet it appears that he was not perfectly conscious of the general implications of what he was writing. Instead Rafael Bombelli, the author of a book named *Algebra* which was published in 1572, consciously introduced a special notation for the number $\sqrt{-1}$ and utilized it in the discussion of solutions to cubic equations.

8.1.3 The Imaginary Numbers froms Descartes to Euler

On the other hand, René Descartes, besides his achievements in philosophy and his invention of analytic geometry, is to be credited for coining the term *imaginary* since

Fig. 8.4 On the left Girolamo Cardano (Pavia 1501–Rome 1576). On the right the front page of *Artis Magnae*, the book published by Cardano in 1545 which contains the solution formulae for the cubic and quartic equations. Cardano's life, illegitimate son of one of Leonardo da Vinci's best friends and collaborators, was quite adventurous and troubled by problems of money that Girolamo usually solved by gambling. He was a medical doctor, a mathematician and a philosopher

he wrote such a sentence[1]: *For any equation one can imagine as many roots as its degree would suggest but in many cases no quantity exists which corresponds to what one imagines*. This is just the *General Theorem of Algebra* in disguise. What was necessary to proceed further in the development of Algebra was to become aware of what is the result of extending the field of real numbers with a new *imaginary* entity.

In the XVIIIth century the inventor of the mathematical notation $\sqrt{-1} = i$ which we still use, is Leonhard Euler, whose contributions to the development of modern mathematics are so extensive and monumental that we do not feel it necessary to mention them in this place. Yet the main point, namely the formulation of the notion of *field* and of *field extension*, of which the inclusion of $i = \sqrt{-1}$ is a primary example, was yet to come.

[1] Some of the historical informations contained in the present section are to be credited to an unpublished note of Orlando Merino available at the following site: http://www.math.uri.edu/~merino/spring06/mth562/ShortHistoryComplexNumbers2006.pdf.

Fig. 8.5 Frontespice of the *General Trattato* of Niccolò Tartaglia, an original copy of which is preserved in the *Biblioteca Peano* of Torino University. Photopicture taken by this author thanks to the courtesy of *Biblioteca Peano*

8.1.4 *Fields, Algebraic Closure and Division Algebras*

The notion of *field* was implicitly used by *Abel* and *Galois*, then the concept developed steadily through the work of *Karl von Staudt*, *Richard Dedekind* who introduced the German denomination *Körper*, *David Hilbert* and finally of *Heinrich Weber* who provided the first axiomatic definition of a *field*.

In the meantime the idea of the field extension and of *algebraic closure* was to be correlated with the other one of *division algebra*.

Let us start with algebraic closure.

A field $\mathbb{K}$ is algebraically closed if and only if it contains a root for every non-constant polynomial $\mathfrak{P}(x) \in \mathbb{K}[x]$, the ring of polynomials in the variable x with coefficients in $\mathbb{K}$.

In other words a field is closed if all algebraic equations can be solved $\mathfrak{P}(x) = 0$ by means of special elements x_ℓ of the same field $\mathbb{K}$ wherefrom the coefficients a_i of the polynomial

$$\mathfrak{P}(x) \equiv \sum_{i=0}^{n} a_i\, x^i$$

are taken, in such a way that $\mathfrak{P}(x_\ell) = 0$ for all $\ell = 1, 2, \ldots, n$. The numbers $x_\ell \in \mathbb{K}$ are named the *roots* of the considered polynomial.

The field of real numbers $\mathbb{R}$ is not closed and just for this reason we have to extend it to the complex number field $\mathbb{C}$ which is closed.

Recalling the definition of algebras we have:

Let $\mathscr{D}$ be an algebra over a field $\mathbb{K}$, and assume that $\mathscr{D}$ does not just consist of its zero element. We call $\mathscr{D}$ a division algebra if $\forall a \in \mathscr{D}$ and for any non-zero element $b \in \mathscr{D}$, $\exists x \in \mathscr{D}$ such that $a = bx$ and $\exists y \in \mathscr{D}$ such that $a = yb$. The name given to this type of algebras is clearly justified by their definition. We can always define the ratio a/b of any two elements a, b of the algebra except for the case where $b = 0$.

Another way of looking at the complex numbers is the following: $\mathbb{C}$ is nothing else but a division algebra of dimension 2 over $\mathbb{R}$, the field of real numbers. This way of thinking is the ultimate understanding of an almost two–thousand–year–long constructive process. Descartes had difficulties to understand imaginary numbers, but he had no problem with the pairs of real numbers (x, y), namely with a two-dimensional vector space, especially taking in due account the fact that he had himself invented such a notion. The key aspect to which Descartes had not given due attention is that pairs of real numbers can not only be summed and subtracted (they form a vector space), but they can also be multiplied among themselves:

$$(x, y) \cdot (u, v) \equiv (xu - yv, xv + yu) \tag{8.1.2}$$

This promotes the complex numbers to an algebra. The marvelous point is that such an algebra is a *division algebra*! Indeed for any pair of real numbers different from $(0, 0)$ we can construct its inverse:

$$(x, y)^{-1} = \left(\frac{x}{x^2 + y^2}, -\frac{y}{x^2 + y^2} \right) \tag{8.1.3}$$

Equation (8.1.3) constitutes a miracle: given a vector $\mathbf{v}$ it makes sense to talk about its inverse $\mathbf{v}^{-1}$ which is clearly a non-sense in most vector spaces.

The algebraic closed field of the complex numbers $\mathbb{C}$ is a division algebra of dimension 2 constructed over the field of real numbers $\mathbb{R}$. It is obtained by introducing one imaginary unit i.

8.1.5 Hamilton and the Quaternions

Sir William Rowan Hamilton (see Fig. 8.6) is probably the greatest Irish scientist of history. He was the fourth of nine children born to Sarah Hutton and Archibald Hamilton who lived in Dublin. Hamilton's father worked as a solicitor and was a joyous fellow, a lover of good wine and of good food. Both of these inclinations Archibald left as a legacy to his son William who developed gout from excessive drinking and overeating and eventually died from it at the age of 60.

Quite early William lost both of his parents and he was looked after and educated by his uncle, the Rev. James Hamilton. This latter was a graduate from Trinity College who was obsessed by the passion of studying all sort of ancient and modern languages. Such a passion he easily communicated to his most talented nephew who learned Latin, Greek, Hebrew, French, Italian and also Persian, Arabic, Sanskrit and other oriental languages, at the pace of a new language every year. It is reported that

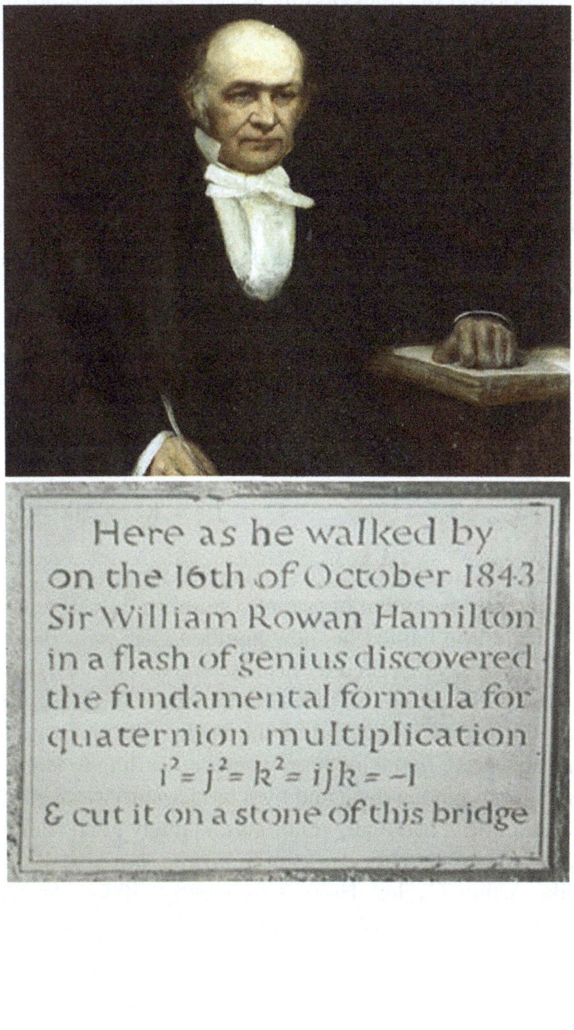

Fig. 8.6 Sir William Rowan Hamilton (August 3rd 1805 – September 2nd 1865)

by thc age of eleven William composed a short welcome poem in Persian for the Persian Ambassador who was visiting Dublin at the time. This almost pathological commitment to language learning might have distracted the young genius from Mathematics and Physics, by which he was also strongly attracted, were it not for a check to his pride when, at the age of 8, William lost a challenge in quick mental calculations against the American calculating prodigy Zerah Colburn who was being exhibited those days in Dublin.

Since then Hamilton concentrated more on science than on languages and shortly he became famous for such contributions to Mechanics and Algebra that place him in the short list of the most distinguished mathematical minds of the XIXth century.

The two greatest achievements of Hamilton are the invention of hamiltonian mechanics, a reformulation of lagrangian analytical mechanics, whose influence on the later development of theoretical physics up to the birth of quantum mechanics and modern field theory are enormous, and the discovery of quaternions.

The obvious curiosity that was obsessing William Rowan Hamilton, was the following: how to construct the next case after the complex numbers, namely a division algebra over the reals of dimension 3? He was trying in all conceivable ways without any success. For this there was a good reason since such a division algebra does not exist, as it was later proved.

One fine October day of 1843 he was strolling with his wife through his city, Dublin, just crossing the Broom Bridge, and he had a fundamental idea. What cannot be done in $d = 3$ has a relatively easy solution in $d = 4$. It suffices to associate one of the four dimensions with the real part of a new *hyper-complex number* and the remaining three with three imaginary units rather than just one. The price one has to pay for that is to assume that the new imaginary units $\mathbf{j}^x$ $(x = 1, 2, 3)$ satisfy the following algebra:

$$\mathbf{j}^x \cdot \mathbf{j}^y = -1 + \varepsilon^{xyz}\,\mathbf{j}^z \tag{8.1.4}$$

where ε^{xyz} is the yet to be invented Levi-Civita epsilon symbol. This is just a condensed modern notation, Hamilton wrote the relations (8.1.4) explicitly one by one [108, 109] as it follows:

$$\mathbf{i}^2 = \mathbf{j}^2 = \mathbf{k}^2 = \mathbf{ijk} = -1 \tag{8.1.5}$$

The consequence of Eq. (8.1.4) is that the division algebra introduced by Hamilton, differently from that of complex numbers is *non-commutative*.

Hamilton named his new numbers, the *quaternions* ($\mathbb{H}$ in current notation) and was extremely excited by his discovery.

According to his own account, using a knife, Hamilton scrabbled formula (8.1.5) on the stones at the end of the bridge, although traces of such a scrabbling have disappeared. In the location where, according to Hamilton, they should have been, a memorial plaque designed by a modern artist was placed in 1958.

Hamilton had many relations and many good friends, among which the jurist and amateur mathematician John T. Graves but also such other giants of Mathematics like Arthur Cayley. He was frequently talking, or corresponding with them. So it happened that the very same year 1843, Graves discovered another division algebra over the fields of real numbers $\mathbb{R}$. This time the dimension was 8 and the number of imaginary units was 7. This division algebra is both *non-commutative* and *non-associative*.

Hamilton reported about the discovery of his friend in a communication [107] to the Irish Academy in 1848 but in the mean-time the same division algebra had been independently discovered by Arthur Cayley [37] and the corresponding numbers, mostly named *octonions* ($\mathbb{O}$ in current notation) are frequently referred to as the *Cayley numbers*.

During his life Hamilton obtained both in the United Kingdom and world-wide a lot of honours and recognitions being member of many Foreign Academies and being also awarded the title of Sir by Queen Victoria.

8.1.6 Frobenius and His Theorem

Ferdinand Georg Frobenius is known for his important contributions to the development of finite group-theory and to some aspects of differential geometry. Born in Charlottenburg he completed his secondary school education in Berlin and then attended the University of Göttingen, where he became Weierstrass'student. After graduation he had a temporary position in Berlin and then for seventeen years he was professor in Switzerland at the ETH of Zürich. There he married, raised his family and did most of his career. In 1891, after Kronecker's death he was suggested by Weierstrass as his successor on the chair of mathematics at the University of Berlin. So in 1893 he made return to Berlin where he lived the last 25 years of his life becoming also member of the Prussian Academy of Science. It was reserved to Ferdinand Georg Frobenius (see Fig. 8.7) to prove by means of a theorem published in 1877 that, up to isomorphism, the only *associative, normed, division algebras* over the reals are $\mathbb{R}$, $\mathbb{C}$ and $\mathbb{H}$, of dimensions 1, 2, 4, respectively [92].

By norm here we mean a quadratic non degenerate form over the algebra $\mathscr{D}$:

$$\begin{aligned} &\mathrm{N} : \mathscr{D} \to \mathbb{R} \\ &\mathrm{N}(\lambda\, a) = \lambda^2\, \mathrm{N}(a) \qquad \forall \lambda \in \mathbb{R}\,,\ \forall a \in \mathscr{D} \\ &\mathrm{N}(a\, b) = \mathrm{N}(a)\, \mathrm{N}(b) \qquad \forall a, b \in \mathscr{D} \end{aligned} \tag{8.1.6}$$

The norm corresponds to the modulus square of the real, complex or quaternionic number.

If we relax the hypothesis of associativity, the landscape is not too much enlarged. Using for instance the very powerful *Bott periodicity theorem* it can be shown that any *real normed division algebra* must be isomorphic to either the real numbers $\mathbb{R}$, the complex numbers $\mathbb{C}$, the quaternions $\mathbb{H}$, or the octonions $\mathbb{O}$.

Fig. 8.7 Ferdinand Georg Frobenius (October 26th 1849 Charlottenburg – August 3rd 1917 Berlin)

8.1.7 Imaginary Units and Geometry

Summarizing the outcome of this long historical excursus we can say that the possible numbers are of four types $\mathbb{R}$, $\mathbb{C}$, $\mathbb{H}$, or $\mathbb{O}$. This is a message for geometry. Keeping the fundamental idea that a geometrical space should be viewed as a manifold, constructed by means of an atlas of open charts, the local coordinates could be chosen not only as real numbers but also as complex, quaternionic or even octonionic numbers. An important lesson, however, is immediately learnt from the previously told story: the other possible numbers are, anyhow, division algebras over the reals, so that the real structure remains the basis for everything. This must be the same also in geometry. Manifolds of complex, quaternionic or octonionic type, if they exist, are, first of all, real manifolds. Their characterization as complex, quaternionic or octonionic must reside in some additional richer structure they are able to support. It is evident that this additional structure are the imaginary units, the same that provide the extensions of the field $\mathbb{R}$ to $\mathbb{C}$, $\mathbb{H}$ or $\mathbb{O}$.

Hence the conceptual path we have to follow starts revealing itself. We have to conceive what the imaginary units might be in the context of differential geometry. The catch is the relation $\mathbf{J}^2 = -\mathbf{1}$. How to reinterpret such a relation? It is rather natural to consider $\mathbf{J}$ as a map, in particular a linear map, and $\mathbf{1}$ as the identity map which always exists. We are almost there, the remaining question is *on which space does* $\mathbf{J}$ *act*? The answer is obvious since for linear maps we need vector spaces and if

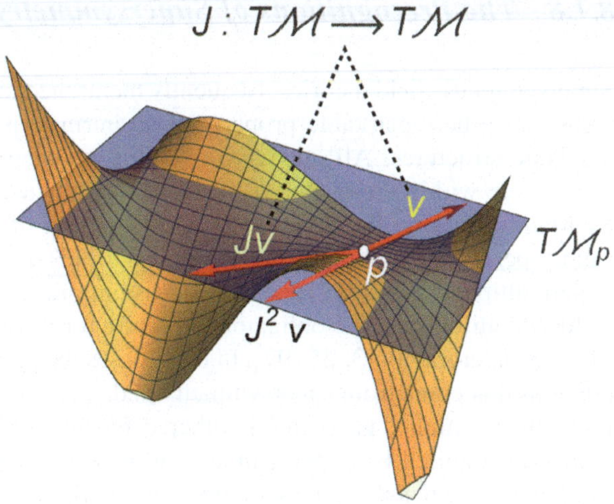

Fig. 8.8 Given a differentiable manifold $\mathcal{M}$, at each of its points $p \in \mathcal{M}$ we can draw the tangent space $T\mathcal{M}_p$. A complex structure $\mathbf{J}$ is a map that sends any vector $\mathbf{v} \in T\mathcal{M}_p$ to its image $\mathbf{Jv}$ which is another vector in the same space. Applying the complex structure twice, the image $\mathbf{Jv}$ is mapped in the vector $-\mathbf{v}$, namely $\mathbf{J}^2\mathbf{v} = -\mathbf{v}$

we want to do things locally, point by point on the manifold, we need *vector bundles*. The universal vector-bundle that it is intrinsically associated with any manifold $\mathcal{M}$ is the tangent bundle $T\mathcal{M} \rightarrow \mathcal{M}$. Hence the imaginary units, that from now on we will name *complex structures*, are linear maps, operating on sections of the tangent bundle, that square to minus one (see Fig. 8.8):

$$\begin{aligned} \mathbf{J} &: T\mathcal{M} \rightarrow T\mathcal{M} \\ \mathbf{J}^2 &= -\mathbf{Id} \end{aligned} \tag{8.1.7}$$

Complex and quaternionic or hyper-complex geometries arise when a manifold admits one or more complex structures satisfying appropriate algebraic relations. This mixture of algebra and geometry leads to new classes of very interesting spaces:

(a) Complex Manifolds
(b) Complex Kähler Manifolds
(c) HyperKähler Manifolds
(d) Quaternionic Kähler Manifolds

It is the mission of the present chapter to define and illustrate such manifolds.

Furthermore when we come to discuss the symmetries of such manifolds, namely their isometries, we discover that the presence of the complex-structures entrains a new very much challenging viewpoint on continuous symmetries. To the Killing vectors, thanks to the symplectic structures implied by the complex-structures we are able to associate *hamiltonian functions*, named *moment maps*. These moment maps open a vast playing ground for new constructions of high relevance both in Physics and Mathematics.

8.1.8 *The Precognitions of Supersymmetry*

Supersymmetric field theories, frequently mentioned in previous pages of the present essay, have the remarkable property of an intrinsic precognition of geometric and algebraic structures. All classes of existing geometries found, in due time, their proper role within the frame of this new type of field theories. For instance Kähler Manifolds describe the most general coupling of scalar multiplets[2] in $\mathcal{N} = 1$ rigid supersymetry, while HyperKähler Manifolds do the same for the rigid $\mathcal{N} = 2$ hypermultiplets[3] (see [117] and the discussions presented in Chap. 11). Quaternionic Kähler Manifolds are the obligatory structure for the coupling of hypermultiplets to $\mathcal{N} = 2$ supergravity [3, 55, 97]. In these cases the precognition resides in algebraic relations that come from supersymmetry and, once duely interpreted, were shown to imply the mentioned geometry. In other, even more spectacular cases, the geometric structures required by supersymmetry were not yet available in the mathematical supermarkets when the corresponding supermultiplets were studied. They were just discovered by the physicists working in supergravity and now constitute new chapters of mathematics. These are the *Special Geometries* whose history is outlined in Chap. 9.

Let us now turn to complex structures and their heritage.

8.2 Fundamental Definitions for Complex Geometry and Its Descendants

Complex, Kähler, HyperKähler and Quaternionic Kähler Manifolds are discussed in ample detail in my parallel more technical book [90]. In this historical essay I confine myself to an abbreviated collection of the most fundamental definitions in order to clarify what the objects of our present discussions actually are.

COMPLEX MANIFOLDS

A 2n-dimensional manifold $\mathcal{M}$ is called almost complex if it has an almost complex structure. An almost complex structure is a linear operator $J : \Gamma(T\mathcal{M}, \mathcal{M}) \to \Gamma(T\mathcal{M}, \mathcal{M})$ which satisfies the following property:

$$J^2 = -\mathbb{1} \tag{8.2.1}$$

[2] Supermultiplet is the name given to a collection of fields that form an irreducible representation of the supersymmetry algebra. The structure of supermultiplets that always involve both bosons and fermions depends on the space-time dimensions D in which we construct our field theory and on the number of supercharges (fermionic Lie algebra generators) that we include in our supersymetry algebra. This latter is usually named $\mathcal{N}$ and in $D = 4$ can range from 1 to 8.

[3] Scalar multiplets and hypermultiplets are multiplets that involve only spin $\frac{1}{2}$ and spin 0 fields. The connection with geometry occurs at level of the scalar fields, that are interpreted, sigma–model like as coordinates of a target manifold with an appropriate geometry (see Sect. 7.6.2).

In every local chart the operator J is represented by a tensor $J^{\alpha}_{\beta}(\phi)$ such that

$$J^{\beta}_{\alpha}(\phi)J^{\gamma}_{\beta}(\phi) = -\delta^{\gamma}_{\alpha} \tag{8.2.2}$$

The tensor

$$T^{\alpha}_{\beta\gamma} = \partial_{[\beta}J^{\alpha}_{\gamma]} - J^{\mu}_{\beta}J^{\nu}_{\gamma}\partial_{[\mu}J^{\alpha}_{\nu]} \tag{8.2.3}$$

is called the *torsion*, or the *Nienhuis tensor*[4] of the almost complex structure J^{α}_{β}. Let $\{x^{\alpha}\}$ be a generic coordinate system and let $w(x)$ be a complex-valued function on the manifold $\mathscr{M}$: we say that w is holomorphic if it satisfies the equation[5]:

$$Jdw = idw \tag{8.2.4}$$

which in the generic coordinate system $\{x^{\alpha}\}$ reads as follows:

$$J^{\beta}_{\alpha}\partial_{\beta}w(x) = i\partial_{\alpha}w(x) \tag{8.2.5}$$

The vanishing of $T^{\alpha}_{\beta\gamma}$ is a necessary condition for the integrability of Eq. (8.2.5). When this latter is integrable, in every patch we can establish a system of complex coordinates z^i, $(i = 1, \ldots, n)$ such that:

$$\begin{aligned} Jdz^i &= idz^i \\ Jdz^{i^*} &= -idz^{i^*} \end{aligned} \tag{8.2.6}$$

and the transition functions between complex well-adapted coordinate in two different patches will be holomorphic. In this way we realize that the manifold supporting an almost complex structure with vanishing Nienhuis tensor is complex. Correspondingly we say that an almost complex structure with vanishing Nienhuis tensor is *a complex structure*.

Holomorphic Vector Bundles

Holomorphic bundles on complex manifolds are defined in complete analogy to fibre-bundles on real manifolds. The essential point is the holomorphicity of the transitions functions. We especially need holomorphic vector bundles.

Let $\mathscr{M}$ be a complex manifold and E be another complex manifold. A holomorphic vector bundle with total space E and base manifold $\mathscr{M}$ is given by a projection map:

$$\pi : E \longrightarrow \mathscr{M} \tag{8.2.7}$$

---

[4] Albert Nijenhuis (November 21, 1926–February 13, 2015) was a Dutch-American mathematician. He wrote his Ph.D. thesis at the University of Amsterdam under the supervision of Jan Arnoldus Schouten.

[5] We defined complex structures as operators acting on sections of the tangent bundle, namely on vector fields. By means of the duality between the tangent bundle and the cotangent bundle, complex structures (or almost complex structures) act equally well on sections of the contangent bundle, namely on differential 1-forms: $\omega(\mathrm{Jv}) \equiv \mathbf{J}\omega(\mathbf{v})$. This is what we use here.

such that

(a) π is a holomorphic map of E onto $\mathcal{M}$
(b) Given any point of the base manifold $p \in \mathcal{M}$, the fibre over p, i.e., $E_p = \pi^{-1}(p)$ is a complex vector space of dimension r. (The number r is called the rank of the vector bundle.) We have an atlas of local trivializations (U_α, h_α) where U_α is a collection of open neighborhoods covering the complex base manifold $\mathcal{M}$ and $h_\alpha : \pi^{-1}(U_\alpha) \to U_\alpha \times \mathbb{C}^r$ is a homeomorphism.
(c) The transition functions between two local trivializations (U_α, h_α) and (U_β, h_β):

$$h_\alpha \circ h_\beta^{-1} : (U_\alpha \cap U_\beta) \otimes \mathbb{C}^r \longrightarrow (U_\alpha \cap U_\beta) \otimes \mathbb{C}^r \quad (8.2.8)$$

induce holomorphic maps:

$$\mathfrak{g}_{\alpha\beta} : (U_\alpha \cap U_\beta) \longrightarrow \mathrm{GL}(r, \mathbb{C}) \quad (8.2.9)$$

In other words the transition function from one local trivialization of the bundle to another one is provided by a non singular $r \times r$ matrix $\mathfrak{g}_{\alpha\beta}(z)$ that depends in a holomorphic way from the complex coordinates z^i of the base manifold in the intersection of the two patches.

CONNECTIONS AND METRICS ON HOLOMORPHIC VECTOR BUNDLES

Let $E \longrightarrow \mathcal{M}$ be a holomorphic vector bundle of rank r and $U \subset \mathcal{M}$ an open subset of the base manifold and consider the concept of fibre bundle sections, illustrated in Fig. 8.9.

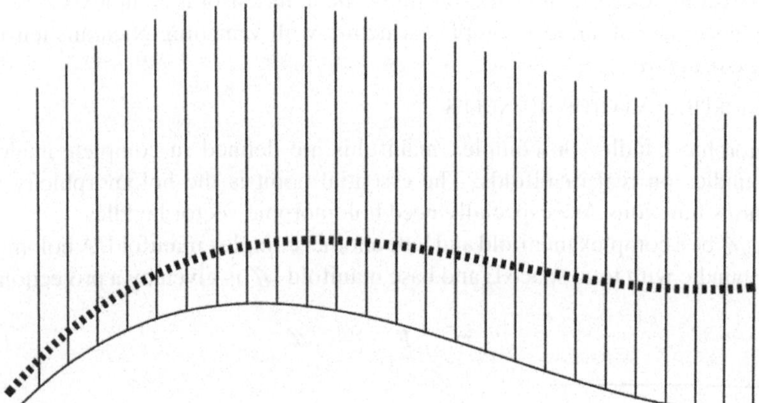

Fig. 8.9 The concept of section of a fibre-bundle is illustrated by the above picture. To every point p of the base manifold a section $\mathfrak{s}$ associates, in a continuous way, a point of the total space $\mathfrak{s}(p) \in P$, that must belong to the fibre over p, namely such that $\pi(\mathfrak{s}(p)) = p$. In the case of vector bundles the section image $\mathfrak{s}(p)$ of a base manifold point p is necessarily an r-dimensional vector, r being the rank of the bundle

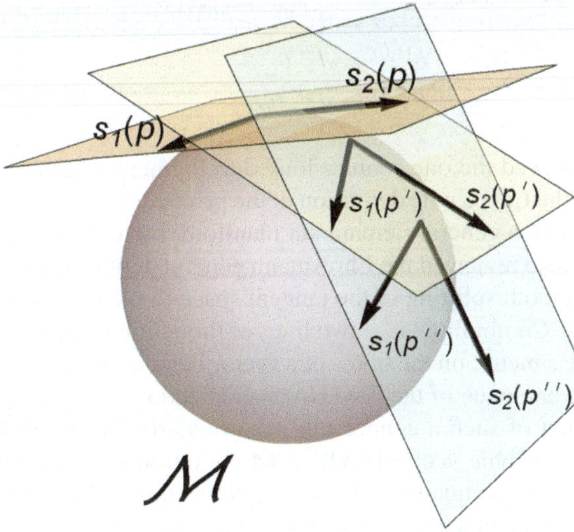

Fig. 8.10 The concept of *a frame of sections* of a vector bundle is illustrated in the above picture using the example of the tangent bundle to a sphere $\mathbb{S}^2$ that, in this case, plays the role of base manifold $\mathcal{M}$. At each point $p \in \mathcal{M}$ of the base manifold we have the *fibre vector space* $\pi^{-1}(p)$ which, in this case, is the tangent plane. The sections $\mathfrak{s}_1(p)$ and $\mathfrak{s}_2(p)$ provide, at each point p, a basis of the tangent plane, namely of the fibre vector space, since they are linearly independent. The system composed by $\mathfrak{s}_{1,2}(p)$ is a frame for this vector bundle

A frame over U is a set of r holomorphic sections $\{\mathfrak{s}_1, \ldots, \mathfrak{s}_r\}$ such that $\{\mathfrak{s}_1(z), \ldots, \mathfrak{s}_r(z)\}$ is a basis for $\pi^{-1}(p)$ for any $p \in U$, having denoted by z^i the complex coordinates labeling the points of the base manifold in the chosen patch (see Fig. 8.10).

Let $f \equiv \{\mathfrak{s}_I(z)\}$ be a frame of holomorphic sections. Any other holomorphic section ξ is described by

$$\xi = \xi^I(z)\,\mathfrak{s}_I \tag{8.2.10}$$

where

$$\bar{\partial}\,\xi^I \equiv d\bar{z}^{j^\star}\,\bar{\partial}_{j^\star}\,\xi^I = 0 \tag{8.2.11}$$

Given a holomorphic bundle with a frame of sections we can discuss metrics connections and curvatures, as we already did for the general case of bundles.

In general a connection θ is defined by introducing the covariant derivative of any section ξ

$$D\,\xi = d\,\xi + \theta\,\xi \tag{8.2.12}$$

where $\theta = \theta^I{}_J$, the connection coefficient, is an $r \times r$ matrix-valued 1-form (see Sect. 7.4.2 and the concept of Ehresmann connection). On a complex manifold this 1-form can be decomposed into its parts respectively of holomorphic type (1,0) and antiholomorphic type (0,1).

$$\begin{aligned} \theta &= \theta^{(1,0)} + \theta^{(0,1)} \\ \theta^{(1,0)} &= dz^i \, \theta_i \\ \theta^{(0,1)} &= d\bar{z}^{i^\star} \, \theta_{i^\star} \end{aligned} \tag{8.2.13}$$

In Chap. 7 we reviewed the one century long development of differential geometry culminating with the Levi Civita definition of the priviledged connection, induced on the tangent bundle to a general riemannian manifold, by the existence of a riemanniann metric. We also reviewed the Ehresmann general definition of a connection on any bundle as a smooth splitting of the tangent space to the bundle in a vertical and a horizontal space. Combining these two lines of thought it is quite obvious that any time we introduce a metric on the fibres of a vector bundle a special connection must emerge that is the analogue of the Levi Civita connection.

The construction of such a connection is particularly elegant and natural when we deal with holomorphic vector bundles and we introduce a *hermitian fibre metric h*. This is a hermitian quadratic form that yields the scalar product of any two holomorphic sections ξ and η at each point of the base manifold:

$$\langle \xi \, , \, \eta \rangle_h \; \equiv \; \bar{\xi}^{I^\star}(\bar{z}) \, \eta^J(z) \, h_{I^\star J}(z, \bar{z}) \; = \; \xi^\dagger \, h \, \eta \tag{8.2.14}$$

As it is evident from the above formula, the metric h is defined by means of the point-dependent hermitian matrix $h_{I^\star J}(z, \bar{z})$, which is requested to transform, from one local trivialization to another, with the inverses of the transition functions $\mathfrak{g}_{\alpha\beta}$ defined in Eq. (8.2.9). This is so because the scalar product $\langle \xi \, , \, \eta \rangle_h$ is by definition an invariant (namely a scalar function globally defined on the manifold).

A hermitian metric for a complex manifold $\mathscr{M}$ is a particular case of the above construction, namely it is a hermitian fibre metric on the tangent bundle $T\mathscr{M}$. In this case the transition functions $\mathfrak{g}_{\alpha\beta}$ are given by the jacobians of the coordinate transformations.

In general h is just a metric on the fibres and the transition functions are different objects from the Jacobian of the coordinate transformations. In any case, as we have emphasized above, given a fibre metric on a holomorphic vector bundle, we can introduce a canonical connection θ associated with it. It is defined by requiring that

$$\begin{aligned} A) \; d \, \langle \xi \, , \, \eta \rangle_h \; &= \; \langle D \xi \, , \, \eta \rangle_h + \langle \xi \, , \, D \eta \rangle_h \\ B) \qquad D^{(0,1)} \xi \; &\equiv \left[\bar{\partial} + \theta^{(0,1)} \right] \xi \; = \; 0 \end{aligned} \tag{8.2.15}$$

namely by demanding that the scalar product be invariant with respect to the parallel transport defined by θ and by requiring that the holomorphic sections be transported into holomorphic sections.

Properties of the Canonical Connection and of Its Curvature

Let f be a holomorphic frame. In this frame the canonical connection is given by

$$\theta(f) \; = \; h(f)^{-1} \, \partial \, h(f) \tag{8.2.16}$$

or, in other words, by

$$\theta^I{}_J = dz^i \, h^{IJ^\star} \, \partial_i \, h_{K^\star J} \tag{8.2.17}$$

In the particular case of a manifold metric, where h is a fibre metric on the tangent bundle $T\mathcal{M}$, the general formula (8.2.17) provides the definition of the Levi-Civita connection:

$$dz^k \, \Gamma^i_{kj} = - g^{il^\star} \, \partial \, g_{l^\star j} \tag{8.2.18}$$

that for complex manifolds with hermitian metrics has a much simpler form as it is evident to the reader from the above formula.

Given a connection we can compute its curvature by means of the standard formula $\Theta = d\theta + \theta \wedge \theta$. In the case of the above-defined canonical connection we obtain

$$\Theta \, (f) = \partial \theta + \bar{\partial} \theta + \theta \wedge \theta = \bar{\partial} \theta \tag{8.2.19}$$

This identity follows from $\partial \theta + \theta \wedge \theta = 0$, which is identically true for the canonical connection (8.2.16). Component-wise the curvature 2-form is given by

$$\Theta^I{}_J = \bar{\partial}_i \left(h^{IK^\star} \, \partial_j \, h_{K^\star J} \right) d\bar{z}^i \wedge dz^j \tag{8.2.20}$$

In the Levi-Civita case, namely when the fibre metric $h = g$ is just a hermitian metric on the tangent bundle, the above formula provides the calculation of the Riemann tensor and of the Ricci tensor appropriate to complex geometry:

$$\begin{aligned} \mathcal{R}^i_j &= \mathcal{R}^i{}_{jk^*\ell} d\bar{z}^{k^*} \wedge dz^\ell \\ \mathcal{R}^i{}_{jk^*\ell} &= \partial_{k^*} \Gamma^i_{j\ell} \end{aligned} \tag{8.2.21}$$

The Ricci tensor has a remarkably simple expression:

$$\mathcal{R}^n_{m^*} = \mathcal{R}^i{}_{m^* n i} = \partial_{m^*} \Gamma^i_{ni} = \partial_{m^*} \partial_n \, ln \, (\sqrt{g}) \tag{8.2.22}$$

where $g = det|g_{\alpha\beta}| = (det|g_{i j^*}|)^2$.

8.2.1 *Kähler Manifolds*

In the previous section we saw the significant simplifications in the codification of *geometry* that occur from the presence of a complex structure. It was the mission of a very original and, under some respects, also extravagant German mathematician of the XXth century to discover an ample class of very much relevant complex manifolds the codification of whose geometry is even more compact since it reduces to the specification of a single real function $\mathcal{K}(z, \bar{z}) = \mathcal{K}^\star(z, \bar{z})$ of the complex variables. The family name of this mathematician is Kähler and the manifolds he

Fig. 8.11 Erich Kähler (1906 Leipzig, Germany - 2000 Wedel, near Hamburg, Germany)

introduced into the modern mathematical landscape are named, after him, Kähler manifods. As we said, Kählerian complex manifolds form a large class, encompassing many important instances of varieties that are relevant both in pure mathematics and in contemporary theoretical physics, in particular in the context of supersymmetric field theories.

Let us first see who Kähler was.

KÄHLER

Born in Leipzig, since the age of 12 Erich Kähler (see Fig. 8.11) developed a strong interest and a true passion for mathematics. Utilizing notes taken at Weierstrass' lectures by one of his school teachers, Kähler became acquainted with Gauss work and with the theory of elliptic functions while he was still a school boy. He entered Leipzig University in 1924 and graduated from it in 1928 writing a thesis entitled *On the existence of equilibrium figures which are derived from certain solutions of the n-body problem*. As the title reveals, the addressed topics were mathematical problems of classical mechanics. After a short term in Königsberg University, he was PrivatDozent in Hamburg where he interacted with Artin. Thanks to a Rockefeller fellowship he spent the academic year 1931–1932 in Italy where he studied with Enriques, Castelnuovo, Levi-Civita, Severi, and Beniamino Segre. This experience was very important in his life, both for the development of his mathematical ideas and for his learning of Italian which he later utilized to write one of his scientific essays.

In 1932 Kähler published a paper entitled *Über eine bemerkenswerte Hermitesche Metrik* [119] in which he introduced the notion of a Kähler metric. This proved to be a major contribution to the development of geometry in the XXth century and the notion of Kähler manifolds has played and still plays a fundamental role not only in Mathematics but also in contemporary Theoretical Physics.

During World War II, Kähler served in the German Navy, was taken prisoner by the French and, while being in a concentration camp of the Allies, he was able to resume his mathematical studies with the help of his several friends among the French mathematicians. After he was released in 1947, Kähler was for a short period in Hamburg and then, in 1948 he accepted a full professorship in Leipzig. After ten years in the gloomy DDR, he managed to escape from the *socialist lager* to the West in 1958, going back to Hamburg where he spent the rest of his life, being professor of Hamburg University until retirement. During his Leipzig years, Kähler wrote in Italian a long essay entitled *Geometria Aritmetica* that was published in 1958 on the journal *Annali di Matematica*. The incipient inclination of this brilliant mathematician towards a somewhat extravagant, almost mystical, reformulation of mathematical lore in a new *philosophical approach* was remarked with the following words by Kähler's affectionate friend André Weil: *This, in more ways than one, is an unusual piece of work. By its size, it is a book; it appears as a volume in a journal. The author is German; the book appears in Italian. The subject combines algebra and geometry, with some arithmetical flavouring; but the author, instead of following in his terminology the accepted usage in either one of those subjects, or adapting it to his purposes, has chosen to borrow his vocabulary from philosophy, so that rings, homomorphisms, factor-rings, ideals, complete local rings appear as "objects", "perceptions", "subjects", "perspectives", "individualities". The book includes altogether new material along with much which turns out to be quite familiar (sometimes to the point of triteness) once it is translated back into more familiar language; but no attempt is made to point out what may be novel and what is not so; there are no historical or bibliographical indications, no "Leitfaden", no introduction apart from a two-page philosophical discourse which ends up with the following statement: "... bibliographical references would probably have obscured the fact that a single philosophical tendency has been the real motive power behind the chain of my reasonings."*

In his last years Kähler, who suffered heavy blows in his personal life from the death in 1966 of his son Reinhard caused by an accident and from the death in 1970 of his wife, caused by leukemia, got more and more involved into philosophical studies. In 1992, evocating F. Nietzsche's most famous piece of work, Kähler wrote a book entitled *Also sprach Ariadne* where he exposed his attempts to bring mathematics and philosophy together. According to a review of the book, written by Doru Stefanescu: *The author considers various mathematical interpretations of some philosophical texts. He especially dwells on the theory of monades of Leibniz and on the work "Also sprach Zarathustra" of F. Nietzsche. His speculative considerations are illustrated by suggestive examples from set theory, mathematical logic, abstract algebra and differential, algebraic and analytic geometry. He considers basic philosophic concepts such as Sein, Schein, monades, transcendental perception, pure reason, Sehkraft, and mathematical objects such as equivalence relations, polynomials, projections. The main thesis of the paper is that algebraic geometry is a prolegomenon to a mathematical theory of monades.*

The Kähler 2-Form and Kähler Metrics

Let $\mathcal{M}$ be a 2n-dimensional manifold with a complex structure J. A metric g on $\mathcal{M}$ is called hermitian with respect to J if, for any pair $\mathbf{u}$, $\mathbf{w}$ of sections of the tangent bundle, i.e. for any pair of vector fields, we have:

$$g(J\mathbf{u}, J\mathbf{w}) = g(\mathbf{u}, \mathbf{w}) \tag{8.2.23}$$

Given a metric g and a complex structure J, let us introduce the following differential 2-form $\mathbf{K}$:

$$\mathbf{K}(\mathbf{u}, \mathbf{w}) = \frac{1}{2\pi} g(J\mathbf{u}, \mathbf{w}) \tag{8.2.24}$$

The components $\mathbf{K}_{\alpha\beta}$ of $\mathbf{K}$ are given by

$$\mathbf{K}_{\alpha\beta} = g_{\gamma\beta} J^{\gamma}_{\alpha} \tag{8.2.25}$$

and by direct computation we can easily verify that g is hermitian if and only if $\mathbf{K}$ is anti-symmetric. A hermitian complex manifold is a complex manifold endowed with a hermitian metric g.

In a well-adapted basis we can write

$$g(u, w) = g_{ij^*} u^i w^{j^*} + g_{i^*j} u^{i^*} w^j \tag{8.2.26}$$

Finally in the well-adapted basis the 2-form $\mathbf{K}$ associated with the hermitian metric g can be written as follows:

$$\mathbf{K} = \frac{i}{2\pi} g_{ij^*} dz^i \wedge d\bar{z}^{j^*} \tag{8.2.27}$$

A hermitian metric on a complex manifold $\mathcal{M}$ is called a Kähler metric if the associated 2-form $\mathbf{K}$ is closed:

$$\mathrm{d}\mathbf{K} = 0 \tag{8.2.28}$$

A hermitian complex manifold endowed with a Kähler metric is called a Kähler manifold.

An Excursus on Cohomology

In order to appreciate the meaning and the conceptual substance of Eq. (8.2.28) my reader should have a minimal familiarity with the fundamental notions of cohomology. For the benefit of readers who might be deprived of that I hereby try to convey some intuitive description of the main concepts regarding differential forms and cohomology; my tools will be the same utilized for similar explanations in previous chapters, namely I will rely on images and on some sketchy example. The addressed topics are anyhow of relevance to the tales told in the next chapter.

Let us begin with Fig. 8.12. The fundamental idea underlying cohomology theory is captured by that image. There is a sequence of *spaces* $\Omega^{[i]}$, whose elements we

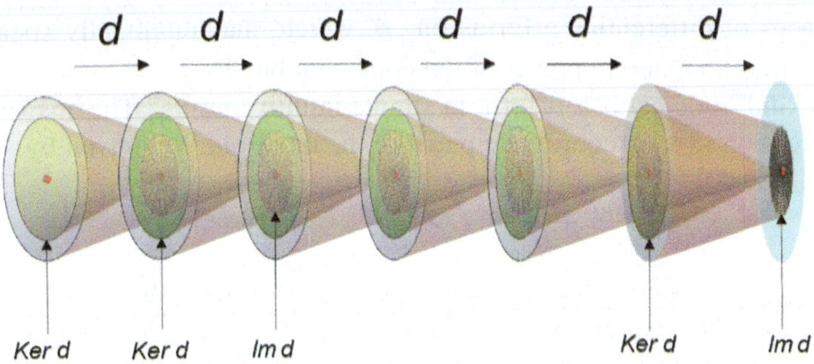

Fig. 8.12 A pictorial view of cohomology. The sequence of spaces $\Omega^{[i]}$ whose elements are named *cochains* are represented as a sequence of large circles. The maps produced by the external differential d are represented by truncated cones. The smaller circles represent the *cocycles*, namely the elements of the kernel of d. The even smaller circles represent the *coboundaries*, namely the elements of the image of d

name the *cochains*[6] and there is a linear operator, named d (the exterior derivative) that provides *non surjective maps* from each space $\Omega^{[i]}$ to the next one $\Omega^{[i+1]}$:

$$\partial_i : \Omega^{[i]} \xrightarrow{d} \Omega^{[i+1]} \quad ; \quad \forall \phi \in \Omega^{[i]} \quad d\phi \in \Omega^{[i+1]} \tag{8.2.29}$$

The fundamental property of the operator d is its nilpotency, namely it squares to zero $d^2 = 0$. In practice this means that the kernel of the map ∂_i, whose elements we name the *cocycles*[7] always contains the image $\mathrm{Im}\partial_{i-1}$ of the previous map ∂_{i-1}, namely the subspace of $\Omega^{[i]}$ formed by all those elements that can be written as $d\phi$ for some ϕ belonging to $\Omega^{[i-1]}$. We name *coboundaries* the elements of $\mathrm{Im}\partial_{i-1}$. In formula one writes:

$$\ker\partial_i \supset \mathrm{Im}\partial_{i-1} \tag{8.2.30}$$

Such a scenario occurs in various mathematical constructions and it is named an *elliptic complex* $\mathscr{C}$. The cohomology groups of the complex, usually denoted $H^{[i]}(\mathscr{C})$ are defined as the set of equivalence classes in which the subspace $\ker\partial_i$ can be partitioned with respect to the following equivalence relation:

$$\forall \omega^{[i]}, \psi^{[i]} \in \ker \partial_i \quad : \quad \omega^{[i]} \sim \psi^{[i]} \;\; \text{iff} \;\; \left(\omega^{[i]} - \psi^{[i]}\right) \in \mathrm{Im}\partial_{i-1} \tag{8.2.31}$$

The standard and best known example of cohomology is *de Rham* cohomology of a differentiable manifold $\mathscr{M}$ (see next chapter). In this case the spaces $\Omega^{[p]}$ are the

[6] We will be not too particular about the algebraic nature of the spaces $\Omega^{[i]}$. What is important is that their elements can be summed and subtracted and that they form an abelian group under addition. In many instances one can take linear combinations of the cochains so that they actually form a vector space over some field, or a module over some ring, but we do not discuss the many subtleties concerning the utilized coefficients.

[7] By definition the kernel of a map $\mu : V \to W$ from a group V to a group W is the subspace of V that is mapped by μ mapped into the neutral element of W that for abelian groups we denote $\mathbf{0}$.

vector spaces of differential p-forms on $\mathcal{M}$ which, mathematically speaking, are sections of the pth external power of the cotangent bundle $T^\star\mathcal{M} \stackrel{\pi}{\longrightarrow} \mathcal{M}$.

Which kind of object is a p-form in simple intuitive terms? The answer is given by its expression in each coordinate patch of the manifold. It is given by:

$$\omega^{[p]} = \omega_{\alpha_1 \dots \alpha_p}(x)\, dx^{\alpha_1} \wedge \dots \wedge dx^{\alpha_p} \tag{8.2.32}$$

where $\omega_{\alpha_1 \dots \alpha_p}(x)$ is a completely antisymmetric tensor depending on the location x in the manifold. The expression (8.2.32) is an object prepared for integration. Given any oriented p-dimensional hypersurface $\Sigma_p \subset \mathcal{M}_n$ of the ambient n-dimensional manifold $\mathcal{M}_n$, the p-form can be integrated on that surface. A simple example can be given in two dimension where the forms can be displayed by plotting. Consider for instance the ambient manifold $\mathcal{M}_n$ to be the xy-plane and consider the following 1-form:

$$\omega^{[1]} = \cos[y]\, dx + \sin[x]\, dy \tag{8.2.33}$$

The components $\omega^{[1]}_\alpha = \{\cos[y], \sin[x]\}$ form a covariant vector field that can be plotted as in Fig. 8.13. The form (8.2.33) is not a cocycle, since its exterior derivative does not vanish. The operator d is just the generalization of the curl operator of three-dimensional vector calculus:

$$d\omega^{[p]} = \partial_{\alpha_1} \omega_{\alpha_2 \dots \alpha_{p+1}}\, dx^{\alpha_1} \wedge \dots \wedge dx^{\alpha_{p+1}} \tag{8.2.34}$$

and in the case of the 1-form (8.2.33) we obtain:

$$\omega^{[2]} \equiv d\omega^{[1]} = -\left(\cos[x] + \sin[y]\right)\, dx \wedge dy \tag{8.2.35}$$

In two dimensions a 2-form is a top form, namely a form of the highest degree that can be integrated over the entire manifold. Its unique component is the function $\omega^{[2]}_{12}(x, y) = -(\cos[x] + \sin[y])$. In this case we can visualize the 2-form by means of a plot like in Fig. 8.14. In the same xy-plane an example of a cocycle, namely of a closed 1-form ($d\psi^{[1]} = 0$), is the following one:

$$\psi^{[1]} = \cos[x]\, dx + \sin[y]\, dy \tag{8.2.36}$$

which is visualized in Fig. 8.15 with the same method utilized in Fig. 8.13.

In the present case the cocycle $\psi^{[1]}$ is also a coboundary, namely it can be represented as the exterior derivative $\psi^{[1]} = d\psi^{[0]}$ of a 0-form, i.e. of a function:

$$\psi^{[0]} = \sin[x] - \cos[y] \tag{8.2.37}$$

Actually it could not be differently. The plane $\mathbb{R}^2$ is a trivial topological space without handles or holes and this triviality implies that all cohomology groups are trivial, namely all closed one-forms are exact. In the case of de Rham cohomology another word for coboundary is exact form!

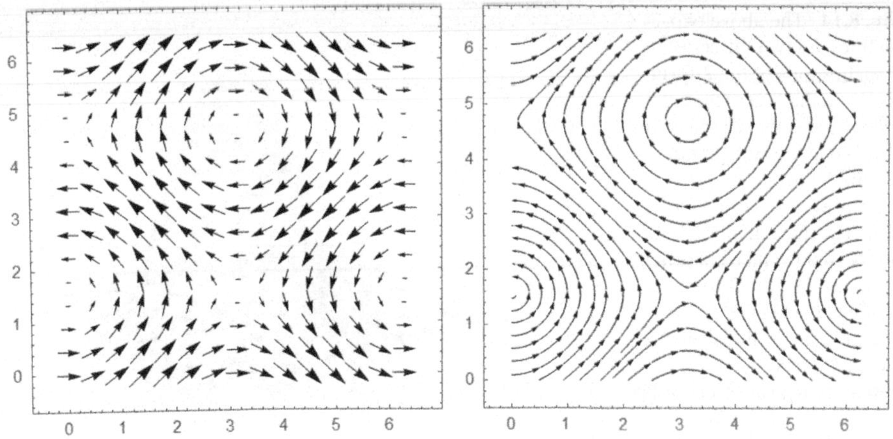

Fig. 8.13 Plotting of the 1-form (8.2.33), as a covariant vector field. On the right we plot the integral curves of this covariant vector field, namely those lines that have at each point the vector field as tangent vector

In this example the reader gets the flavor of what algebraic topology is about. The calculation of cohomology is a way of probing the topology of manifolds.

After these clarifications we come back to the Kähler manifolds, condition (8.2.28) states that the Kähler two form should be a cocycle.

THE KÄHLER POTENTIAL

Seen from a different view point, Eq. (8.2.28) is a differential equation for the metric g_{ij^*}, whose general solution in any local chart can be found since, locally, every closed form is also exact; the non-triviality of a cocycle appears only while gluing together its representations in different charts. The solution for the hermitian metric is given by the following expression:

$$g_{ij^*} = \partial_i \partial_{j^*} \mathcal{K} \tag{8.2.38}$$

where $\mathcal{K} = \mathcal{K}^* = \mathcal{K}(z, z^*)$ is a real function of z^i, z^{i^*}. The function $\mathcal{K}$ is called the Kähler potential and it is defined only up to the real part of a holomorphic function $f(z)$. Indeed one sees that

$$\mathcal{K}'(z, z^{i^*}) = \mathcal{K}(z, z^{i^*}) + f(z) + f^*(z^*) \tag{8.2.39}$$

gives rise to the same metric g_{ij^*} as $\mathcal{K}$. The transformation (8.2.39) is called a Kähler transformation.[8] The differential geometry of a Kähler manifold is described by Eq. (8.2.21) with g_{ij^*} given by (8.2.38). Kähler geometry is that implied by $\mathcal{N} = 1$ supersymmetry for the scalar multiplets [10].

[8] The non triviality of the Kähler 2-form manifests itself in the Kähler transformations that are required to connect the Kähler potential as given in one-chart with the Kähler potential as given in another one.

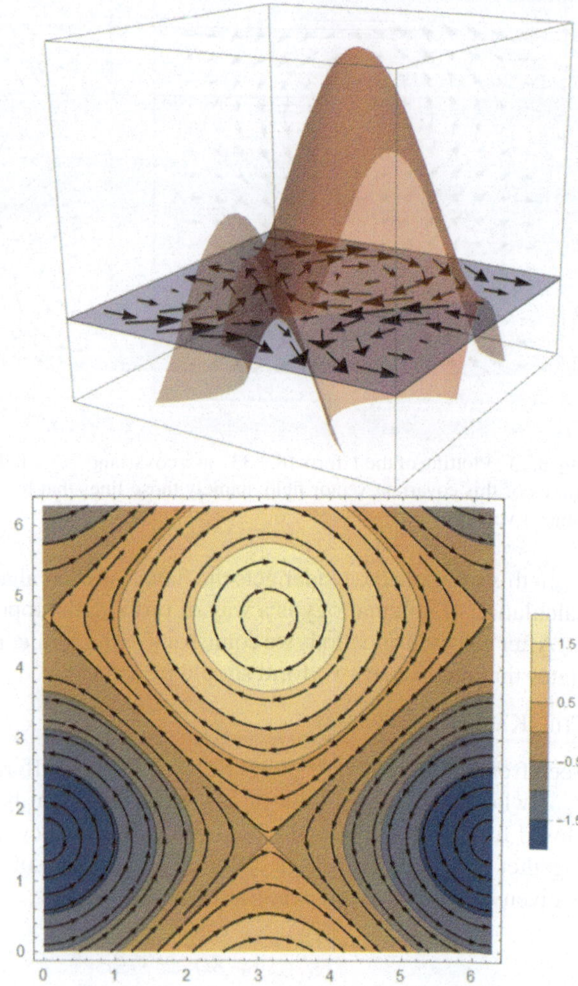

Fig. 8.14 The above two pictures are two different visualizations of the 2-form $d\omega^{[1]}$. In the first picture the blue plane is the xy space on which one sees the plot of the the 1-form $\omega^{[1]}$, displayed as a vector field plot. The third dimension z has been introduced in order to diplay also the 2-form $\omega^{[2]} = d\omega[1]$. The z-coordinate of the displayed surface points (in yellow-brownish color) is the value of $\omega^{[2]}_{12}(x, y)$. In the second planar diagram the value of $\omega^{[2]}_{12}(x, y)$ is codified by colors according to the plotlegend. As a reference, on the same plot we display also the integral curve of the one-form $\omega^{[1]}$

8.2.2 *Quaternionic Kähler, Versus HyperKähler Manifolds*

We saw the wealth of new features and, at the same time, the stricter structure of a manifold geometry when $\mathcal{M}$ admits one complex structure and is, therefore, complex. Following Hamilton, it is natural to inquiry what it might happen if a manifold admitted not just one, rather three complex structures, fulfilling the relations scribbled 170 years ago by the Irish genious on the walls of his Dublin bridge. Such manifolds do indeed exist and they happen to be of the highest relevance for supersymmetric field theories in the context of the Supergravity/Superstring world. Actually these geometries fall in two ample classes that we presently introduce and are respectively named that of *Quaternionic Kähler manifolds* and that of *HyperKähler manifolds*.

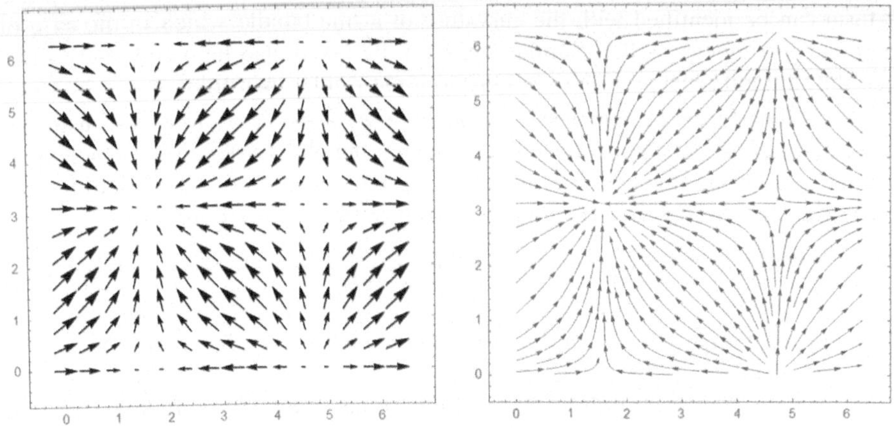

Fig. 8.15 Plotting of the 1-form (8.2.36), as a covariant vector field. On the right we plot the integral curves of this covariant vector field, namely those lines that have at each point the vector field as tangent vector

Both a Quaternionic Kähler or a HyperKähler manifold $\mathcal{QM}$ is a $4m$-dimensional real manifold endowed with a metric h:

$$ds^2 = h_{uv}(q)dq^u \otimes dq^v \quad ; \quad u, v = 1, \ldots, 4m \tag{8.2.40}$$

and three complex structures

$$(J^x) \; : \; T(\mathcal{QM}) \longrightarrow T(\mathcal{QM}) \qquad (x = 1, 2, 3) \tag{8.2.41}$$

that satisfy the quaternionic algebra

$$J^x J^y = -\delta^{xy} \, \mathbb{1} + \varepsilon^{xyz} J^z \tag{8.2.42}$$

and respect to which the metric is hermitian:

$$\forall \mathbf{X}, \mathbf{Y} \in T\mathcal{QM} \; : \quad h\left(J^x\mathbf{X}, J^x\mathbf{Y}\right) = h\left(\mathbf{X}, \mathbf{Y}\right) \qquad (x = 1, 2, 3) \tag{8.2.43}$$

From Eq. (8.2.43) it follows that one can introduce a triplet of 2-forms

$$\mathbf{K}^x = K^x_{uv} dq^u \wedge dq^v \; ; \; K^x_{uv} = h_{uw}(J^x)^w_v \tag{8.2.44}$$

that provide the generalization of the concept of Kähler form occurring in the complex case. The triplet $\mathbf{K}^x$ is named the *HyperKähler* form. It is an SU(2) Lie-algebra valued 2-form in the same way as the Kähler form is a U(1) Lie-algebra valued 2-form. In the complex case the definition of Kähler manifold involves the statement that the Kähler 2-form is closed. At the same time in Hodge–Kähler manifolds the Kähler

2-form can be identified with the curvature of a line-bundle which in the case of rigid supersymmetry is flat. Similar steps can be taken also here and lead to two possibilities: either HyperKähler or Quaternionic Kähler manifolds.

Let us introduce a principal SU(2)-bundle $\mathcal{SU} \xrightarrow{\pi} \mathcal{QM}$ over the considered manifold. Let ω^x denote a connection on such a bundle. To obtain either a HyperKähler or a Quaternionic Kähler manifold we must impose the condition that the HyperKähler 2-form is covariantly closed with respect to the connection ω^x:

$$\nabla \mathbf{K}^x \equiv d\mathbf{K}^x + \varepsilon^{xyz}\omega^y \wedge \mathbf{K}^z = 0 \tag{8.2.45}$$

The only difference between the two kinds of geometries resides in the structure of the $\mathcal{SU}$-bundle.

A HyperKähler manifold is a $4m$-dimensional manifold with the structure described above and such that the $\mathcal{SU}$-bundle is **flat**. Defining the $\mathcal{SU}$-curvature by:

$$\boldsymbol{\Omega}^x \equiv d\omega^x + \frac{1}{2}\varepsilon^{xyz}\omega^y \wedge \omega^z \tag{8.2.46}$$

in the HyperKähler case we have:

$$\boldsymbol{\Omega}^x = 0 \tag{8.2.47}$$

Viceversa a quaternionic Kähler manifold is a $4m$-dimensional manifold with the structure described above and such that the curvature of the $\mathcal{SU}$-bundle is proportional to the HyperKähler 2-form. Hence, in the quaternionic case we can write:

$$\boldsymbol{\Omega}^x = \lambda\, K^x \tag{8.2.48}$$

where λ is a non vanishing real number.

As a consequence of the above structure the manifold $\mathcal{QM}$ has a holonomy group of the following type:

$$\begin{aligned} \mathrm{Hol}(\mathcal{QM}) &= \mathrm{SU}(2) \otimes \mathrm{H} \quad \text{(Quaternionic Kähler)} \\ \mathrm{Hol}(\mathcal{QM}) &= \mathbb{1} \otimes \mathrm{H} \quad \text{(HyperKähler)} \\ \mathrm{H} &\subset \mathrm{Sp}(2\mathrm{m}, \mathbb{R}) \end{aligned} \tag{8.2.49}$$

Let us briefly comment on the notion of holonomy group. For any differentiable manifold, using Cartan's formulation of *moving frames* that leads to the vielbein and the spin connection (see Sect. 7.3.5), the curvature 2-form $\mathfrak{R}^{ab}$ is an $\mathfrak{so}(n)$ Lie Algebra valued object if n is the real dimension of $\mathcal{M}$. When the geometry of $\mathcal{M}$ is not generic, rather restricted as it is the case of complex, Kähler and quaternionic Kähler or HyperKähler manifolds, the curvature 2-form is forced to take values in subalgebras of $\mathfrak{so}(n)$ and these subalgebras are named the holonomy Lie algebras of the manifold. The corresponding Lie groups are the *holonomy groups*. This is the modern declination of Klein's Erlangen programme: the holonomy groups are the main algebraic classifiers of the different types of geometries.

8.3 L'Ésprit de Géométrie

..Olibri, habile géomètre, et grand physicien fonda la secte de vorticoses. Circino, habile physicien et grand géomètre fut le premier attractionnaire. ...On entre sans préparation dans l'école dOlibri; tout le monde en a la clef. Celle de Circino n'est ouverte qu'aux premiers géomètres.

In his philosophical novel *Les Bijoux Indiscrets*, published in 1748, Diderot (see Fig. 8.16) described with such words the antithesis of Descartes' mechanicistic vortex theory, that denies vacuum existence and long distance interactions, with the Newtonian theory of central forces (see Fig. 8.17). As we know, the latter won the historic competition for preminence and became the final basis of Classical Physics namely of the *Système du Monde*, to use the title of Laplace's monumental work, which of Newtonian Physics constitutes the apotheosis.

Fig. 8.16 Denis Diderot (Langres, 1713 - Paris, 1784) and his philosophical novel *Les Bijoux Indiscrets*

Fig. 8.17 Olibri = René Descartes (La Haye en Touraine 1596 - Stockholm 1650) and Circino = Isaac Newton (Woolsthorpe-by-Colsterworth, 1642 - London 1727)

Diderot's contraposition of Réné Descartes (Olibri) qualified *habile géomètre, et grand physicien* against Newton (Circino), described as *habile physicien et grand géomètre*, looks, at first sight, rather curious. Indeed it seems strange that, according to the Encyclopedist Philosopher's viepoint, the founder of Rationalist Thought and the inventor of Analytic Geometry, should miss the *Ésprit de Géométrie* theorized in the so entitled treatise that Blaise Pascal (see Fig. 8.18), great scholar in Projective Geometry, wrote in 1657, seven years after the death of the *Cogito, ergo sum* author.

According to Pascal, the geometric spirit corresponds to human mind's hability to possess, just by intuition and without any formal definition, such fundamental notions as those of point, line, surface and space. After having accepted such intuitive concepts, that constitute the bricks of the Euclidian building, Geometry is constructed with the use of logical deductive faculties, in line with the *Discourse on the Method*. Hence the *cartesian method* is fully accepted by Pascal, yet Diderot was probably right when, in the Descartes' *horror vacui* and in the mechanicistic transmission of the motion by means of vortices he identified the very negation of Pascal's *Ésprit de Géométrie*.

Olibri et Circino se proposèrent l'un et l'autre d'expliquer la nature. So Diderot says and continues: *Les principes d'Olibri ont au premier coup d'oeil une simplicité qui séduit: ils satisfont en gros aux principaux phénomènes; mais ils se démentent dans les détails. Quant à Circino, il semble partir d'une absurdité; mais il n'y a que le premier pas qui lui coute. Les détail minutiex qui ruinent le système d'Olibri affermissent le sien. Il suit une route obscure à l'entrée, mais qui s'éclaire à mesure qu'on avance. Celle au contraire, d'Olibri, claire à l'entrée, va toujours en s'obscurcissant.*

The attentive reader of Chap. 6 can compare these words of Diderot with Weyl's conception of the mathematical way of thinking when he says that …*we forget about what the symbols stand for. The mathematician is concerned with the catalogue alone; he is like the man in the catalogue room who does not care what books or pieces of an intuitevely given manifold the symbols of his catalogue denote. He need not be idle; there are many operations which he may carry out with these symbols, without ever having to look at the things they stand for.*

Fig. 8.18 Blaise Pascal (Clermont Ferrand, 1623 - Paris, 1662)

Pascal's characterization of the *Ésprit de Géométrie* as the hability to know a priori what points, lines, surfaces and space are, is probably the first step towards that conceptual elaboration that led Kant to elevate Euclidian Geometry to the fundament of any sensitive perception. As we extensively discussed, this secular sanctification of Euclid's postulates, including the fifth, was, in the early XIXth century, the main barrier to the development of non-euclidian geometries. Notwithstanding this obstruction, Bolyai, Lobachevsky and the same Gauss succeeded to introduce their new geometrical conceptions that start from different axioms, yet utilize the same cartesian method in the following deductive phase. Thus the screams of the Beotians were not so loud as Gauss feared and the road was paved to the Revolution of Geometry pursued in different fashions by Riemann, Klein, Beltrami, Lie, Helmholtz, Poincaré, culminating, on the mathematical side, with the tensor calculus introduced by Ricci and Levi-Civita and on the physical side with Einstein's General Relativity.

How can we characterize the reformulation of the *Ésprit de Géométrie* after XIXth century conceptual revolutions? The path is clear. Once the Kantian apriorism was removed, dramatically flunk by the explicit construction of non-euclidian models, once the notion of intrinsic curvature was introduced, first by means of Gauss' 1828 *Disquisitiones Generales supra Superficies Curvas*, then through the concept of *differentiable manifold* advanced by Riemann in his 1854 Habilationschrift *Ueber die Hypothesen, welche der Geometrie zu grunde liegen*, mathematical-philosophical attention shifted to the problem of determining the True Geometry of *Space* by means of physics, thus anticipating Einstein's philosophical foundations of General Relativity. Already Riemann, in his *Ueber die Hypothesen*, wrote: *the propositions of*

Geometry cannot be deduced from the general notion of manifold (merfach asgedehnter Grosse, in the original German text), rather the properties which distinguish our Space from other three-dimensional varieties can be assessed only on the basis of experiment.

Following Klein's 1872 Erlangen Programme, history witnessed a vast and variagated collection of conceptual elaborations aimed at the axiomatization of our three-dimensional physical space. The standard philosophical vice, aimed at an *a priori determination* of *existing reality*, came back into the game in new ways. Extremely interesting and paradigmatic is the 1868 article written Hermann Helmholtz with the title *Ueber di Tatschen, die der Geometrie zum Grunde liegen*, namely *The Facts that lie at the foundations of Geometry*. On the basis of what Helmholtz said in this essay I am sure that Diderot would have qualified him *habile géomètre* and, quite duely, *grand physicien.*

Joining the notion of manifold with a vision similar to the one that four year later was advocated by Klein in his Erlangen Programme, Helmholtz ventured into an axiomatic definition of *Physical Space* which was articulated into five axioms. The fifth axiom establishes that the dimension of Space is $d = 3$. In modern mathematical parlance, the remaining axioms, that are formulated in terms of the existence and mobility of rigid bodies, can be summarized, together with the fifth one, just in one sentence: *Space is a Riemaniann three-dimensional manifold of constant curvature.* In this way only three choices survive $\kappa = 1, 0, -1$, having denoted by κ the sign of the constant curvature. The first choice was immediately dismissed by Helmholtz since it would lead to a compact and hence finite Universe (a three-sphere). That the Universe was infinite was accepted as a dogma in the middle XIXth century, notwithstanding Olbers paradox. Among the Euclid Geometry ($\kappa = 0$) and that of Lobachevsky ($\kappa = -1$) the new Olibri–Helmholtz was not able to choose and his essay remained inconclusive.

Looking at these matters from our XXI century standpoint, educated by the impressive advances of Observational Cosmology in the last fifteen years, Helmholtz discussion is much less extravagant than it might look at first sight. Indeed the question at stake is about the spatial curvature of the Universe which, by means of Einstein equations, is determined by its overall energy-matter contents, whether larger, equal or less than the so called critical density. On the other hand, that the Universe has a large scale constant curvature is something established by the Cosmological Principle which is an axiomatization of an observational fact, namely the homogeneity and the isotropy of what we see.

In conclusion *L'Esprit de Géométrie* in the second half of the XIXth century was ambiguous: one hand the idea that a priori intuitive notions of *Space* and of its subvarieties did exist was dismissed and the task of choosing the *True Geometry* was assigned to Nature. On the other hand efforts were abundant aimed at predicting on philosophical grounds Nature's choice.

Two lines of developments slowly changed and completely reversed this attitude. On one side the heritage of General Relativity shew that Geometry is not only revealed through physical experiments rather it can be *dynamically* predicted by the development of Einstein equations, starting from given initial data that vary from case to case. On the other hand the impressive advances in the very formulation of geometrical questions and the new visions, whose development we have outlined in the present chapter shifted the attention of mathematicians to deeper characterizations of the available spaces that started to be organized in a vast and quite articulated *bestiary*. In the same way as in *Zoology* no one ventures to predict the unique existing animal in the same way in modern geometry no one aims anymore at singling out the unique God given True Geometry.

Further visions were to be established and developed in the XXth century that we address in the next chapter.

Chapter 9
Geometry Becomes Special

La géométrie...est une science née à propos de l'expérience...nous avons créé l'espace qu'elle etudie, mais en l'adaptant au monde où nous vivons. Nous avons choisie l'espace le plus commode...

Henri Poincaré

9.1 The Evolution of Geometry in the First Half of the XX Century

Let us begin with a quotation from a talk given in Cambridge by William Hodge

The last thirty years (1925–1955) have seen an enormous improvement in the position of geometry as a branch of mathematics, or, rather, have seen the re-integration of geometry into the main fabric of mathematics. Indeed, one can go further and say that with the restoration of geometry to its rightful place in the mathematical scheme the process of fragmentation which had been doing so much harm to mathematics has been reversed, and we may look forward to the day in which there are no longer analysts, algebraists, geometers and so on, but simply mathematicians. Mathematical research has two aspects, motivation and technique, and when the latter gains control the result is apt to be excessive specialization. The revolution of geometrical thought, and the reinstatement of geometry as one of the major mathematical disciplines, have helped to bring about a unification of mathematics which we may justly regard as one of the major contributions of the last quarter century to the subject.

HODGE

Born in Scotland, William Hodge, (see Fig. 9.1) studied Mathematics first in Edinburgh, then in Cambridge and graduated under the supervision of Edmund Whittaker. In 1936 he was appointed as Lowndean Professor of Astronomy and Geometry in Cambridge, a position that he held up to retirement. Prior to that he was Lecturer in Bristol University and held temporary positions in Princeton. Hodge has given outstanding contributions in the field of differential forms, topological invariants, harmonic integrals and complex analysis. He invented Hodge duality. Sir Michael Atiyah has been one of Hodge' students.

P. G. Fré, *A Conceptual History of Space and Symmetry*,
https://doi.org/10.1007/978-3-319-98023-2_9

Fig. 9.1 Sir William Vallance Douglas Hodge (Edinburgh 1903 – Cambridge 1975)

In 1925, Hodge, whose biography is shortly summarized in the above lines, was a student in Cambridge, dividing a room in St. John's college with Alan Broadbent who would become his brother in law by marrying his sister Janet. In 1955, as we have recalled he was a distinguished Cambridge Professor of Mathematics sitting on the chair named after Lowndean. Four years later he was to be knighted by Queen Elizabeth. Just in the middle of those thirty years that, according to him, had reintegrated geometry into the main fabrics of mathematics, Hodge had published a paper on the *Theory of Harmonic Integrals* that won him the 1937 Adams Prize and then, in 1941 the book *The Theory and applications of harmonic integrals* whose content was described by Hermann Weyl as:

...*one of the great landmarks in the history of science in the present century.*

Certainly William Hodge had himself given an outstanding contribution to the conceptual revolution he refers to in his quoted words. By means of the integrals of harmonic forms on Riemannian manifolds, he had brought together geometry, topology and analysis into a solid unity that, since then, would last in the successive history of mathematics and physics. Incidentally the conserved quantities of physical theories (the electric charge, for example) are instances of harmonic integrals and the link between topology and physics is by them best exemplified.

However Hodge's contribution was not the only one. Other high class mathematicians were active in those thirty years and gave extremely relevant contributions. A look at the second chapter of Hodge's book helps our orientation. It is entitled *Integrals and their Periods* and it includes as section 23, 24 and 25 the exposition and proof of de Rham theorems.

Fig. 9.2 Georges de Rham (1903 Roche, Switzerland – 1990 Lausanne, Switzerland)

DE RHAM

Born in the small town of Roche at the foot of the Jura, de Rham made his first studies there. In 1919 his family moved to the city of Lausanne, where Georges lived most of his life up to his death in 1990. Since 1924 he attended the University of Lausanne but in 1931 he wrote his doctoral thesis *Sur l'Analysis situs des variétés a n dimensions* under the supervision of Élie Cartan and defended it in Paris, obtaining his doctorate in mathematical sciences from the Faculty of Science, University of Paris. In the French capital, where he lived at his own expenses, he attended courses from the great mathematicians of the time Gaston Julia, Arnaud Denjoy, Émile Picard and met with André Weil. He spent also a short period in Göttingen where he met with Richard Courant, Charles Ehresmann, Andrey Kolmogorov, Emmy Noether and Hermann Weyl. Since 1943 to his retirement in 1971 he was professor in Lausanne and he also held a chair at the University of Geneva. The academic year 1957–1958 he was visiting professor in Princeton. His greatest achievement in mathematics is the theorem that bears his name and states the isomorphism of de Rham cohomology of differential forms with the dual of singular homology based on simplexes (Fig. 9.2).

In 1926 and in 1928, as we recalled above, he had been in Paris renting a cheap room at his own expenses and following courses at the Sorbonne where he was particularly impressed by the lectures of Jacques Hadamard and Henri Lebesgue. But he also read Cartan's book *Sur les nombres de Betti des espaces de groupes clos*. Topology, geometry and symmetry were approaching each other from many sides in those years! The first seeds of the *de Rham theorem* were implanted in the mind of the young Swiss mathematician by such readings and by his personal contacts with the magic circle of contemporary French mathematicians, at whose center, stood the great personality of Cartan. Indeed Georges de Rham's doctoral thesis is essentially the first exposition of his theorem.

Another great mathematician, Raoul Bott assessed the impact of de Rham theorem in the following way:

In some sense the famous theorem that bears his name dominated his mathematical life, as indeed it dominates so much of the mathematical life of this whole century. When I met de Rham in 1949 at the Institute in Princeton he was lecturing on the Hodge theory in the context of his currents. These are the natural extensions to manifolds of the distributions which had been introduced a few years earlier by Laurent Schwartz and of course it is only in this extended setting that both the de Rham theorem and the Hodge theory become especially complete. The original theorem of de Rham was most probably believed to be true by Poincaré and was certainly conjectured (and even used!) in 1928 by Élie Cartan. But in 1931 de Rham set out to give a rigorous proof. The technical problems were considerable at the time, as both the general theory of manifolds and the 'singular theory' were in their early formative stages.

Three Things

Three things should have become manifest to the reader from the facts and from the evaluations that we have collected above.

First a historical political note. In the middle thirties of the XXth century, while the ominous shadows of nazism, antisemitism and stalinism were growing, projecting threats of new wars, the world centers of Mathematics, that once upon a time had been Paris, Berlin and Göttingen, underwent a shift, Princeton replacing the German centers from which the greatest scientists of Jewish origin were fleeing away. Great mathematicians, philosophers and physicists from other countries, will soon replenish Princeton ranks in the eves of World War II.

The second notable thing is the deep influence of Cartan, which is felt in all respects. Looking at all new developments we invariably discover a direct root in the thought and in the teaching of this great man.

Thirdly we see that what was in the process of formation in the thirties was a new vision of geometry where the global properties of manifolds were coming to prominence and new instruments to measure these properties were introduced. Integrals, being extended to the whole space, probe its structure at large and this was the viewpoint introduced by Hodge. On the other hand, through *isomorphism theorems* like de Rham's theorem, the result of Hodge's harmonic integrals can be largely predicted a priori, since these integrals are associated with algebraic structures, like cohomologies (see Sect. 8.2.1), which allow for a formal abstract description. Groups were once again coming into prominence but in a new capacity, that of classifiers of topology.

The Emergence of Fibre-Bundles

In this context the notion of *fibre-bundle* was slowly emerging as the appropriate conception that encompasses all geometrical spaces of interest both in mathematics and in physics. Slowly it was becoming clear that the key point is the *locality* of the *product structure*. Indeed fibre-bundles are just those manifolds that look like the direct product of two spaces at the local level but are not a direct product in globality.

Fig. 9.3 On the left Shiing-shen Chern (1911 in Chia-hsing, China – 2004 Tianjin, China)

The lesson of Riemannian geometry had thought mathematicians that the deviation from some property which, in a class of objects, is valid only for a subclass, is measured by appropriate *intrinsic indicators*, such as the curvature tensor. In a flat manifold the parallel transport of vectors is independent from the transport path. In a curved manifold, this is not true and the curvature tensor measures the deviation from absolute parallelism. Trivial fibre-bundles are direct-product of spaces. Non trivial fibre-bundles are such because they globally deviate from a direct product. What is it, analogous to the curvature, that indicator, or set of indicators, that measures the deviation from globality of the product? The answer is provided by the so called *characteristic classes*.

The theory of characteristic classes was founded by the Chinese-American mathematician *Shiing-shen Chern*.

CHERN

Born in China, son to a classical Confucian scholar, Chern (see Fig. 9.3) studied there mathematics until 1932, when he left his country for Europe, ending up first in Hamburg and then in Paris. In Hamburg he met with Kähler who introduced him to Cartan's works and he graduated from that University in 1936. The same year he reached Paris where he interacted with Cartan himself and met with André Weil. In 1937 he left Paris and went back to China to take a professorship of mathematics at the Tsing Hua University. He was trapped in China by the break-up of the Chinese-Japanese war until 1942, when he received an invitation to the Institute of Advanced Study of Princeton which he reached by means of an adventurous trip on board of a US military plane that took him from the remote inland regions of China to destination, crossing through India, Africa and South America. In the US he met again with Weil who, at the time, was teaching in Pennsylvania. After the war Chern went back to China, but in 1949 he fled once again from his country to US, escaping from Mao Tse Dung's communist revolution. In the States he became US citizen and he was professor in Chicago and at Berkeley, until his retirement. In the last years of his life he went once again back to China. Chern's contributions to mathematics are very ample and deep. As stated above he has been the founder of the theory of characteristic classes and of the modern vision of fibre-bundles.

From the above short biographical note we know that Chern's first studies were in Peking. In 1932 a visit to the Chinese capital of the Austrian mathematician Blashke was at the origin of a turning point in Chern's life. He received a scholarship from Tsing Hua University to study in the United States but he asked that it might be rather used to go to the University of Hamburg. He was convinced that the mathematics he was interested in was done more in Europe than in the States. Chern wrote: *It was professor Blaschke whose influence on me cannot be overstated. In 1932 he visited Peking as part of his world tour. I was a young college student in his audience. I was immediately impressed by his fresh ideas and his insistence on mathematics being a lively and intelligible subject. This contact with him was instrumental in making me to decide to come to Hamburg as a student.*

In 1934 Chern arrived in Hamburg and there he met with the young Kähler who had just written a book describing Cartan's mathematics. The conversations with Kähler were very influential in determining Chern to spend in Paris the third year of his fellowship, after his graduation from Hamburg that took only two years.

So in 1936 Chern went to Paris and there he absorbed Cartan's viewpoint on mathematics that has been described in this way:

There is a tendency in mathematics to be abstract and have everything defined, whereas Cartan approached mathematics more intuitively. That is, he approached mathematics from evidence and the phenomena which arise from special cases rather than from a general and abstract viewpoint.

Not surprisingly, with such an attitude, Cartan has been not only a mathematician but also a theoretical physicist, in the sense Herman Weyl also was. Indeed, as we have explained in Chap. 7, Cartan introduced the vielbein or *réperes mobiles* formalism for Gravity, and in that he proved to be more right than Einstein himself. Looking at it a posteriori Cartan's entire work has been a monumental investigation of the mathematical structure of what we name *Space* and in doing that he used those concrete spaces that he had the venture to classify exhaustively, namely symmetric spaces.

Speaking of Cartan's ideas Chern said:

Without the notation and terminology of fibre bundles, it was difficult to explain these concepts in a satisfactory way.

Chern also gave an alive description of his interaction with Élie Cartan:

Usually the day after meeting with Cartan I would get a letter from him. He would say, After you left, I thought more about your questions ... - he had some results, and some more questions, and so on. He knew all these papers on simple Lie groups, Lie algebras, all by heart. When you saw him on the street, when a certain issue would

Fig. 9.4 André Abraham Weil (1906 Paris, France – 1998 Princeton, USA)

come up, he would pull out some old envelope and write something and give you the answer. And sometimes it took me hours or even days to get the same answer. I saw him about once every two weeks, and clearly I had to work very hard.

THE STATUS OF GEOMETRY IN THE EVES OF WORLD WAR II

We can summarize the status of geometry in 1937–38 as follows.

Since a few years de Rham had put the systematic use within topology of differential forms and of their cohomology on firm grounds. In 1937 Hodge, by means of harmonic differential forms and of their integrals, had given to Poincaré duality between homology an cohomology its solid basis. The notion of fibre-bundle was slowly coming into being and Chern and André Weil were about to provide their proper characterizations by means of characteristic classes. The great examples of Cartan's symmetric spaces were in everyone's mind, providing a precious, guiding principle.

ANDRÉ WEIL

The life of André Weil (see Fig. 9.4), one of the greatest mathematician of the XXth century, has also been somewhat adventurous with several shifts and turning points on the dramatic background of World War II. Born in Paris in a Jewish family that had escaped from Alsace after the 1870 annexation to the German Empire, he graduated there in 1928 under the supervision of Hadamard. Strongly interested in classical languages and ancient cultures he had an experience as a teacher at a Muslim University in India and, once he was back to France, he interacted with Cartan's son Henri in Strasbourg, where he became involved with the famous group of mathematicians writing under the name Nicolas Bourbaki. In 1939 he was in Finland when the Finnish-Soviet war broke up. He was arrested as a spy. Released, he went back to France to be arrested once again as a renitent to the military service. Released, he took part in the 1940 campaign and in the May Debacle. In January 1941 he fled from France to the US with his entire family. During the war he survived teaching in a minor university in Pennsylvania. After the war he went for a few years to Brazil and from there he made return to the US, where he became member of Princeton Institute for Advanced Studies in 1958. In Princeton he spent the rest of his life. Weil's contributions to mathematics are extremely important in many fields, algebraic geometry and number theory, in particular. The proof of the fundamental theorem stating the Chern–Weil homomorphism was independently developed by Weil and Chern in the 1940s.

Indeed, when Chern arrived in Princeton in 1943, after his adventurous air-trip across three continents, he found there such outstanding mathematical personalities as Hermann Weyl, Claude Chevalley and Solomon Lefshetz. Einstein was also in the group. Without any doubt in such an environment Chern's ideas found a fertile humus where to grow. However he had not forgotten André Weil with whom he had met seven years before in Paris and who was teaching in Pennsylvania, few tens of miles away from Princeton. The two met several times, talking about Cartan's mathematics and certainly such conversations are responsible for their almost simultaneous but independent discovery of the Chern–Weil homorphism. With great generosity, in all of his later lectures and books Chern always referred to Chern–Weil homomorphism as to the Weil homomorphism. On the other hand recalling these war-time encounters with Chern, Weil wrote:

...we seemed to share a common attitude towards such subjects, or towards mathematics in general; we were both striving to strike at the root of each question while freeing our minds from preconceived notions about what others might have regarded as the right or the wrong way of dealing with it.

SUMMARIZING...

So the theory of fibre-bundles and characteristic classes was developed in those years while Ehresman definitely fixed the notion of a connection on a Principal Bundle as we have explained in Chap. 7 (see Sect. 7.4). The appropriate mathematical language of modern gauge theories in which the Standard Model of non gravitational interactions could be properly formulated was essentially ready by the mid fifties of the XXth century when Hodge wrote his sentence, quoted at the beginning of this chapter. Yet it took half a century before theoretical physicists became fully aware of the mathematical nature of those objects with which they were playing, striving to describe the Fundamental Forces of Nature.

9.1.1 Complex Geometry Rises to Prominence

On the purely mathematical front in the years from 1953 to 1955, Pierre Dolbeault (see Fig. 9.5) introduced a new very important mathematical instrument: the $\bar{\partial}$-cohomology of the differential forms defined on complex analytic manifolds,

Fig. 9.5 Pierre Dolbeault (1924–2015). Dolbeault graduated at the École Normale Supérieure with a thesis written under Henri Cartan's supervision. He held positions at several French universities, such as Montpellier and Bordeaux; in 1960, he became a professor at the University of Poitiers and finally moved to Paris VI (Université Pierre et Marie Curie) in 1972, where he stayed until his retirement in 1992. Dolbeault has given outstanding contributions to complex geometry and analysis and he is specially known for the creation of Dolbeault cohomology and the theorems associated with it

namely the holomorphic analogue of de Rham cohomology defined on real manifolds. The essence of Dolbeault cohomology is the topic of Dolbeault's thesis, prepared by him under the direction of Henri Cartan, Élie's son and one of the closest friends of André Weil. The thesis was defended in Paris in 1955.

Complex Geometry and, within it Kähler Geometry, arose to high prominence in the three decades from 1950 to 1980. The language of fibre-bundles and characteristic classes was combined with the notion of holomorphicity and line-bundles, namely Principal Bundles whose structural group is the group of non vanishing complex numbers $\mathbb{C}^\star$, became ubiquitous in the discussion of complex manifolds.

A new innovative conception developed in this context, namely that of characterizing the geometry of base manifolds $\mathcal{M}$ by means of statements on the characteristic classes of bundles defined over them.

The first example, which plays an important role in the sequel, is that of *Hodge–Kähler manifolds* that are Kähler manifolds $\mathcal{M}$ characterized by the existence of a line bundle $\mathcal{L} \to \mathcal{M}$, such that its first Chern Class coincides with the cohomology class of the Kähler 2-form: $c_1(\mathcal{L}) = [K]$.

CALABI YAU n-FOLDS

Another important example is provided by Calabi–Yau n-folds. These latter were introduced by Eugenio Calabi (see Fig. 9.6) in 1964 with the definition of complex n-dimensional algebraic varieties $\mathcal{M}_n$, the first Chern class of whose tangent bundle

Fig. 9.6 On the left Eugenio Calabi (Milano, Italy 1923). On the right Shing-Tung Yau (Shantou, China 1949). Born Italian, Calabi is an American citizen. He graduated in 1946 from MIT and obtained his Ph.D. from Princeton in 1950. He held temporary positions in Minnesota and in Princeton, then since 1967 to retirement he was Full Professor of Mathematics at the University of Pennsylvania, successor of Hans Rademacher. He came to the definition of Calabi–Yau n-folds while exploring the geometry of complex manifolds that support harmonic spinors. Born in China, Yau studied first at Hong Kong University, then he went to the USA where he got his Ph.D. in 1971 from Berkeley under the supervision of Chern. Post-doctoral fellow in Princeton and in Stony Brook, he became Professor in Stanford. Since 1987 he is Professor of Mathematics at Harvard University. Yau's proof of Calabi 1964 conjecture was published in 1977

vanishes: $c_1\,(T\mathscr{M}_n) = 0$. Later, the American-Chinese mathematician Shin-Tung Yau (see Fig. 9.6) proved the theorem that for Calabi–Yau n-folds, every $(1,1)$ Dolbeault cohomology class contains a representative that can be identified with the Kähler 2-form of a Ricci flat Kähler metric: the Calabi–Yau metric.

9.1.2 On the Way to Special Geometries

Other notable examples of this way of thinking, applying both to complex and to real geometry are the *manifolds of restricted holonomy*. One considers Riemannian manifolds $\mathscr{M}_n$ in dimension n and their *spin bundles*, namely the principal bundles on which their spin connections ω^{ab} are defined as Ehresman connections. Generically such bundles have, as structural group, Spin(n), which is the double covering of SO(n), yet it may happen that ω^{ab} is Lie algebra–valued in a proper subalgebra $\mathbb{G} \subset \mathfrak{so}(n)$.[1] Choosing algebras $\mathbb{G}$ for which this might happen and imposing that it should happen is a strong constraint on the geometry of the manifold $\mathscr{M}_n$.

[1]We have already described the notion of holonomy at the end of Sect. 8.2.2.

Research on manifolds of restricted holonomy went on in the 1980s and 1990s in the mathematical community but, not too surprisingly, it was heavily stimulated by issues in theoretical physics and particularly in Superstring/Supergravity theory.

It is easy to understand why. The main input in Superstring/Supergravity is Supersymmetry, a generalization of Lie algebras where spinor representations and vector representations of groups SO(n) are transformed one into the other by new symmetry operators Q^α, dubbed the *supercharges*, that are themselves spinors. At the level of field theories we work with fibre-bundles and the fields we consider are sections of such bundles. Field theories can be supersymmetric if the supercharges Q^α find a field-theoretic realization which is a symmetry of the action, leaving the door open for its desired spontaneous breaking. It is quite intuitive that such a realization of the supercharges requires special restrictions on the bundles and this reflects into heavy constraints on the geometry of the base manifolds.

The above simple reasoning reveals what, in the opinion of this author, is the main conceptual contribution of supergravity theories to the development of geometrical thought and, eventually, of physical thought, provisionally assuming that geometry and physics are, once properly interpreted, the same thing. Supersymmetry tackles with one of the most fundamental and so far unexplained pillars of physics, namely the separation of the physical world into bosons and fermions and the spin-statistics theorem. The distinction between vector and spinor representations is at the basis of all that and it is a distinctive property of the $\mathfrak{so}(n)$ Lie algebras, unexisting for the other simple Lie algebras. On the other hand the reduction of the tangent-bundle to an $\mathfrak{so}(n)$-bundle is the same thing as the existence of a metric and can be interpreted as gravity. Special Geometries arise because of supersymmetry, in order to allow the mixing of boson and fermions. It is the mathematical investigation of *Space* from this new viewpoint the new quality of geometrical studies inspired by supergravity. Before telling such a story we need to recall another mathematical conception, that was developed independently from Superstring/Supergravity yet found its most ample and fertile applications in the supersymmetric context.

9.1.3 The Geometry of Geometries

Let us recall Hermann Weyl's discussion of the ellipses, used by him to introduce his conception of mathematical thinking and reported by us in Sect. 6.1.2. The coefficients a, b, c of the quadratic form quoted by Weyl are the first example of *moduli* and the portion of $\mathbb{R}^3$ where they are allowed to take values is the first example of a *moduli-space*. In complex algebraic geometry one considers loci of some projective space $\mathbb{P}_n(\mathbb{C})$ cut out by some homogeneous polynomial constraint of degree m:

$$0 = \mathscr{W}(a, X) = \sum_{i_1 \dots i_m} a_{i_1 \dots i_m} X^{i_1} \dots X^{i_m} \tag{9.1.1}$$

imposed on the $n+1$ homogeneous coordinates X^i $(i = 1, \ldots, n+1)$. The complex coefficients $a_{i_1 \ldots i_m}$ are also *moduli* and fill some complex manifold $\mathcal{M}$. If we consider the following constraint imposed on the metric tensor of some Riemannian manifold $\mathcal{M}_n$:

$$R_{\mu\nu}[g] = \lambda \, g_{\mu\nu} \tag{9.1.2}$$

where $R_{\mu\nu}[g]$ is the Ricci tensor and λ some constant, we actually write a set of differential equations for the metric tensor $g_{\mu\nu}$, which, on the manifold $\mathcal{M}_n$, admit a solution depending on a set of parameters $\{p_1, \ldots p_r\}$, among which λ is included. Also these are moduli and they fill a space named the *the moduli space of Einstein metrics* on $\mathcal{M}_n$.

Several other examples can be made of manifolds $\mathcal{M}_{mod}$ whose points correspond to the specification of a particular geometry within a class, for instance the moduli ρ^i of an instanton parameterize the solution of the self duality constraint:

$$F^{\Lambda}_{\mu\nu}(\rho, x) = \tfrac{1}{2} \, \varepsilon_{\mu\nu\lambda\sigma} \, F^{\Lambda}_{\lambda\sigma}(\rho, x) \tag{9.1.3}$$

imposed on the field strength of a connection on a principal fibre bundle $P(\mathrm{G}, \mathcal{M}_4)$.

A new mathematical idea that is of outmost relevance both for physics and for mathematics is encoded in the following almost obvious argument. Being a manifold, the moduli space $\mathcal{M}_{mod}$ can support such geometrical structures like a metric, like a complex structure, or a fibration. We call this the *geometry of geometries*. There are several mathematical constructions, dictated by the mathematical nature of the objects of which we consider the moduli, that single out a canonical determination of the *geometry of geometries*, yet it is precisely at this level that the interaction between physics and mathematics becomes most profound and fertile. Indeed the geometry of geometries is typically what enters the supergravity lagrangians under the form of sigma-models for scalar fields that on one side are the spin zero members of supersymmetry multiplets, while on the other side they are *moduli* of some manifold, for a example a Calabi–Yau threefold, on which the superstring has been compactified.

This evenience produces a double check on the geometry of geometries. Its use in supersymmetric lagrangians, imposes strong constraints on the geometry of the scalar fields that, in many cases, have a recognizable solution in terms of known geometrical categories, in other cases it leads to the definition of new types of restricted geometries, generically dubbed *special geometries*. It is particularly rewarding that the *special geometries* selected by supersymmetry are just those apt to accomodate *the moduli spaces* of such mathematical structures as *the complex structures* or the *Kähler structures* of a compactification manifold like a Calabi–Yau threefold.

Altogether, a really new chapter has been written in the two decades from 1990 to 2010 in the history of geometry, where the distinction between physics and mathematics has become somewhat obsolete, ideas from one field compenetrating the other in an essential way.

9.1.4 The Advent of Special Geometries

The first instance of a special geometry was found by brute force, immediately after the discovery in 1976 by Sergio Ferrara, Daniel Freedman and Peter van Nieuwenhuizen of $\mathcal{N} = 1, d = 4$ supergravity (see Fig. 9.7). The next year, considering the coupling of a scalar multiplet to the newly found gravitational theory, the three supergravity founders, together with Breitenlohner, Gliozzi and Scherk, constructed a rather impressive and cumbersome lagrangian, depending on an arbitrary real function $G(A, B)$ of a scalar A and a pseudoscalar B and on all its derivatives up to the fourth one [73]. It was Bruno Zumino (see Fig. 9.8) who, in 1979, decoded the meaning of this monster, showing that $G(A, B)$ is just the Kähler potential of a Kähler metric, all of the introduced derivatives obtaining their adequate interpretation as metric, connection and curvature of the Kählerian manifold [171]. In this way the generalization to several scalar multiplets was singled out: it suffices to utilize an n-dimensional Kähler manifold.

Shortly after, the so named holomorphic superpotential introduced by physicists to describe fermion–scalar interactions and to produce a scalar potential consistent

Fig. 9.7 From left to right the three founders of Supergravity Theory, Daniel Freedman (1939), Sergio Ferrara (1945), Peter van Nieuwenhuizen (1938). Dan Freedman was born in the USA, graduated from Wisconsin University. He has been professor at Stony Brook University and he is currently full-professor at MIT. Sergio Ferrara born in Rome in 1945 graduated from la Sapienza University under the supervision of Raoul Gatto. Permanent Member of the CERN Theoretical Division for many years, he is also professor of physics at UCLA. Peter van Nieuwenhuizen born in Holland in 1938, graduated in Utrecht under the supervision of Veltman, held various positions in the United States and since the middle 1980s he is full-professor of physics at Stony Brook University. The paper containing the lagrangian and the transformation rules of $\mathcal{N} = 1, d = 4$ supergravity was published by the three founders of the theory in 1976. Since then all the three have contributed extensively and in various different directions to the development of supergravity. Sergio Ferrara among the three has largely contributed to the development of special geometries

Fig. 9.8 Bruno Zumino (1923–2014). Born in 1923 in Rome, he graduated from the University La Sapienza in 1945. He died in 2014 in California, where he was emeritus professor of Berkeley University. For many years he was permanent member of the Theoretical Division at CERN. Zumino has given many important contributions to Theoretical Physics in several directions: supersymmetry, anomalies, conformal field theories, quantum groups

with supersymmetry, was also interpreted geometrically. The superpotential is just a holomorphic section of the Hodge line-bundle over the Kähler manifold.

In this way the firstly found special geometry was a known one, namely Hodge-Kähler geometry. This is not so for the next case.

At the beginning of the 1980s the next obvious case was the coupling of vector multiplets to $\mathcal{N} = 2, d = 4$ supergravity. Each multiplet contains a complex scalar field and the question was what is the geometry of the scalar manifold $\mathcal{M}_{scalar}$ in the case of several such multiplets. Certainly $\mathcal{M}_{scalar}$ had to be Kähler, since $\mathcal{N} = 2$ is in particular $\mathcal{N} = 1$. Yet the stronger supersymmetry imposes additional constraints so that $\mathcal{M}_{scalar}$ had to be a *special Kähler manifold*. A pioneering work on this problem was conducted in several different combinations by a group of French, Belgian, Dutch, Swiss and Italian theoretical physicists in the papers mentioned in [46, 59, 60]. Using a special set of complex coordinates, the special Kähler manifolds that can accomodate the scalar fields of $\mathcal{N} = 2$ vector multiplets were described as those where the Kähler potential is obtained from a holomorphic prepotential according to a specific formula.

Once this was established, a natural question arose whether among so defined *special Kähler manifolds* there were symmetric spaces G/H. The answer to this question was given in Paris in 1985 by Eugene Cremmer and Antoine Van Proeyen (see Fig. 9.9) who, in a beautiful paper absolutely worth of Cartan's tradition [47], provided the exhaustive classification shown in the first column of Table 9.1. As one sees, exceptional Lie groups make their appearance in such a list through peculiar real forms. This was no longer a surprise for supergravity researchers since, four years before, the same Eugene Cremmer, in collaboration with Bernard Julia (see Fig. 9.10),

Fig. 9.9 On the left Antoine Van Proeyen (1953 Belgium), on the right Eugene Cremmer (Paris 1942). Antoine Van Proeyen graduated from KU Leuven and worked in several Laboratories and Universities, among which the École Normale of Paris, CERN Theoretical Division and Torino University, before becoming full-professor in Leuven. He is currently the Head of the Theoretical Physics Section at the K.U. Leuven. Since 1979, he has been involved in the construction of various supergravity theories, the resulting special geometries and their applications to phenomenology and cosmology. Cremmer is *directeur de recherche* of the CNRS working at the École Normale Supérieure of Paris. In 1978, together with Bernard Julia and Joël Scherk, he derived the space-time formulation of 11 dimensional supergravity theory, regarded today as the low energy limit of the so far mysterious M-theory. In the following few years, Cremmer, together with Bernard Julia, constructed the dimensional reductions of $d = 11$ supergravity, arriving in $d = 4$ at the maximal extended $\mathcal{N} = 8$ theory, whose structure is completely determined by the non-compact coset $\frac{E_{7(7)}}{SU(8)}$ accomodating the 70 scalars of the gravitational multiplet. Active research is going on at the present time to demonstrate that $\mathcal{N} = 8$ supergravity is a finite quantum field theory

Table 9.1 List of special Kähler symmetric spaces with their Quaternionic Kähler c-map images. The number n denotes the complex dimension of the Special Kähler preimage. On the other hand $4n + 4$ is the real dimension of the Quaternionic Kähler c-map image. The c-map is a central item in the mathematical technical exposition of special geometries contained in the parallel book [90]

$\mathcal{SK}_n$ Special Kähler manifold	$\mathcal{QM}_{4n+4}$ Quaternionic Kähler manifold	$\dim \mathcal{SK}_n = n$
$\frac{SU(1,1)}{U(1)}$	$\frac{G_{2(2)}}{SU(2)\times SU(2)}$	$n = 1$
$\frac{Sp(6,R)}{SU(3)\times U(1)}$	$\frac{F_{4(4)}}{USp(6)\times SU(2)}$	$n = 6$
$\frac{SU(3,3)}{SU(3)\times SU(3)\times U(1)}$	$\frac{E_{6(2)}}{SU(6)\times SU(2)}$	$n = 9$
$\frac{SO^{\star}(12)}{SU(6)\times U(1)}$	$\frac{E_{7(-5)}}{SO(12)\times SU(2)}$	$n = 15$
$\frac{E_{7(-25)}}{E_{6(-78)}\times U(1)}$	$\frac{E_{8(-24)}}{E_{7(-133)}\times SU(2)}$	$n = 27$
$\frac{SL(2,\mathbb{R})}{SO(2)} \times \frac{SO(2,2+p)}{SO(2)\times SO(2+p)}$	$\frac{SO(4,4+p)}{SO(4)\times SO(4+p)}$	$n = 3 + p$
$\frac{SU(p+1,1)}{SU(p+1)\times U(1)}$	$\frac{SU(p+2,2)}{SU(p+2)\times SU(2)}$	$n = p + 1$

Fig. 9.10 Bernard Julia (Paris 1952). He graduated from Université de Paris-Sud in 1978, and he is *directeur de recherche* of the CNRS working at the *École Normale Supŕieure*. In 1978, together with Eugene Cremmer and Joël Scherk, he constructed 11-dimensional supergravity. Shortly afterwards, Cremmer and Julia constructed the classical Lagrangian of four-dimensional $\mathcal{N} = 8$ supergravity by dimensional reduction from the 11-dimensional theory

had shown that the dimensional reduction of maximally extended supergravity from $D = 11$ down to $D = 10, D = 9, \ldots, D = 4, D = 3$ produces, as scalar manifolds, the following maximally split symmetric spaces:

$$\mathrm{M}_D = \frac{\mathrm{E}_{11-\mathrm{D}(11-\mathrm{D})}}{\mathrm{H}_\mathrm{c}} \tag{9.1.4}$$

where:

$$\begin{aligned} E_{5(5)} &\simeq D_{5(5)} \simeq \mathrm{SO}(5,5) \\ E_{4(4)} &\simeq A_{4(4)} \simeq \mathrm{SL}(5,\mathbb{R}) \\ E_{3(3)} &\simeq A_{1(1)} \times A_{2(2)} \simeq \mathrm{SL}(2,\mathbb{R}) \otimes \mathrm{SL}(3,\mathbb{R}) \\ E_{2(2)} &\simeq A_{1(1)} \times A_{1(1)} \simeq \mathrm{SL}(2,\mathbb{R}) \otimes \mathrm{SL}(2,\mathbb{R}) \end{aligned} \tag{9.1.5}$$

So exceptional Lie groups that had been regarded for long time as mathematical curiosities were brought to prominence by supergravity and in parallel also by superstring theory.

The fact that all such results were obtained in the *École Normale Supérieure de Paris* demonstrates the far reaching influence of Élie Cartan's tradition.

At the end of the eighties the intrinsic definition of *special Kähler geometry*, free from the use of special coordinates, was independently obtained with two different strategies by Andrew Strominger (see Fig. 9.12) and by Leonardo Castellani, Riccardo D'Auria and Sergio Ferrara (see Fig. 9.11).

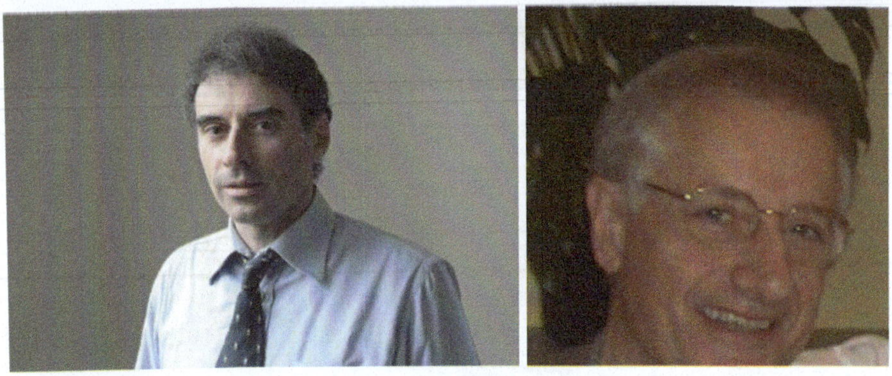

Fig. 9.11 On the left Leonardo Castellani (born 1953 in Freiburg, Switzerland). On the right Riccardo D'Auria (born 1940 in Rome). Leonardo Castellani studied physics at the University of Florence in Italy and obtained his Ph.D. from Stony Brook University in the US, with a thesis written under the supervision of van Nieuwenhuizen. He had post-doctoral positions at Caltech and at CERN, then he became permanent Researcher in the Torino section of the National Institute of Nuclear Research (INFN) and in 1993 he was appointed full-professor of Theoretical Physics at the University of Eastern Piedmont, position that he holds at the present time. He is especially known for his contributions, together with D'Auria and Fré to the rheonomic formulation of supersymmetric theories, for his derivation together with Larry Romans of the list of G/H compactifications of $d = 11$ supergravity and more recently for developments in quantum group theories and, together with P.A. Grassi and R. Catenacci for the extension of Hodge theory to supermanifolds. Riccardo D'Auria studied at the University of Torino and graduated there with a thesis written under the supervision of Tullio Regge. He was for several years Associate Professor at the University of Torino, in 1987 he was appointed full-professor of Theoretical Physics at the University of Padua. Few years later he was offered a full professor chair at the Politecnico of Torino where he concluded his academic career becoming emeritus professor in 2011. D'Auria, together with Fré has been the founder of the rheonomic formulation of supergravity and also with Fré he introduced the notion of super Free Differential Algebras, that were singled out as the algebraic basis of all supergravity theories in dimension higher than four. In particular in 1982, D'Auria and Fré obtained the FDA formulation of $d = 11$ supergravity. D'Auria has given many more contributions to supergravity theory in particular in connection with special geometries, with the classification of black-hole solutions, with duality rotations, with the various formulations of the $d = 6$ theories and with several other aspects of the superworld

While Strominger derived his definition from the properties of Calabi–Yau moduli spaces [160], Castellani, D'Auria and Ferrara [35, 36] (and later D'Auria Ferrara and Fré [55]) derived their own definition from the constraints imposed by supersymmetry on the curvature tensor of the Kählerian manifold. With some labour they also showed the full equivalence of the two definitions.

In the same years, Antoine Van Proeyen and Bernard de Wit (see Fig. 9.12), in some publications together with a younger collaborator, established a full classification of *homogeneous special geometries*, namely of special manifolds that admit a solvable transitive group of isometries [61, 62, 64]. They also explored the relation [62] between *special Kähler geometries* and quaternionic geometries that can be obtained from them by means of a very interesting map, originally discovered by Cecotti [40]

Fig. 9.12 On the left Bernard Quirinus Petrus Joseph de Wit (born 1945 in the Netherlands). On the right Andrew Eben Strominger (born 1955 in the USA). Bernard de Wit studied theoretical physics at Utrecht University, where he got his Ph.D. under the supervision of the Nobel Prize laureate Martinus Veltman in 1973. He held postdoc positions in Stony Brook, Utrecht and Leiden. He became a staff member at the National Institute for Nuclear and High Energy Physics (NIKHEF) in 1978, where he became head of the theory group in 1981. In 1984 he was appointed professor of theoretical physics at Utrecht University where he has stayed for the rest of his career. Bernard de Wit has given important contributions to the development of supergravity theory building, in collaboration mainly with Van Proeyen, the so named *conformal tensor calculus*. Together with Herman Nicolai he constructed the $\mathfrak{so}(8)$-gauged version of $\mathcal{N} = 8$ supergravity that has provided the paradigmatic example for all supergravity gaugings. Andrew Strominger completed his undergraduate studies at Harvard in 1977 before attending the University of California, Berkeley for his Master diploma. He received his Ph.D. from MIT in 1982 under the supervision of Roman Jackiw. Prior to joining Harvard as a professor in 1997, he held a faculty position at the University of California, Santa Barbara. Strominger is especially known for introducing, together with Cumrun Vafa the string theory explanations of the microscopic origin of black hole entropy, originally calculated thermodynamically by Stephen Hawking and Jacob Bekenstein. Strominger, together with Philippe Candelas, Gary Horowitz and Edward Witten was the first proposer of Calabi–Yau threefolds as compactification manifolds for superstrings and supergravities in $d = 10$

and further developed by Ferrara et al. in [56, 77]. So doing they came in touch with the classification of quaternionic manifolds with a transitive solvable group of motion that had been performed several years before by Alekseevsky [1, 45].

The map mentioned above is named the c-map and can be given a modern compact definition exhibited in [88]. Furthermore the c-map has a non euclidian analogue, the $c^{\star}$-map that plays an important role in the discussion of supergravity based black-holes, another instance of geometry that is extensively discussed in the parallel book [90].

Since these constructions involve an extensive use of advanced mathematical techniques and a lot of intermediate steps, just in the same spirit as that adopted in previous chapters, I confine myself to quote without development the two basic definitions of special Kähler geometry and of c-map, referring the reader to [90] for all further mathematical information.

9.1.5 Special Kähler Geometry

In this section we present the definition of Special Kähler Geometry. Let us begin by summarizing some relevant concepts and definitions that are propaedeutical to the main one.

HODGE–KÄHLER MANIFOLDS

Consider a *line bundle* $\mathscr{L} \xrightarrow{\pi} \mathscr{M}$ over a Kähler manifold $\mathscr{M}$. By definition this is a holomorphic vector bundle of rank $r = 1$. For such bundles the only available Chern class is the first:

$$c_1(\mathscr{L}) = \frac{i}{2}\bar{\partial}\left(h^{-1}\,\partial\, h\right) = \frac{i}{2}\bar{\partial}\,\partial \log h \tag{9.1.6}$$

where the 1-component real function $h(z, \bar{z})$ is some hermitian fibre metric on $\mathscr{L}$. Let $\xi(z)$ be a holomorphic section of the line bundle $\mathscr{L}$: noting that under the action of the operator $\bar{\partial}\,\partial$ the term $\log\left(\bar{\xi}(\bar{z})\,\xi(z)\right)$ yields a vanishing contribution, we conclude that the formula in Eq. (9.1.6) for the first Chern class can be re-expressed as follows:

$$c_1(\mathscr{L}) = \frac{i}{2}\bar{\partial}\,\partial \log \parallel \xi(z) \parallel^2 \tag{9.1.7}$$

where $\parallel \xi(z) \parallel^2 = h(z, \bar{z})\,\bar{\xi}(\bar{z})\,\xi(z)$ denotes the norm of the holomorphic section $\xi(z)$.

Equation (9.1.7) is the starting point for the definition of Hodge–Kähler manifolds. A Kähler manifold $\mathscr{M}$ is a Hodge manifold if and only if there exists a line bundle $\mathscr{L} \xrightarrow{\pi} \mathscr{M}$ such that its first Chern class equals the cohomology class of the Kähler two-form K:

$$c_1(\mathscr{L}) = [\,\mathrm{K}\,] \tag{9.1.8}$$

In local terms this means that there is a holomorphic section $\xi(z)$ such that we can write

$$\mathrm{K} = \frac{i}{2}\, g_{ij^\star}\, dz^i \wedge d\bar{z}^{j^\star} = \frac{i}{2}\bar{\partial}\,\partial \log \parallel \xi(z) \parallel^2 \tag{9.1.9}$$

Recalling the local expression of the Kähler metric in terms of the Kähler potential $g_{ij^\star} = \partial_i\,\partial_{j^\star}\mathscr{K}(z, \bar{z})$, it follows from Eq. (9.1.9) that if the manifold $\mathscr{M}$ is a Hodge manifold, then the exponential of the Kähler potential can be interpreted as the metric $h(z, \bar{z}) = \exp\left(\mathscr{K}(z, \bar{z})\right)$ on an appropriate line bundle $\mathscr{L}$.

CONNECTION ON THE LINE BUNDLE

On any complex line bundle $\mathscr{L}$ there is a canonical hermitian connection (see Eq. 8.2.16) defined as:

$$\theta \equiv h^{-1}\,\partial\, h = \tfrac{1}{h}\,\partial_i h\, dz^i \;;\; \bar{\theta} \equiv h^{-1}\,\bar{\partial}\, h = \tfrac{1}{h}\,\partial_{i^\star} h\, d\bar{z}^{i^\star} \tag{9.1.10}$$

For the line-bundle advocated by the Hodge-Kähler structure we have

$$\left[\bar{\partial}\,\theta\right] = c_1(\mathscr{L}) = [\mathrm{K}] \qquad (9.1.11)$$

and since the fibre metric h can be identified with the exponential of the Kähler potential we obtain:

$$\theta = \partial\,\mathscr{K} = \partial_i \mathscr{K}\, dz^i \; ; \; \bar{\theta} = \bar{\partial}\,\mathscr{K} = \partial_{i^\star} \mathscr{K}\, d\bar{z}^{i^\star} \qquad (9.1.12)$$

To define special Kähler geometry, in addition to the afore-mentioned line–bundle $\mathscr{L}$ we need a flat holomorphic vector bundle $\mathscr{S}V \longrightarrow \mathscr{M}$ whose sections play an important role in the construction of the supergravity Lagrangians. For reasons intrinsic to such constructions the rank of the vector bundle $\mathscr{S}V$ must be $2\,n_V$ where n_V is the total number of vector fields in the theory. If we have n-vector multiplets the total number of vectors is $n_V = n + 1$ since, in addition to the vectors of the vector multiplets, we always have the graviphoton sitting in the graviton multiplet. On the other hand the total number of scalars is $2n$. Suitably paired into n-complex fields z^i, these scalars span the n complex dimensions of the base manifold $\mathscr{M}$ to the rank $2n + 2$ bundle $\mathscr{S}V \longrightarrow \mathscr{M}$.

SPECIAL KÄHLER MANIFOLDS

We are now ready to give the first of two equivalent definitions of special Kähler manifolds. We can formulate it as follows. A Hodge Kähler manifold is **Special Kähler (of the local type)** if there exists a completely symmetric holomorphic 3-index section W_{ijk} of $(T^\star\mathscr{M})^3 \otimes \mathscr{L}^2$ (and its antiholomorphic conjugate $W_{i^\star j^\star k^\star}$) such that the following identities are satisfied and the Riemann tensor of the Levi-Civita connection can be written as follows:

$$\begin{aligned}
\partial_{m^\star} W_{ijk} &= 0 \quad \partial_m W_{i^\star j^\star k^\star} = 0 \\
\nabla_{[m} W_{i]jk} &= 0 \quad \nabla_{[m} W_{i^\star]j^\star k^\star} = 0 \\
\mathscr{R}_{i^\star j \ell^\star k} &= g_{\ell^\star j} g_{k i^\star} + g_{\ell^\star k} g_{j i^\star} - e^{2\mathscr{K}}\, W_{i^\star \ell^\star s^\star} W_{t k j} g^{s^\star t}
\end{aligned} \qquad (9.1.13)$$

In the above equations ∇ denotes the covariant derivative with respect to both the Levi-Civita and the U(1) holomorphic connection of the above mentioned Hodge line-bundle. In the case of W_{ijk}, the U(1) weight is $p = 2$.

Out of the W_{ijk} we can construct covariantly holomorphic sections of weight 2 and - 2 by setting:

$$C_{ijk} = W_{ijk}\, e^{\mathscr{K}} \quad ; \quad C_{i^\star j^\star k^\star} = W_{i^\star j^\star k^\star}\, e^{\mathscr{K}} \qquad (9.1.14)$$

The flat bundle mentioned in the previous subsection apparently does not appear in this definition of special geometry. Yet it is there. It is indeed the essential ingredient in the second definition whose equivalence to the first we shall shortly outline.

Let $\mathcal{L} \xrightarrow{\pi} \mathcal{M}$ denote the complex line bundle whose first Chern class equals the cohomology class of the Kähler form K of an n-dimensional Hodge–Kähler manifold $\mathcal{M}$. Let $\mathcal{S}V \longrightarrow \mathcal{M}$ denote a holomorphic flat vector bundle of rank $2n+2$ with structural group $\mathrm{Sp}(2n+2, \mathbb{R})$. Consider tensor bundles of the type $\mathcal{H} = \mathcal{S}V \otimes \mathcal{L}$. A typical holomorphic section of such a bundle will be denoted by Ω and will have the following structure:

$$\Omega = \begin{pmatrix} X^\Lambda \\ F_\Sigma \end{pmatrix} \quad \Lambda, \Sigma = 0, 1, \ldots, n$$

By definition the transition functions between two local trivializations $U_i \subset \mathcal{M}$ and $U_j \subset \mathcal{M}$ of the bundle $\mathcal{H}$ have the following form:

$$\begin{pmatrix} X \\ F \end{pmatrix}_i = e^{f_{ij}} M_{ij} \begin{pmatrix} X \\ F \end{pmatrix}_j$$

where f_{ij} are holomorphic maps $U_i \cap U_j \rightarrow \mathbb{C}$ while M_{ij} is a constant $\mathrm{Sp}(2n+2, \mathbb{R})$ matrix. For a consistent definition of the bundle the transition functions are obviously subject to the cocycle condition on a triple overlap: $e^{f_{ij}+f_{jk}+f_{ki}} = 1$ and $M_{ij} M_{jk} M_{ki} = 1$.

Let $\mathrm{i}\langle\ |\ \rangle$ be the compatible hermitian metric on $\mathcal{H}$

$$\mathrm{i}\langle \Omega \mid \bar{\Omega} \rangle \equiv -\mathrm{i}\Omega^T \begin{pmatrix} 0 & \mathbb{1} \\ -\mathbb{1} & 0 \end{pmatrix} \bar{\Omega}$$

Given these preliminaries we formulate the second definition of special Kähler geometry as follows. We say that a Hodge–Kähler manifold $\mathcal{M}$ is **special Kähler** if there exists a bundle $\mathcal{H}$ of the type described above such that for some section $\Omega \in \Gamma(\mathcal{H}, \mathcal{M})$ the Kähler two form is given by:

$$\mathrm{K} = \frac{\mathrm{i}}{2} \partial\bar{\partial} \log \left(\mathrm{i}\langle \Omega \mid \bar{\Omega} \rangle \right) = \frac{i}{2}\, g_{i,j^*}\, dz^i \wedge d\bar{z}^{j^*} \,.. \tag{9.1.15}$$

From the point of view of local properties, Eq. (9.1.15) implies that we have an expression for the Kähler potential in terms of the holomorphic section Ω:

$$\mathcal{K} = -\log \left(\mathrm{i}\langle \Omega \mid \bar{\Omega} \rangle \right) = -\log \left[\mathrm{i} \left(\bar{X}^\Lambda F_\Lambda - \bar{F}_\Sigma X^\Sigma \right) \right] \tag{9.1.16}$$

The relation between the two definitions of special manifolds is obtained by introducing a non–holomorphic section of the bundle $\mathcal{H}$ according to:

$$V = \begin{pmatrix} L^\Lambda \\ M_\Sigma \end{pmatrix} \equiv e^{\mathcal{K}/2} \Omega = e^{\mathcal{K}/2} \begin{pmatrix} X^\Lambda \\ F_\Sigma \end{pmatrix}$$

so that Eq. (9.1.16) becomes:

$$1 = \mathrm{i}\langle V \mid \bar{V}\rangle = \mathrm{i}\left(\bar{L}^{\Lambda} M_{\Lambda} - \bar{M}_{\Sigma} L^{\Sigma}\right) \tag{9.1.17}$$

Since V is related to a holomorphic section by Eq. (9.1.5) it immediately follows that:

$$\nabla_{i^\star} V = \left(\partial_{i^\star} - \frac{1}{2}\partial_{i^\star}\mathscr{K}\right) V = 0 \tag{9.1.18}$$

On the other hand, from Eq. (9.1.17), defining:

$$U_i = \nabla_i V = \left(\partial_i + \frac{1}{2}\partial_i \mathscr{K}\right) V \equiv \begin{pmatrix} f_i^{\Lambda} \\ h_{\Sigma|i} \end{pmatrix}$$

$$\bar{U}_{i^\star} = \nabla_{i^\star} \bar{V} = \left(\partial_{i^\star} + \frac{1}{2}\partial_{i^\star}\mathscr{K}\right) \bar{V} \equiv \begin{pmatrix} \bar{f}_{i^\star}^{\Lambda} \\ \bar{h}_{\Sigma|i^\star} \end{pmatrix}$$

it follows that:

$$\nabla_i U_j = \mathrm{i} C_{ijk}\, g^{k\ell^\star}\, \bar{U}_{\ell^\star} \tag{9.1.19}$$

where ∇_i denotes the covariant derivative containing both the Levi-Civita connection on the bundle $\mathscr{T}M$ and the canonical connection θ on the line bundle $\mathscr{L}$. In this way we reveal the existence of the completely symmetric tensor C_{ijk} and all the other identities follow. For further details we refer the reader to [90].

THE VECTOR KINETIC MATRIX $\mathscr{N}_{\Lambda\Sigma}$ IN SPECIAL GEOMETRY

In the construction of supergravity actions another essential item is the complex symmetric matrix $\mathscr{N}_{\Lambda\Sigma}$ whose real and imaginary parts are necessary in order to write the kinetic terms of the vector fields. The matrix $\mathscr{N}_{\Lambda\Sigma}$ constitutes an integral part of the Special Geometry set up and we provide its general definition in the following lines. Explicitly $\mathscr{N}_{\Lambda\Sigma}$ which, in relation to its interpretation in the case of Calabi–Yau threefolds, is named the *period matrix*, is defined by means of the following relations:

$$\bar{M}_{\Lambda} = \mathscr{N}_{\Lambda\Sigma}\bar{L}^{\Sigma} \quad ; \quad h_{\Sigma|i} = \mathscr{N}_{\Lambda\Sigma} f_i^{\Sigma} \tag{9.1.20}$$

which can be solved introducing the two $(n+1)\times(n+1)$ vectors

$$f_I^{\Lambda} = \begin{pmatrix} f_i^{\Lambda} \\ \bar{L}^{\Lambda} \end{pmatrix} \quad ; \quad h_{\Lambda|I} = \begin{pmatrix} h_{\Lambda|i} \\ \bar{M}_{\Lambda} \end{pmatrix}$$

and setting:

$$\mathscr{N}_{\Lambda\Sigma} = h_{\Lambda|I} \circ \left(f^{-1}\right)^{I}{}_{\Sigma} \tag{9.1.21}$$

9.1.6 The Quaternionic Kähler Geometry in the Image of the c-Map

A very rich class of geometries is made by those Quaternionic Kähler manifolds that are in the image of the c-map.[2] This latter

$$\text{c-map} \; : \; \mathcal{SK}_n \Longrightarrow \mathcal{QM}_{4n+4} \tag{9.1.22}$$

is a universal construction that starting from an arbitrary Special Kähler manifold $\mathcal{SK}_n$ of complex dimension n, irrespectively whether it is homogeneous or not, leads to a unique Quaternionic Kähler manifold $\mathcal{QM}_{4n+4}$ of real dimension $4n+4$ which contains $\mathcal{SK}_n$ as a submanifold. The precise modern definition of the c-map, originally introduced in [56, 77], is provided below.

Let $\mathcal{SK}_n$ be a special Kähler manifold whose complex coordinates we denote by z^i and whose Kähler metric we denote by $g_{ij^\star}$. Let moreover $\mathcal{N}_{\Lambda\Sigma}(z,\bar{z})$ be the symmetric period matrix defined by Eq. (9.1.21), introduce the following set of $4n+4$ coordinates:

$$\{q^u\} \equiv \underbrace{\{U,a\}}_{\text{2 real}} \bigcup \underbrace{\underbrace{\{z^i\}}_{\text{n complex}}}_{\text{2n real}} \bigcup \underbrace{\mathbf{Z} = \{Z^\Lambda \, , \, Z_\Sigma\}}_{\text{(2n+2) real}} \tag{9.1.23}$$

Let us further introduce the following $(2n+2) \times (2n+2)$ matrix $\mathcal{M}_4^{-1}$:

$$\mathcal{M}_4^{-1} = \left(\begin{array}{c|c} \mathrm{Im}\mathcal{N} + \mathrm{Re}\mathcal{N}\,\mathrm{Im}\mathcal{N}^{-1}\,\mathrm{Re}\mathcal{N} & -\mathrm{Re}\mathcal{N}\,\mathrm{Im}\,\mathcal{N}^{-1} \\ \hline -\mathrm{Im}\mathcal{N}^{-1}\,\mathrm{Re}\mathcal{N} & \mathrm{Im}\mathcal{N}^{-1} \end{array}\right) \tag{9.1.24}$$

which depends only on the coordinate of the Special Kähler manifold. The c-map image of $\mathcal{SK}_n$ is the unique Quaternionic Kähler manifold $\mathcal{QM}_{4n+4}$ whose coordinates are the q^u defined in (9.1.23) and whose metric is given by the following universal formula

$$ds^2_{\mathcal{QM}} = \frac{1}{4}\left(dU^2 + 4g_{ij^\star}\,dz^j\,d\bar{z}^{j^\star} + e^{-2U}\,(da + \mathbf{Z}^T\mathbb{C}d\mathbf{Z})^2 - 2\,e^{-U}\,d\mathbf{Z}^T\,\mathcal{M}_4^{-1}\,d\mathbf{Z}\right) \tag{9.1.25}$$

---

[2]Not all non-compact, homogeneous Quaternionic Kähler manifolds which are relevant to supergravity (which are *normal*, i.e. exhibiting a solvable group of isometries having a free and transitive action on it) are in the image of the c-map, the only exception being the quaternionic projective spaces [40, 64].

The metric (9.1.25) has the following positive definite signature

$$\text{sign}\left[ds^2_{\mathscr{Q}\mathscr{M}}\right] = \left(\underbrace{+,\cdots,+}_{4+4n}\right) \tag{9.1.26}$$

since the matrix $\mathscr{M}_4^{-1}$ is negative definite.

In the case the Special Kähler pre-image is a symmetric space $\mathrm{U}_{\mathscr{S}\mathscr{K}}/\mathrm{H}_{\mathscr{S}\mathscr{K}}$, the manifold $\mathscr{Q}\mathscr{M}$ turns out to be a symmetric space as well, $\mathrm{U}_Q/\mathrm{H}_Q$. In any case it is Quaternionic Kähler since there are general formulas for the three complex structures and for the $\mathfrak{su}(2)$ connection with the necessary properties. We refer to [90] for all details.

9.2 The Geometric Role of Solvable Algebras and the Tits Satake Projection

In the last section of the present chapter we sketch the history of a relatively new development in Geometry which occurred over the last twenty years under the crucial influence of supergravity. This development is based on a mathematical theory pertaining to Lie Algebras which, although it has its roots in mathematical work of the 1960s [20, 154, 163], contributed by two great algebrists, Jacques Tits and Ichiro Satake (see Fig. 9.13), yet fully revealed its profound significance for Geometry and Physics only much later, by the end of the XXth century, and, as anticipated, within the context of supergravity.

The addressed topics is the Tits-Satake projection, a construction which, according to certain rules, from a class of homogeneous manifolds, extracts a single representative of the entire class. What is extremely surprising and inspiring is that such a projection, invented long before the advent of supergravity *special geometries*, has very nice properties with respect to *special structures*. Indeed it maps *special Kähler manifolds* into *special Kähler manifolds*, *quaternionic Kähler* into *quaternionic Kähler* and commutes with the c-map discussed in the previous section. Actually it also commutes with another map, the $c^\star$-map, which is relevant for the construction of supergravity black-hole solutions and is thoroughly illustrated in [90].

A conceptual procedure specially cheered by theoretical physicists is that of *Universality Classes*. Considering complex phenomena like, for instance, phase-transitions one looks for universal features that are the same for entire classes of such phenomena. After grouping the multitude of cases into *universality classes*, one tries to construct a theoretical model of the behavior shared by all elements of each class. A mathematical well founded projection is likely to provide a powerful weapon to this effect. Indeed one might expect that there are *universal features* shared by all cases that have the same projection and that the theoretical model of this shared behavior is encoded in the algebraic structure of the projection image. In

Fig. 9.13 On the left J. Tits (1930 Uccle, Belgium). On the right Ichiro Satake (1927 Yamaguchi Japan - 2014 Tokyo Japan). Jacques Tits was born in Uccle, on the southern outskirts of Brussels. He graduated from the Free University of Brussels in 1950 with a dissertation *Généralisation des groupes projectifs basés sur la notion de transitivité*. From 1956 to 1962 Tits was an assistant at the University of Brussels. He became professor there in 1962 and remained in this role for two years before accepting a professorship at the University of Bonn in 1964. In 1973 he was offered the *Chair of Group Theory* at the *College de France* which he occupied until his retirement in 2000 being naturalised French citizen since 1974. Jacques Tits has given very prominent contributions to the advancement of Group Theory in many directions and he is especially known for the *Theory of Buildings*, which he founded, and for the *Tits alternative*, a theorem on the structure of finitely generated groups. After his retirement from the College the France, a special Vallée-Poussin Chair was created for him at the University of Louvain. Ichiro Satake was born in the Province of Yamaguchi in Japan and graduated from the University of Tokyo in 1959. He held various academic positions in the USA and since 1968 to his retirement in 1983 he was Full Professor of Mathematics at the University of California, Berkeley. He is specially known for his contributions to the theory of algebraic groups and for the Satake diagrams that classify the real forms of a complex Lie algebra

[90] it is shown that this is precisely what happens with the Tits-Satake projection that captures *universal geometrical features* of supergravity models.

Since the interplay between Mathematics and Theoretical Physics has been essential in the development of this new chapter of *homogeneous space geometry* we briefly recall the key facts of this short but intellectually intense history.

(1) In the early 1990s, as we have already reported, B. de Wit, A. Van Proeyen, F. Vanderseypen studied the classification of homogeneous special manifolds admitting a solvable transitive group of isometries [63, 64]. This work extended and completed the results obtained several years before by Alekseevsky in relation with the classification of quaternionic manifolds also admitting a transitive solvable group of isometries [2].
(2) In 1996–1998, L. Andrianopoli, R. D'Auria, S. Ferrara, P. Fré and M. Trigiante explored the general role of *solvable Lie algebras* in supergravity [4, 5, 78, 165],

pointing out that, since all homogenous scalar manifolds of all supergravity models are of the non-compact type, they all admit a description in terms of a solvable group manifold. The solvable representation of the scalar geometry was shown to be particularly valuable in connection with the description of BPS black hole solutions of various supergravity models.

(3) In the years 1999–2005 Thibaut Damour, Marc Henneaux, Hermann Nicolai, Bernard Julia, F. Englert, P. Spindel and other collaborators, elaborating on old ideas of V.A. Belinsky, I.M. Khalatnikov, E.M. Lifshitz [11, 12, 72], introduced the conception of *rigid cosmic billiards* [28, 50–54, 58, 65–67, 71, 114, 123, 124]. According to this conception the various dimensions of a higher dimensional gravitational theory are identified with the generators of the Cartan Subalgebra $\mathscr{H}$ of a supergravity motivated Lie algebra and cosmic evolution takes place in a Weyl chamber of $\mathscr{H}$. Considering the Cartan scalar fields as the coordinate of a fictitious ball, during cosmic evolution such a ball scatters on the walls of the Weyl chambers and this pictorial image of the phenomenon is at the origin of its denomination *cosmic billiard*. In this context the distinction between compact and non-compact directions of the Cartan subalgebra appeared essential and this brought the Tits Satake projection into the game.

(4) In 2003–2005 F. Gargiulo, K. Rulik, P. Fré, A.S. Sorin and M. Trigiante developed the conception of *soft cosmic billiards* [81, 86, 87], corresponding to exact, purely time dependent solutions of supergravity, including not only the Cartan fields but also those associated with roots which dynamically construct the Weyl chamber walls advocated by *rigid cosmic billiards*.

(5) In 2005, Fré, Gargiulo and Rulik constructed explicit examples of soft cosmic billiards in the case of a *non maximally split symmetric manifold*. In that context they analyzed the role of the Tits Satake projection and introduced the new mathematical concept of *Paint Group* [80].

(6) In 2007, P. Fré, F. Gargiulo, J. Rosseel, K. Rulik, M. Trigiante and A. Van Proeyen [79] axiomatized the Tits Satake projection for all homogeneous special geometries. They based their formulation of the projection on the intrinsic definition of the *Paint Group* as the group of outer automorphisms of the solvable transitive group of motion of the homogeneous manifold. This is the theory that is thoroughly explained in an appropriate chapter of [90]. Up to the knowledge of this author, this theory was never previously developed in the mathematical literature.

(7) In the years 2009–2011 the integration algorithm utilized in the framework of soft cosmic billiards was extended by P. Fré, A.S. Sorin and M. Trigiante to the case of spherical symmetric black-holes for manifolds in the image of the $c^\star$-map [42, 82, 83].

(8) In 2011, P. Fré, A.S. Sorin and M. Trigiante demonstrated that the classification of nilpotent orbits for a non maximally split Lie algebra depends only on its Tits-Satake projection and it is a property of the Tits-Satake universality class (see Chap. 10).

Through the above sketched historical course, which unfolded in about a decade, the theory of the Tits-Satake projection has acquired a quite solid and ramified profile, intertwined with the c and $c^{\star}$ maps that opens new viewpoints and provides new classification tools in the geometry of homogeneous manifolds and symmetric spaces. Although the theory is distinctively algebraic and geometric, yet it is poorly known in the mathematical community due to its supergravity driven origins. Hopefully its full-fledged exposition included in [90] will improve its status in the mathematical club.

Chapter 10
Black Holes: The Physics of Geometry

Deep into that darkness peering, long I stood there, wondering, fearing, doubting, dreaming dreams no mortal ever dared to dream before

Edgar Allan Poe

10.1 An Exciting Historical Moment

When on September 14th 2015 the gravitational wave signal emitted 1.5 billion year ago by two coalescing black stars was detected at LIGO I and LIGO II, we not only obtained a new spectacular confirmation of General Relativity but we actually saw the dynamical process of formation of the most intriguing objects populating the Universe, namely black holes (Fig. 10.1).

Black Holes are on one side physical objects capable of interacting with the emission of enormous quantities of energy, on the other side they are just pure geometries. Indeed a classical black-hole is nothing else but a solution of Einstein equations which are just geometrical statements on the curvature tensor.

10.2 A Short History of Black Holes

It seems that the first to conceive the idea of what we call nowadays a black-hole was the English Natural Philosopher and Geologist John Michell (1724–1793). As early as 1783, Michell, member of the Royal Society, had invented a device to measure Newton's gravitational constant, namely the torsion balance that he built independently from its co-inventor Charles Augustin de Coulomb. He did not live long enough to put into use his apparatus which was inherited by Cavendish. In 1784 in a letter addressed precisely to Cavendish, John Michell advanced the hypothesis that there could exist heavenly bodies so massive that even light could not escape from their gravitational attraction. This letter surfaced back to the attention of contempo-

P. G. Fré, *A Conceptual History of Space and Symmetry*,
https://doi.org/10.1007/978-3-319-98023-2_10

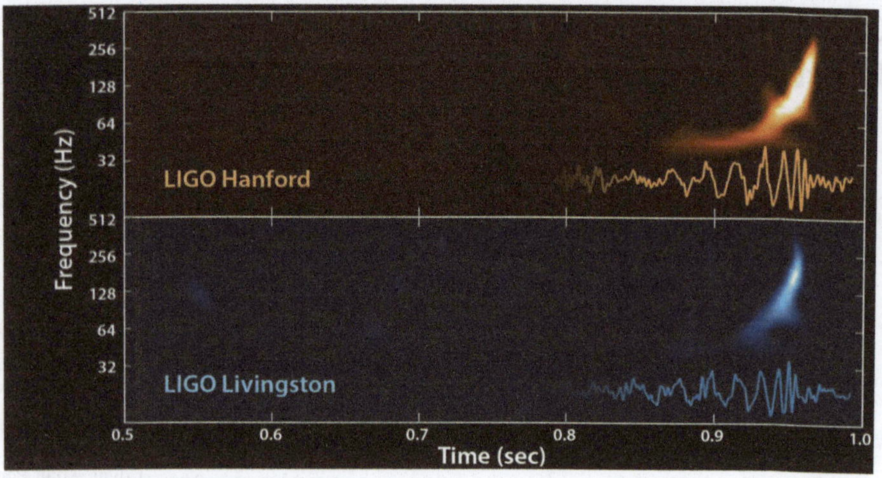

Fig. 10.1 The gravitational wave signal emitted in the coalescence of two black holes which occurred 1.5 billion of years ago was simultaneously detected september 14th 2015 by the two interferometers LIGO I and LIGO II

rary scientists only in the later seventies of the XXth century. Before that finding, credited to be the first inventor of black-holes was Pierre Simon Laplace.

LAPLACE

Pierre Simon Laplace (see Fig. 10.2) was born in Beaumont en Auge in Normandy in the family of a poor farmer. He could study thanks to the generous help of some neighbors. Later with a recommendation letter of d'Alembert he entered the military school of Paris where he became a teacher of mathematics. There he started his monumental and original research activity in Mathematics and Astronomy that made him one of the most prominent scientists of his time and qualified him to the rank of founder of modern differential calculus, his work being a pillar of XIXth century Mathematical Physics. A large part of his work on Astronomy was still done under the Ancien Regime and dates back to the period 1771–1787. He proved the stability of the Solar System and developed all the mathematical tools for the systematic calculus of orbits in Newtonian Physics. His results were summarized in the two fundamental books *Mecanique Cèleste* and *Exposition du Système du Monde*. Besides introducing the first idea of what we call nowadays a black-hole, Laplace was also the first to advance the hypothesis that the Solar System had formed through the cooling of a globular–shaped, rotating, cluster of very hot gas (a nebula). In later years of his career Laplace gave fundamental and framing contributions to the mathematical theory of probability. His name is attached to numberless corners of differential analysis and function theory. He received many honors both in France and abroad. He was member of all most distinguished Academies of Europe. He also attempted the political career serving as Minister of Interiors in one of the first Napoleonic Cabinets, yet he was soon dismissed by the First Consul as a person not qualified for that administrative

Fig. 10.2 Pierre Simon laplace (1749–1827)

job notwithstanding Napoleon's recognition that he was a great scientist. Politically Laplace was rather cynic and ready to change his opinions and allegiance in order to follow the blowing wind. Count of the First French Empire, after the fall of Napoleon he came on good terms with the Bourbon Restoration and was compensated by the King with the title of marquis.

In the 1796 edition of his monumental book *Exposition du Système du Monde* he presented exactly the same argument put forward in Michell's letter, developing it with his usual mathematical rigor. All historical data support the evidence that Michell and Laplace came to the same hypothesis independently. Indeed the idea was quite mature for the physics of that time, once the concept of escape velocity v_e had been fully understood.

THE ESCAPE VELOCITY

Consider a spherical celestial body of mass M and radius R and let us pose the question what is the minimum initial vertical velocity that a point–like object located on its surface, for instance a rocket, should have in order to be able to escape to infinite distance from the center of gravitational attraction. Energy conservation provides the immediate answer to such a problem. At the initial moment $t = t_0$ the energy of the missile is:

$$E = \frac{1}{2} m_m v_e^2 - \frac{G M m_m}{R} \qquad (10.2.1)$$

where G is Newton's constant. At a very late time, when the missile has reached $R = \infty$ with a final vanishing velocity its energy is just $0 + 0 = 0$. Hence E

vanished also at the beginning, which yields:

$$v_e = \sqrt{2\frac{G M}{R}} \tag{10.2.2}$$

If we assume that light travels at a finite velocity c, then there could exist heavenly bodies so dense that:

$$\sqrt{2\frac{G M}{R}} > c \tag{10.2.3}$$

In that case not even light could escape from the gravitational field of that body and no-one on the surface of the latter could send any luminous signal that distant observers could perceive. In other words by no means distant observers could see the surface of that super-massive object and even less what might be in its interior.

Obviously neither Michell nor Laplace had a clear perception that the speed of light c is always the same in every reference frame, since Special Relativity had to wait its own discovery for another century. Yet Laplace's argument was the following: let us assume that the velocity of light is some constant number a on the surface of the considered celestial body. Then he proceeded to an estimate of the speed of light on the surface of the Sun, which he could do using the annual light aberration in the Earth-Sun system. The implicit, although unjustified, assumption was that light velocity is unaffected, or weakly affected, by gravity. Analyzing such an assumption in full-depth it becomes clear that it was an anticipation of Relativity in disguise.

Actually condition (10.2.3) has an exact intrinsic meaning in General Relativity. Squaring this equation we can rewrite it as follows:

$$R > r_S \equiv 2\frac{GM}{c^2} \equiv 2\,m \tag{10.2.4}$$

where r_S is the Schwarzschild radius of a body of mass M, namely the unique parameter which appears in the Schwarzschild solution of Einstein Equations.

So massive bodies are visible and behave qualitatively according to human common sense as long as their dimensions are much larger then their Schwarzschild radius. Due to the smallness of Newton's constant and to the hugeness of the speed of light, this latter is typically extremely small. Just of the order of a kilometer for a star, and about 10^{-23} cm for a human body. Nevertheless, sooner or later all stars collapse and regions of space–time with outrageously large energy–densities do indeed form, whose typical linear size becomes comparable to r_S. The question of what happens if it is smaller than r_S is not empty, on the contrary it is a fundamental one, related with the appropriate interpretation of what lies behind the apparent singularity of the Schwarzschild metric at $r = r_S$.

The Event Horizon

Any singularity is just the signal of some kind of criticality. At the singular point a certain description of physical reality breaks down and it must be replaced by a different one: for instance there is a phase–transition and the degrees of freedom that capture most of the energy in an ordered phase become negligible with respect to other degrees of freedom that are dominating in a disordered phase. What is the criticality signaled by the singularity $r = r_S$ of the Schwarzschild metric? Is it a special feature of this particular solution of Einstein Equations or it is just an instance of a more general phenomenon intrinsic to the laws of gravity as stated by General Relativity? The answer to the first question is encoded in the wording *event horizon*. The answer to the second question is that event horizons are a generic feature of static solutions of Einstein equations.

An event-horizon $\mathfrak{H}$ is a hypersurface in a pseudo-Riemannian manifold $(\mathcal{M}, g)$ which separates two sub-manifolds, one $\mathfrak{E} \subset \mathcal{M}$, named the exterior, can communicate with infinity by sending signals to distant observers, the other BH $\subset \mathcal{M}$, named the black-hole, is causally disconnected from infinity, since no signal produced in BH can reach the outside region $\mathfrak{E}$. The black-hole is the region deemed by Michell and Laplace where the escape velocity is larger than the speed of light.

In order to give a precise mathematical sense to the above explanation of event-horizons a lot of things have to be defined and interpreted. First of all what is infinity and is it unique? Secondly which kind of hypersurface is an event-horizon? Thirdly can we eliminate the horizon singularity by means of a suitable analytic extension of the apparently singular manifold? Finally, how do we define causal relations in a curved Lorentzian space-time?

The answers were found in the course of the XXth century and constitute the principal milestones in the history of black-holes.

SCHWARZSCHILD

Karl Schwarzschield (see Fig. 10.3) was born in Frankurt am Mein in a well to do Jewish family. Very young he determined the orbits of binary stars. Since 1900 he was Director of the Astronomical Observatory of Göttingen, the hottest point of the world

Fig. 10.3 Karl Schwarzschield (1873–1916)

for Physics and Mathematics at that time and in subsequent years. Famous scientist and member of the Prussian Academy of Sciences in 1914 he enrolled as a volunteer in the German Army and went to war first on the western and then on the eastern front against Russia. At the front in 1916 he wrote two papers. One containing the quantization rules discovered by him independently from Sommerfeld. The second containing the Schwarzschild solution of General Relativity. At the front he had learnt General Relativity just two months before, by reading Einstein's paper of November 1915. He sent his paper to Einstein who wrote in reply : *...I did not expect that one could formulate the exact solution of the problem in such a simple way*. Few months later Schwarschild died from an infection taken at the front. Although Schwarzschild metric [156] was discovered in 1916 (see Fig. 10.3), less than six months after the publication of General Relativity, its analytic extension, that opened the way to a robust mathematical theory of black-holes, was found only forty-five years later, six after Einstein's death.

KRUSKAL

Student of the University of Chicago, Martin Kruskal (see Fig. 10.4) obtained his Ph.D. from New York University and was for many years professor at Princeton University. In 1989 he joined Rutgers University where he remained the rest of his life. Mathematician and Physicist, Martin Kruskal gave very relevant contributions in theoretical plasma physics and in several areas of non-linear science. He discovered exact integrability of some non-linear differential equations and is reported to be the inventor of the concept of *solitons*. In 1960, Kruskal found a one-to-many coordinate transformation [132] that allowed him to represent Schwarzschild space-time as a portion of a larger space-time where the locus $r = r_S$ is non-singular, rather it is a well-defined light-like hypersurface constituting precisely the event-horizon. A similar coordinate change was independently proposed the same year also by the Australian-Hungarian mathematician Georges Szekeres (see Fig. 10.4).

SZEKERES

Born in Budapest, Szekeres graduated from Budapest University in Chemistry. As a Jewish he had to escape from Nazi persecution and he fled with his family to China where he remained under Japanese occupation till the beginning of the Communist Revolution. In 1948 he was offered a position at the University of Adelaide in Australia. In this country he remained the rest of his life. Notwithstanding his degree in chemistry Szekeres was a Mathematician and he gave relevant contributions in various of its branches. He is among the founders of combinatorial geometry.

THE GRAVITATIONAL COLLAPSE

These mathematical results provided a solid framework for the description of the final state in the gravitational collapse of those stars that are too massive to stop at the stage of white–dwarfs or neutron–stars. Robert Openheimer and H. Snyder in their 1939 paper, wrote:*When all thermonuclear sources of energy are exhausted, a sufficiently heavy star will collapse. Unless something can somehow reduce the star's mass to the order of that of the sun, this contraction will continue indefinitely...past white*

Fig. 10.4 Martin David Kruskal (1925–2006) on the left and George Szekeres (1911–2005) on the right

dwarfs, past neutron stars, to an object cut off from communication with the rest of the universe. Such an object, could be identified with the interior of the event horizon in the newly found Kruskal space-time. Yet, since the Kruskal–Schwarzschild metric is spherical symmetric such identification made sense only in the case the parent star had vanishing angular momentum, namely was not rotating at all. This is quite rare since most stars rotate.

KERR AND PETROV

In 1963 the New Zealand physicist Roy Kerr, working at the University of Texas (see Fig. 10.5), found the long sought for generalization of the Schwarzschild metric that could describe the end-point equilibrium state in the gravitational collapse of a rotating star. Kerr metric [122] introduced the third missing parameter characterizing a black-hole, namely the angular momentum J. The first is the mass M, known since Schwarzschild's pioneering work, the second, the charge Q (electric, magnetic or both) had been introduced already in the first two years of life of General Relativity. Indeed the Reissner–Nordström metric, which solves coupled Einstein–Maxwell equations for a charged spherical body, dates back to 1916–1918 (see Fig. 10.6).

The long time delay separating the early finding of the spherical symmetric solutions and the construction of the axial symmetric Kerr metric is explained by the high degree of algebraic complexity one immediately encounters when spherical symmetry is abandoned. Kerr's achievement would have been impossible without the previous monumental work of the young Russian theoretician A.Z. Petrov. Educated in the same University of Kazan where, at the beginning of the XIXth century Lobachevsky had first invented non-euclidian geometry, in his 1954 doctoral dissertation, Petrov conceived a classification of Lorentzian metrics based on the

Fig. 10.5 Born in 1934 in New Zealand, where later he was professor until his retirement, Kerr worked for several years in the U.S.A. and in particular in Texas. There, in 1963, he found what was sought for a long time by many people, i.e. the generalization of Schwarzschild solution with cylindrical symmetry and angular momentum: The rotating black hole!

properties of the corresponding Weyl tensor. This leads to the concept of principal null-directions. According to Petrov there are exactly six types of Lorentzian metrics and, in current nomenclature, Schwarzschild and Reissner Nodström metrics are of Petrov type D. This means that they have two double principal null directions. Kerr made the hypothesis that the metric of a rotating black-hole should also be of Petrov type D and searching in that class he found it.

From 1964 to 1974

The decade from 1964 to 1974 witnessed a vigorous development of the mathematical theory of black-holes. Brandon Carter solved the geodesic equations for the Kerr-metric, discovering a fourth hidden first integral which reduces these differential equations to quadratures. In the same time through the work of Stephen Hawking, George Ellis, Roger Penrose and several others, general analytic methods were established to discuss, represent and classify the causal structure of space-times. Slowly a new picture emerged. Similarly to soliton solutions of other non-linear differential equations, black-holes have the characteristic features of a new kind of particles, mass, charge and angular momentum being their unique and defining attributes. Indeed it was proved that, irrespectively from all the details of its initial structure, a gravitational collapsing body sets down to a final equilibrium state parameterized only by (M, J, Q) and described by the so called Kerr–Newman metric, the generalization of the Kerr solution which includes also the Reissner Nordström charges.

Fig. 10.6 Hans Jacob Reissner (1874–1967) was a German aeronautical engineer with a passion for mathematical physics. He was the first to solve Einstein's field equations with a charged electric source and he did that already in 1916. Emigrated to the United States in 1938 he taught at the Illinois Institute of Technology and later at the Polytechnic Institute of Brooklyn. Reissner's solution was retrieved and refined in 1918 by Gunnar Nordström (1881–1923) a Finnish theoretical physicist who was the first to propose an extension of space-time to higher dimensions. Independently from Kaluza and Klein and as early as 1914 he introduced a fifth dimension in order to construct a unified theory of gravitation and electromagnetism. His theory was, at the time, a competitor of Einstein's theory. Working at the University of Leiden in the Netherlands with Paul Ehrenfest, in 1918 he solved Einstein field equations for a spherically symmetric charged body thus extending the Hans Reissner's results for a point charge

The Information Loss

This introduced the theoretical puzzle of *information loss*. Through gravitational evolution, a supposedly coherent quantum state, containing a detailed fine structure, can evolve to a new state where all such information is unaccessible, being hidden behind the event horizon. The information loss paradox became even more severe when Hawking on one side demonstrated that black-holes can evaporate through a quantum generated thermic radiation and on the other side, in collaboration with Bekenstein, he established, that the horizon has the same properties of an entropy and obeys a theorem similar to the second principle of thermodynamics.

Hence from the theoretical view-point black-holes appear to be much more profound structures than just a particular type of classical solutions of Einstein's field equations. Indeed they provide a challenging clue into the mysterious realm of quantum gravity where causality is put to severe tests and needs to be profoundly revised. For this reason the study of black-holes and of their higher dimensional analogues within the framework of such candidates to a Unified Quantum Theory of all Interactions as Superstring Theory is currently a very active stream of research.

Laplace's Demon

Ironically such a Revolution in Human Thought about the Laws of Causality, whose settlement is not yet firmly acquired, was initiated two century ago by the observations of Laplace, whose unshakable faith in determinism is well described by the following quotation from the *Essai philosophique sur les probabilités*. In that book he wrote: *We may regard the present state of the universe as the effect of its past and the cause of its future. An intellect which at a certain moment would know all forces that set nature in motion, and all positions of all items of which nature is composed, if this intellect were also vast enough to submit these data to analysis, it would embrace in a single formula the movements of the greatest bodies of the universe and those of the tiniest atom; for such an intellect nothing would be uncertain and the future just like the past would be present before its eyes.* The vast intellect advocated by Pierre Simon and sometimes named the Laplace demon might find some problems in reconstructing the past structure of a star that had collapsed into a black hole even if that intellect had knowledge of all the conditions of the Universe at that very instant of time.

From the astronomical view-point the existence of black-holes of stellar mass has been established through many overwhelming evidences, the best being provided by binary systems where a visible normal star orbits around an invisible companion which drags matter from its mate. Giant black-holes of millions of stellar masses have also been indirectly revealed in the core of active galactic nuclei and also at the center of our Milky Way a black hole is accredited.

Gravitational Wave Evidence

As we recalled at the beginning, in the last two years the detection of gravitational events emitted in the coalescence of two compact stars, has provided new dramatic evidence on the existence of black holes and on their formation. A spectacular almost theatrical event which might be used as a convenient temporary conclusion of this short history of black-holes, has been in the current 2017 year the direct detection, both gravitational and electromagnetic, of the formation of a new black hole in the coalescence of two neutron stars. Clearly this is just the conclusion of a chapter in the story. A new exciting age has just started and we are going to learn much more about these intriguing manifestations of space-time geometry that hide many of the most profound secrets of quantum physics.

10.3 Black Holes in Supergravity and Superstrings

A new season of research in Black Hole theory started in the middle nineties of the XXth century with the contributions of Sergio Ferrara, Renata Kallosh, Andrew Strominger and Cumrun Vafa, that are described in the following short summary:

1. In 1995 R. Kallosh, S. Ferrara and A. Strominger considered black holes in the context of $\mathcal{N} = 2$ supergravity and introduced the notion of attractors [75, 76].
2. In 1996 S. Ferrara (see Fig. 9.7) and R. Kallosh (see Fig. 10.7) formalized the attractor mechanism for supergravity black holes [75, 76].
3. In 1996 A. Strominger (see Fig. 9.12) and C. Vafa (see Fig. 10.8) showed that an extremal BPS black hole in d = 5 has a horizon area that exactly counts the number of string microstates it corresponds to [162].[1]
4. In the years 1997–2000 the horizon area of BPS supergravity black holes was interpreted in terms of a symplectic invariant constructed with the black hole electromagnetic charges (for a review containing also an extensive bibliography see [57]).
5. In the years 2006–2009 new insights extended the attractor mechanism to non BPS black-holes [6, 13, 14, 21–25, 41, 103, 104, 120, 121, 166].
6. Since 2010 new exact integration techniques for Sugra Black Holes were found by A. Sorin, P. Fré, M. Trigiante and their younger collaborators [42, 81–87].

10.4 The Black Holes Mathematically Discussed in the Twin Book *Advances in Geometry and Lie Algebras from Supergravity*

The intriguing relation between Geometry and Physics arises at several levels, the most profound and challenging being provided by the identification of the *horizon area* with the *statistical entropy* of the mysterious dynamical system which is encoded in a *classical black hole solution*.

Neither here, nor in the twin volume [90] we touch upon the physics of black holes[2] and on the exciting question of their interpretation in terms of *microstates*, yet we can not avoid discussing their several nested geometrical aspects, glimpses of which we provide next.

In the context of supergravity a black hole solution of Einstein equations comes equipped with other associated geometrical data, namely those encoded in a set of electromagnetic fields that are connections on suitable bundles and those encoded in scalar fields that describe a map from 4-dimensional space-time $\mathcal{M}_4$ to some internal manifold whose geometry is dictated by supersymmmetry. Indeed the general form of a bosonic supergravity lagrangian in $D = 4$ is the following one:

$$\mathcal{L}^{(4)} = \sqrt{|\det g|}\left[\frac{R[g]}{2} - \frac{1}{4}\partial_\mu\phi^a\partial^\mu\phi^b h_{ab}(\phi) + \mathrm{Im}\mathcal{N}_{\Lambda\Sigma}\, F^\Lambda_{\mu\nu} F^{\Sigma|\mu\nu}\right] + \frac{1}{2}\mathrm{Re}\mathcal{N}_{\Lambda\Sigma}\, F^\Lambda_{\mu\nu} F^\Sigma_{\rho\sigma}\varepsilon^{\mu\nu\rho\sigma}\,, \tag{10.4.1}$$

[1] There followed a vast literature some items of which are are quoted in [16, 49, 100, 101, 133, 141, 161]

[2] For a short but comprehensive exposition of basic black-hole physics we refer the reader to Chaps. 2 and 3 in the second volume of [89] by the present author.

Fig. 10.7 Renata Kallosh (on the left) born in Moscow in 1943 completed her Bachelor's from Moscow State University in 1966 and obtained her Ph.D. from Lebedev Physical Institute, Moscow in 1968. She then held a position, as professor, at the same institute, before moving to CERN for a year in 1989. Kallosh joined Stanford University in 1990 and continues to work there. She is married with the famous cosmologist Andrei Linde. Renata Kallosh is renowned for her pioneering contributions with Ferrara to the attractor mechanism in supergravity black holes, for her studies in supergravity cosmology and for her early work with A. Van Proeyen on the AdS/CFT correspondence. Indeed Kallosh and Van Proeyen were the first to propose the interpretation of the anti de Sitter group as the conformal group on a brane boundary. Anna Ceresole (on the right), born 1961 in Torino, graduated from Torino University in 1984 with a thesis on Kaluza Klein supergravity written under the supervision of Hermann Nicolai and the author of this book. In 1989 she obtained her Ph.D. from Stony Brook University under the supervision of Peter van Nieuwenhuizen. Post doctoral fellow at Caltech for two years she was Assistant Professor at the Politecnico di Torino for several years. Then she became senior research scientist of INFN and joined the Torino University String Group. Anna Ceresole has given many important contributions to the development of supergravity, in particular in relation with special Kähler Geometry and black hole charges, duality transformations, gaugings and inflaton potentials. She has worked both with younger students and post-doc and, in different combinations, with all the main actors in the development of supergravity theory

The fields included in the theory are the metric $g_{\mu\nu}(x)$, n_v abelian gauge fields A^Λ_ν, whose field strengths (or curvatures) we have denoted by $F^\Lambda_{\mu\nu} \equiv (\partial_\mu A^\Lambda_\nu - \partial_\nu A^\Lambda_\mu)/2$ and n_s scalar fields ϕ^a that parameterize a scalar manifold $\mathscr{M}^{D=4}_{scalar}$ that, for supersymmetry $\mathscr{N} > 2$, is necessarily a coset manifold:

$$\mathscr{M}^{D=4}_{scalar} = \frac{\mathrm{U}_{D=4}}{\mathrm{H}_c} \tag{10.4.2}$$

$\mathrm{U}_{D=4}$ being a non-compact real form of a semi-simple Lie group, essentially fixed by supersymmetry and H_c its maximal compact subgroup. For $\mathscr{N} = 2$ Eq. (10.4.2) is not obligatory yet it is possible: a well determined class of symmetric homogeneous

Fig. 10.8 Cumrun Vafa (on the left) was born in Tehran, Iran in 1960. He graduated from Alborz High School and went to the US in 1977. He got his undergraduate degree from the Massachusetts Institute of Technology with a double major in physics and mathematics. He received his Ph.D. from Princeton University in 1985 under the supervision of Edward Witten. He then became a junior fellow at Harvard, where he later got a junior faculty position. In 1989 he was offered a senior faculty position, and he has been there ever since. Currently, he is the Donner Professor of science at Harvard University. Vafa's most relevant achievement is, together with Strominger, the first example of interpretation of the Bekenstein Hawking black hole entropy in terms of superstring microstates. He has also given pioneering contributions to topological strings, F-theory and to the the general vision named *geometric engineering of quantum field theories*, which is a programme aimed at decoding quantum field theories in terms of algebraic geometry constructions. Dieter Luest (on the right) born in Chicago in 1956, graduated from the Ludvig Maximillian University in Muenchen in 1985. He was postdoctoral fellow in Caltech, Pasadena, in the Max Planck Institute in Muenchen and at CERN in Geneva. From 1993 to 2004 he was full professor of quantum field theory at the von Humboldt University in Berlin. Since 2004 he made return to Muenchen where he is both full professor at the Ludwig Maximilan University and research director at the Max Planck Institute. Dieter Luest has given very important contributions in a large variety of topics connected with string theory and supergravity, in particular in relation with black hole solutions, D-brane engineering, Calabi Yau compactifications, double geometries, flux compactifications and string cosmology

manifolds that are special Kähler manifolds fall into the set up of the present general discussion.

Hence we see that we are dealing with geometries at three levels:

1. We deal with the geometry of space-time $\mathscr{M}_4^{st}$, encoded in its metric $g_{\mu\nu}$ which is dynamical, in the sense that we have to determine it through the solution of field equations, many possibilities being available, among which we have black-hole geometries with event horizons and all the rest.
2. We deal with connections on a fibre bundle $P\left(\mathscr{G}, \mathscr{M}_4^{st}\right)$, whose base manifold is the dynamically determined space-time $\mathscr{M}_4^{st}$ and whose structural group is an abelian group $\mathscr{G}$ of dimension equal to the number n_v of involved gauge fields. These connections are also dynamical in the sense that they have to be determined as solutions of the coupled field equations.

3. We deal with a fixed Riemannian geometry encoded in the target manifold (10.4.2) of which the scalar fields ϕ^a are local coordinates. Any solution of the coupled field equations defines a map

$$\phi \ : \ \mathscr{M}_4^{st} \to \mathscr{M}^{D=4}_{scalar} \tag{10.4.3}$$

of space-time into the scalar manifold.

There is still encoded into the lagrangian (10.4.1) another geometrical datum of utmost relevance. Let us describe it. Considering the n_v vector fields A^Λ_μ let

$$\mathscr{F}^{\pm|\Lambda}_{\mu\nu} \equiv \tfrac{1}{2}\left[F^\Lambda_{\mu\nu} \mp \mathrm{i}\,\frac{\sqrt{|\det g|}}{2}\varepsilon_{\mu\nu\rho\sigma}\,F^{\rho\sigma}\right] \tag{10.4.4}$$

denote the self-dual (respectively antiself-dual) parts of the field-strengths. As displayed in Eq. (10.4.1) they are non minimally coupled to the scalars via the symmetric complex matrix

$$\mathscr{N}_{\Lambda\Sigma}(\phi) = i\,\mathrm{Im}\mathscr{N}_{\Lambda\Sigma} + \mathrm{Re}\mathscr{N}_{\Lambda\Sigma} \tag{10.4.5}$$

The key point is that the isometry group $\mathrm{U}_{D=4}$ of the scalar manifold (10.4.2) is promoted to a symmetry of the entire lagrangian through the projective transformations of $\mathscr{N}_{\Lambda\Sigma}$ under the group action.[3]

Indeed the field strengths $\mathscr{F}^{\pm|\Lambda}_{\mu\nu}$ plus their magnetic duals:

$$G_{\Lambda|\mu\nu} \equiv \tfrac{1}{2}\,\varepsilon_{\mu\nu}{}^{\rho\sigma}\,\frac{\delta\mathscr{L}^{(4)}}{\delta F^\Lambda_{\rho\sigma}} \tag{10.4.6}$$

fill up a $2\,n_v$–dimensional symplectic representation of $\mathbb{U}_{D=4}$ which we call by the name of $\mathbf{W}$.

We rephrase the above statements by asserting that there is always a symplectic embedding of the duality group $\mathrm{U}_{D=4}$,

$$\mathrm{U}_{D=4} \mapsto \mathrm{Sp}(2\mathrm{n_v},\mathbb{R}) \quad ; \quad n_v \equiv \#\text{ of vector fields} \tag{10.4.7}$$

so that for each element $\xi \in \mathrm{U}_{D=4}$ we have its representation by means of a suitable real symplectic matrix:

$$\xi \mapsto \Lambda_\xi \equiv \begin{pmatrix} A_\xi & B_\xi \\ C_\xi & D_\xi \end{pmatrix} \tag{10.4.8}$$

satisfying the defining relation:

$$\Lambda^T_\xi \begin{pmatrix} \mathbf{0}_{n\times n} & \mathbf{1}_{n\times n} \\ -\mathbf{1}_{n\times n} & \mathbf{0}_{n\times n} \end{pmatrix} \Lambda_\xi = \begin{pmatrix} \mathbf{0}_{n\times n} & \mathbf{1}_{n\times n} \\ -\mathbf{1}_{n\times n} & \mathbf{0}_{n\times n} \end{pmatrix} \tag{10.4.9}$$

[3] See Sect. 9.1.5 for the interpretation of $\mathscr{N}_{\Lambda\Sigma}$ in the context of Special Kähler Geometry.

Under an element of the duality group the field strengths transform as follows:

$$\begin{pmatrix} \mathscr{F}^+ \\ \mathscr{G}^+ \end{pmatrix}' = \begin{pmatrix} A_\xi & B_\xi \\ C_\xi & D_\xi \end{pmatrix} \begin{pmatrix} \mathscr{F}^+ \\ \mathscr{G}^+ \end{pmatrix} \quad ; \quad \begin{pmatrix} \mathscr{F}^- \\ \mathscr{G}^- \end{pmatrix}' = \begin{pmatrix} A_\xi & B_\xi \\ C_\xi & D_\xi \end{pmatrix} \begin{pmatrix} \mathscr{F}^- \\ \mathscr{G}^- \end{pmatrix} \tag{10.4.10}$$

where, by their own definitions we get:

$$\mathscr{G}^+ = \mathscr{N}\,\mathscr{F}^+ \quad ; \quad \mathscr{G}^- = \overline{\mathscr{N}}\,\mathscr{F}^- \tag{10.4.11}$$

and the complex symmetric matrix $\mathscr{N}$ should transform as follows:

$$\mathscr{N}' = \left(C_\xi + D_\xi\,\mathscr{N}\right)\left(A_\xi + B_\xi\,\mathscr{N}\right)^{-1} \tag{10.4.12}$$

Choose a parametrization of the coset $\mathbb{L}(\phi) \in \mathrm{U}_{\mathrm{D}=4}$, which assigns a definite group element to every coset point identified by the scalar fields. Through the symplectic embedding (10.4.8) this produces a definite ϕ-dependent symplectic matrix

$$\begin{pmatrix} A(\phi) & B(\phi) \\ C(\phi) & D(\phi) \end{pmatrix} \tag{10.4.13}$$

in the $\mathbf{W}$-representation of $\mathrm{U}_{\mathrm{D}=4}$. In terms of its blocks the kinetic matrix $\mathscr{N}(\phi)$ is explicitly given by a formula that was found at the beginning of the 1980.s by Gaillard–Zumino [95]:

$$\mathscr{N}(\phi) = \left[C(\phi) - i\,D(\phi)\right]\left[A(\phi) - i\,B(\phi)\right]^{-1}, \tag{10.4.14}$$

As we have already remarked, the matrix $\mathscr{N}$ is the same which appears in the definition of special Kähler geometry and it transforms according to Eq. (10.4.12).

Summarizing the geometrical structure of the bosonic supergravity lagrangian is essentially encoded in two data. The duality-isometry group $\mathrm{U}_{\mathrm{D}=4}$ and its symplectic representation $\mathbf{W}$ that corresponds to the embedding (10.4.7).

A brilliant discovery that occurred in the first two decades of the XXIst century can be dubbed the $D = 3$ approach to supergravity black-holes. Mainly originating from the contributions included in the following papers [15, 24, 26, 96, 105, 106, 145], it consists of the following.

The radial dependence of all the relevant functions parameterizing the supergravity solution can be viewed as the field equations of another one-dimensional σ-model where the evolution parameter τ is actually a monotonic function of the radial variable r and where the target manifold is a *pseudo-quaternionic manifold* $\mathscr{Q}^\star_{(4n+4)}$ related to the quaternionic manifold $\mathscr{Q}_{(4n+4)}$ in the following way.[4] The coordinates of $\mathscr{Q}^\star_{(4n+4)}$ are the same as those of $\mathscr{Q}_{(4n+4)}$, while the two metrics differ only by a change of sign. Indeed we have

[4]Compare with Sect. 9.1.6.

$$ds^2_{\mathcal{Q}} = \frac{1}{4}\left[dU^2 + 2\,g_{ij^\star}\,dz^i\,d\bar{z}^{j^\star} + \mathrm{e}^{-2U}\,(da + \mathbf{Z}^T\mathbb{C}d\mathbf{Z})^2 \;-\; 2\,e^{-U}\,d\mathbf{Z}^T\,\mathcal{M}_4(z,\bar{z})\,d\mathbf{Z}\right]$$

$$\Downarrow \text{ Wick rot.} \qquad (10.4.15)$$

$$ds^2_{\mathcal{Q}^\star} = \frac{1}{4}\left[dU^2 + 2\,g_{ij^\star}\,dz^i\,d\bar{z}^{j^\star} + \mathrm{e}^{-2U}\,(da + \mathbf{Z}^T\mathbb{C}d\mathbf{Z})^2 \;+\; 2\,e^{-U}\,d\mathbf{Z}^T\,\mathcal{M}_4(z,\bar{z})\,d\mathbf{Z}\right] \qquad (10.4.16)$$

In Eqs. (10.4.15), (10.4.16), $\mathbb{C}$ denotes the $(2n+2)\times(2n+2)$ antisymmetric matrix defined over the fibres of the symplectic bundle characterizing special geometry, while the *negative definite*, $(2n+2)\times(2n+2)$ matrix $\mathcal{M}_4(z,\bar{z})$ is the one already introduced in Eq. (9.1.24). The pseudo-quaternionic metric is non-Euclidean and it has the following signature:

$$\mathrm{sign}\left(ds^2_{\mathcal{Q}^\star}\right) = \left(\underbrace{+,\ldots,+}_{2n+2},\underbrace{-,\ldots,-}_{2n+2}\right) \qquad (10.4.17)$$

In this way we arrive at a *Geometry of the Geometries*. As solutions of the σ-model defined by the metric (10.4.16), all spherically symmetric black-holes correspond to geodesics and consequently a geodetic in the manifold $\mathcal{Q}^\star$ encodes all the geometrical structures listed below:

(a) A spherical black-hole metric,
(b) a spherical symmetric connection on the fibre bundle $P\left(\mathcal{G},\mathcal{M}_4^{st}\right)$
(c) a spherical symmetric map from $\mathcal{M}_4^{st}$ into the manifold (10.4.2)

The indefinite signature (10.4.17) introduces a clear-cut distinction between non-extremal and extremal black-holes: the non-extremal ones correspond to time-like geodesics, while the extremal black-holes are associated with light-like ones. Space-like geodesics produce supergravity solutions with naked singularities [26].

In those cases where the Special Manifold $\mathcal{SK}_n$ is a symmetric space $\frac{\mathrm{U_{D=4}}}{\mathrm{H_{D=4}}}$ also the quaternionic manifold defined by the metric (10.4.15) is a symmetric coset manifold:

$$\frac{\mathrm{U_{D=3}}}{\mathrm{H_{D=3}}} \qquad (10.4.18)$$

where $\mathrm{H_{D=3}} \subset \mathrm{U_{D=3}}$ is the *maximal compact subgroup* of the U-duality group, in three dimensions $\mathrm{U_{D=3}}$. The change of sign in the metric (10.4.17) simply turns the coset (10.4.18) into a new one:

$$\frac{\mathrm{U_{D=3}}}{\mathrm{H^\star_{D=3}}} \qquad (10.4.19)$$

where $\mathrm{H_{D=3}}^\star \subset \mathrm{U_{D=3}}$ is another *non-compact maximal subgroup* of the U-duality group whose Lie algebra $\mathbb{H}^\star$ happens to be a different real form of the complexification of the Lie algebra $\mathbb{H}$ of $\mathrm{H_{D=3}}$. That such a different real form always exists within $\mathrm{U_{D=3}}$ is one of the group theoretical miracles of supergravity.

Chapter 6 of the twin book [90] contains a detailed analysis of the new very rich *geometric lore* which emerges from the issue of black–hole constructions within the σ-model approach, considering also multicenter and non spherical symmetric solutions that correspond to maps, satisfying certain general conditions, of $\mathbb{R}^3$ into the target manifold $\mathcal{M}_s = \frac{\mathrm{U}_{D=3}}{\mathrm{H}^\star_{D=3}}$. In such constructions all the issues discussed in previous chapters of the present conceptual history enter the game in an essential way:

1. Special Kähler Geometry,
2. Lie Algebra invariants,
3. $c^\star$ map,
4. Tits Satake projection and its universality classes,
5. Weyl Group and its extensions,
6. Classification of nilpotent orbits.

In the quoted chapter of the twin book [90], at the end of my mathematical exposition I arrive at some conclusions on the upgrading of the *episteme* contributed by such recent developments that I report here *in toto*, since they strongly pertain to the conceptual history of symmetry.

10.5 Upgrading of the Episteme from the Supergravity Approach to Black-Holes

Although the inspiring motivations for the quite recent research results I sketched above come from *Supergravity*, yet the presented constructions are of genuine algebraic and geometrical character; indeed they might be understood and treated within the scope of pure Mathematics. As usual, the role of supersymmetry is just that of directing our choices, leading us to focus on *special manifolds* endowed with *special geometries*.

Actually the methods and the constructions described above are general and might be dealt with no knowledge of supermultiplets and supercharges. Additional inspiration coming from *Supergravity* is encoded in the strategic attention paid to the *Tits–Satake projection* and to *Tits–Satake universality classes*, which, however, are purely mathematical phenomena, self-contained in Lie algebra theory.

Even the very final physical motivation of constructing *extremal black-hole solutions* might be forgotten once, in the spirit of *the geometry of geometries*, a physical–geometrical problem has been mapped into another purely geometrical one.

Thus let us summarize into a list of points the mathematical logic of what we have been discussing above.

(A) The problem of constructing extremal black-hole solutions is reduced to the construction and classification of mappings:

$$\Phi \; : \; \mathbb{R}^3 \Longrightarrow \mathcal{M}_s \tag{10.5.1}$$

where $(\mathcal{M}_s, g)$ is a pseudo-Riemmannian manifold and the map Φ satisfies both the σ-model equations of motion and the stress-tensor vanishing condition:

$$\partial_i \left(\frac{\partial \Phi^\mu}{\partial x^i} \nabla_\mu \Phi^\nu \right) = 0 \ ; \quad g_{\mu\nu}(\Phi)\, \partial_i \Phi^\mu \, \partial_j \Phi^\nu = 0 \qquad (10.5.2)$$

(B) The geometrical problem posed in (A) can be considered for any Lorentzian-manifold $\mathcal{M}_s$ but, instructed by supersymmetry, we localize it on the homogeneous manifolds:

$$\mathcal{M}_s = \frac{\mathrm{U_{D=3}}}{\mathrm{H}^\star} \qquad (10.5.3)$$

that are in the image of the $c^\star$-map.

(C) For the reasons discussed at length in previous sections and chapters we are actually interested only in those maps of the type (10.5.1) where:

$$\Phi\left[\mathbb{R}^3\right] \subset \frac{\mathrm{U^{TS}_{D=3}}}{\mathrm{H^\star_{TS}}} \subset \frac{\mathrm{U_{D=3}}}{\mathrm{H}^\star} \qquad (10.5.4)$$

namely where the image of the three-dimensional space $\mathbb{R}^3$ lies entirely inside the Tits-Satake submanifold.

(D) The $\mathrm{H}^\star$–orbits of solutions can be classified and explicitly constructed thanks to an algorithm, thoroughly explained in [90], that associates such solutions to each $\mathrm{H}^\star$–orbit of nilpotent operators $X \in \mathbb{K}$, where $\mathbb{K}$ is the orthogonal complement of the subalgebra $\mathbb{H}^\star \subset \mathbb{U}$. The classification of U-nilpotent orbits is a frontier topic in Mathematics and, further specialized to $\mathrm{H}^\star \subset \mathrm{U}$ orbits, involves items and techniques generically not yet available in the mathematical supermarket, like the generalized Weyl group $\mathcal{GW}$ and the H-Weyl subgroup $\mathcal{W}_\mathrm{H}$.

(E) Within the class of manifolds in the image of the $c^\star$-map, the problem of $\mathrm{H}^\star$ nilpotent orbits acquires very special features because of the special nature of the subgroup $\mathrm{H}^\star$. These special features are ultimately related with the algebraic structure of special geometries.

(F) The association of the considered mathematical problem with extremal black-holes provides the features pointed out in (E) with physical interpretations in terms of electromagnetic charges, horizon areas and fixed scalars. Yet we might complete ignore such interpretations and ask ourself the question of what is the abstract, purely mathematical meaning of such relations as that between $\mathrm{U_{D=4}}$–orbits in the **W**-representation and $\mathrm{H}^\star$ nilpotent orbits. Such a study has not yet been performed but might be the source of new precious insights.

Generally speaking the problem considered in this chapter unveils new very profound aspects of Special Geometries pertaining both to the scope of Geometry and of Lie Algebra Theory. As we tried to emphasize in point (F) of the above list a mathematical reformulation of all the mechanisms spotted in this context might be of great moment. We might find clues to some generalization of the golden splitting that goes beyond both supersymmetry and even homogeneous spaces and opens some new direction in differential and algebraic geometry. Inspiring clues come probably from a careful analysis of Weyl subgroups and the characterization among them of those that can be regarded as H-subgroups.

In this context an inspiring observation appears to be the one highlighted in previous pages that regular finite horizon black-holes can be regarded as bound-states of small or very small black-holes. An in depth investigation of the proper mathematics lurking behind this feature is potentially capable of revealing new exciting perspectives both in geometry and physics. In view of the deep relation between quantum physics and geometry encapsulated into black-holes it is to be expected that all the intriguing geometrical relations listed above are the tip of an iceberg of theoretical knowledge yet to be uncovered.

Ultimately we can state the following. Black holes have been recognized to be the most intriguing arena where geometry and quantum physics are entangled. Within the scope of Supergravity black-holes display a much richer geometrical structure than in pure General Relativity and that geometrical structure is based on the most advanced aspects of Lie algebra theory. It is to be expected that interrogating this advanced mathematics in the proper way we can make further non trivial steps in our physical conceptions.

Chapter 11
Modern Manifolds from Ancient Polyhedra

Quotiens bella non ineunt, non multum venatibus, plus per otium transigunt, dediti somno ciboque

Tacitus, Germania, XV

11.1 Historical Introduction

In this chapter which is the last we turn to the analysis of important developments in complex geometry which took place in the 1980s–1990s, directly motivated by supersymmetry and supergravity and completely inconceivable outside such a framework. Notwithstanding their roots in the theoretical physics of the superworld, such developments constitute, by now, the basis of some of the most innovative and alive research directions of contemporary geometry.

As we already remarked in passing, an entire new life was contributed to Geometry by the problems posed by the coupling of matter multiplets to supergravity or by the description of their self-interaction in rigid supersymmetry. This was the cradle of *special geometries* whose theory gained momentum by the end of the 1980s and the beginning of the 1990s. In connection with supersymmetry a basic problem which was to reveal his deep geometrical implications is that of *gauging*: namely how to promote global symmetries of supersymmetric lagrangians to *local gauge* ones. In that context one crucial geometrical item happens to be the moment-map. Indeed the hamiltonian functions $\Sigma_A(\phi)$ associated with the generators T_A of any Lie isometry group play a distinctive role in supersymmetric field theories: they are the on-shell value taken by the so named *auxiliary fields* and appear in the supersymmetry transformation rules of the fermion members of the supermultiplets: spin $\frac{1}{2}$ or spin $\frac{3}{2}$ fields. Furthermore, according with a general scheme, these hamiltonian functions, or *moment maps*, are also the building blocks of the *scalar potential* generated by the *gauging*.

By the end of the 1980s the geometrical characterization of the scalar manifolds appearing in $\mathcal{N} = 2$ field theories in $D = 4$ or $\mathcal{N} = 4$ in $D = 3$ was universally

P. G. Fré, *A Conceptual History of Space and Symmetry*,
https://doi.org/10.1007/978-3-319-98023-2_11

Fig. 11.1 The first picture dating 1979 is the historical one taken during the first international conference on Supergravity, held at Stony Brook ITP. The second picture dating 1982 shows Peter van Nieuwenhuizen, the present author and Riccardo D' Auria in front of the Stony Brook house they were sharing during a one month stay of the two Italians for collaboration with van Nieuwenhuizen. The third and the fourth pictures were taken in November 2001 during the conference *Supergravity at 25* held in Stony Brook ITP. In the second picture one sees Leonardo Castellani, the present author, Peter van Nieuwenhuizen and Alberto Lerda. The last picture is the group photo of all participants to the workshop. In the 1980s the scientific relations between Torino University and Stony Brook were particularly intense and fruitful. Equally important were the relations of Stony Brook with Leuven in Belgium, Utrecht in the Netherlands and the École Normale Superiéure in Paris

clear and the notion of *HyperKähler manifolds*, well established both in Theoretical Physics and in Mathematics, was attracting a lot of interest in both communities. The prototype of compact *HyperKähler manifolds* were the torus T^4 and the Kummer surface K3, largely utilized in supergravity and string compactifications. From the mathematical point of view the main interest was focused on the identification and on the construction of new examples, compact or non compact of *HyperKählerian spaces*: supersymmetry came to aid.

In the 1980s, with the presence of Peter van Nieuwenhuizen, one of the three founders of supergravity, and the contiguity to a Department of Mathematics of very high level, the Institute of Theoretical Physics (ITP) of New York State University at Stony Brook had become a very prominent center of Mathematical Physics, particularly active in those geometrical directions that are more closely related to supersymmetry. Several young researchers from Europe who extensively contributed to the topics outlined in this essay and mathematically explained in [90], spent research stages in Stony Brook in various capacities, either as post-doctoral fellows or as visiting scientists (see Fig. 11.1).

In 1987 a milestone paper for the history of HyperKähler geometry was written by four authors, three of which were or had been associated with Stony Brook (see Fig. 11.2). The mentioned paper, entitled *HyperKähler metrics and supersymmetry* authored by Anders Karlhede, Nigel Hitchin, Ulf Lindstrom and Martin Roček [117] grew out from two different cultural traditions turning out to be extremely influential both in Physics and in Mathematics.

The British author Hitchin, former student of Sir Michael Atiyah and presently his successor on the Savilian Chair of Geometry in Oxford, brought in the distin-

Fig. 11.2 From the left to the right: Martin Roček, Anders Karlhede (1952), Nigel J. Hitchin (1946), Ulf Lindstrom (1947), finally a view of the campus of New York University at Stony Brook. Martin Roček is currently Professor of Theoretical Physics at Stony Brook and a member of the C. N. Yang Institute for Theoretical Physics. He received A.B. and Ph.D. degrees from Harvard University in 1975 and 1979, respectively. He did post-doctoral research at the University of Cambridge and Caltech before becoming a professor at Stony Brook. Anders Karlhede is currently Vice Rector of Stockholm University and a member of the Swedish Academy of Sciences. Nigel Hitchin is currently Savilian Professor of Geometry, Oxford, a position previously held by his doctoral supervisor (and later research collaborator) Sir Michael Atiyah. Hitchin is responsible, together with Atiyah for the index theorem and for the ADHM construction of instantons. Ulf Lindstrom is currently chairman of the theoretical physics department at the University of Upsala. He originally graduated from Stockholm University. Lindstrom and Hitchin have both contributed to the development of the notion of generalized complex geometry. In 1987 when their fundamental paper on HyperKähler quotients was written, three of the above four authors (Karlhede, Lindstrom and Roček) were working at the ITP of Stony Brook

guished geometrical and topological tradition of the Cambridge school, whose roots can be traced back to Hodge and which is responsible for such other milestones as, for instance, the index theorem. Martin Roček, Anders Karlhede and Ulf Lindstrom, together with Marc Grisaru and Jim Gates, were among the early founders of the superspace formalism for supersymmetric theories and had a deep working knowledge of the latter. From the inbreeding of these two traditions arose a quite powerful new mathematical vision, that of *HyperKähler quotient*.

The guiding line was provided by the lagrangian realization of a supersymmetric field theory encompassing hypermultiplets that span a flat HyperKähler manifold $\mathcal{S}$ and are coupled to gauge vector multiplets which promote a group $\mathcal{G}$ of global isometries of the space $\mathcal{S}$ to local symmetries of the lagrangian. If the kinetic terms of these vector multiplets $\mathbf{V}$ are omitted, the latter can be integrated away by means of a gaussian integration. The result of this functional integration yields, as a remnant, a set of constraints. The systematic solution of such constraints provides the geometrical construction of a new non trivial, yet smaller, HyperKähler manifold, namely the HyperKähler quotient $\mathcal{S}//\mathcal{G}$.

The great value of paper [117] was the clear cut axiomatization of this procedure which, extracted from field theory, was recast in pure mathematical terms as a self contained mathematical construction.

In the following years the HyperKähler quotient was adopted by mathematicians as a preferred constructive algorithm for new HyperKähler manifolds.

A very important instance of such constructions was provided a couple of years after the publication of [117] by Kronheimer, who succeeded in showing that all asymptotically flat gravitational instantons, the so named ALE manifolds, can be realized as HyperKähler quotients [130, 131]. The classification of ALE manifolds is a new incarnation of the ADE classification of simply laced Lie algebras, finite subgroups of SU(2) and of singularities. It clearly encodes a very deep connection between fundamental issues of Geometry and Physics.

Many current research lines in geometry related with manifolds of *restricted holonomy*, *spin(7) manifolds* and the like are intimately related with the idea of the HyperKähler quotient or of its smaller version, namely Kähler quotient. Similarly *quiver* constructions in brane physics and most of the geometrical constructions in the CFT/gauge correspondence are off-springs of the HyperKähler/Kähler quotient algorithm.

11.2 The Ideology of the Kähler/HyperKähler Quotient

Any Kähler manifold $\mathcal{M}$ is symplectic, the symplectic two-form being provided precisely by the Kähler two-form. Henceforth if $\mathcal{M}$ admits a Lie group $\mathcal{G}$ of isometries one can introduce the moment map:

$$\mathcal{P} : \mathcal{M} \longrightarrow \mathbb{R} \otimes \mathbb{G}^* \tag{11.2.1}$$

where $\mathbb{G}^*$ denotes the dual of the Lie algebra $\mathbb{G}$ of the group $\mathcal{G}$, i.e. the space of linear functionals on $\mathbb{G}$.

The moment-map is defined by the following properties that have to be satisfied by the functions $\mathcal{P}_{\mathbf{X}}$ associated with each holomorphic Killing vector field $\mathbf{X}$ belonging to $\mathbb{G}$:

$$-\,d\mathcal{P}_{\mathbf{X}} = i_{\mathbf{X}}\mathbf{K} \tag{11.2.2}$$

$$\mathbf{X}\mathcal{P}_{\mathbf{Y}} = \mathcal{P}_{[\mathbf{X},\mathbf{Y}]} \tag{11.2.3}$$

where $\mathbf{K}$ is the Kähler two-form and $i_{\mathbf{X}}$ denotes the contraction with the mentioned vector field. Equation (11.2.3) is named the equivariance condition of the moment-map.

HyperKähler manifolds are characterized by the presence of three complex structures and of three corresponding Kähler forms $\mathbf{K}^x$ $(x = 1, 2, 3)$.[1] This allows to introduce a tri-holomorphic moment-map:

$$\mathcal{P} : \mathcal{M} \longrightarrow \mathbb{R}^3 \otimes \mathbb{G}^* \tag{11.2.4}$$

where the triplet of functions $\mathcal{P}^x_{\mathbf{X}}$ associated with each holomorphic Killing vector field $\mathbf{X}$ belonging to $\mathbb{G}$ satisfy the conditions:

$$-\,d\mathcal{P}^x_{\mathbf{X}} = i_{\mathbf{X}}\mathbf{K}^x \tag{11.2.5}$$

$$\mathbf{X}\mathcal{P}^x_{\mathbf{Y}} = \mathcal{P}^x_{[\mathbf{X},\mathbf{Y}]} \tag{11.2.6}$$

Given a HyperKähler manifold $\mathcal{S}$ which admits a Lie group $\mathcal{G}$ of triholomorphic isometries, the *HyperKähler quotient* [117] is a procedure that provides a way to construct from $\mathcal{S}$ a lower-dimensional HyperKähler manifold $\mathcal{M}$, as follows. Let $\mathfrak{Z}^* \subset \mathbb{G}^*$ be the dual of the center of the Lie algebra $\mathbb{G}$. For each $\zeta \in \mathbb{R}^3 \otimes \mathfrak{Z}^*$ the level set of the momentum map

$$\mathcal{N} \equiv \bigcap_{\mathbf{X}} \mathcal{P}^{-1}_{\mathbf{X}}(\zeta^x) \subset \mathcal{S}, \tag{11.2.7}$$

which has dimension $\dim\,\mathcal{N} = \dim\,\mathcal{S} - 3\,\dim\,\mathcal{G}$, is invariant under the action of $\mathcal{G}$, due to the equivariance of the moment map $\mathcal{P}$. Thus one can take the quotient:

$$\mathcal{M} = \mathcal{N}/\mathcal{G} \tag{11.2.8}$$

The manifold $\mathcal{M}$ is smooth of dimension $\dim\mathcal{M} = \dim\mathcal{S} - 4\dim\mathcal{G}$ as long as the action of $\mathcal{G}$ on $\mathcal{N}$ has no fixed points. The three two-forms κ^x on $\mathcal{M}$, defined via the restriction to $\mathcal{N} \subset \mathcal{S}$ of the three Kähler forms $\mathbf{K}^x$ on $\mathcal{S}$ are closed and satisfy the quaternionic algebra thus providing $\mathcal{M}$ with a HyperKähler structure.

[1] Compare with Sect. 8.2.2.

In view of this fundamental property, the HyperKähler quotient offers a natural way to construct a $\mathcal{N} = 2$, $\mathrm{D} = 4$ or $\mathcal{N} = 4$, $\mathrm{D} = 2$ σ-model on a non-trivial manifold $\mathcal{M}$ starting from a free σ-model on a trivial flat-manifold $\mathcal{S} = \mathbf{H}^n$. It suffices to gauge appropriate triholomorphic isometries by means of non-propagating gauge multiplets. Omitting the kinetic term of these gauge multiplets and performing the gaussian integration of the corresponding fields one realizes the HyperKähler quotient in a Lagrangian way. In the four-dimensional case, this fact was fully exploited, by Hitchin, Kärlhede, Lindstrom and Roček in their seminal paper [117], was further discussed by Galicki [97] and was applied, in the context of string theory by Ferrara, Girardello, Kounnas and Porrati [74]. Actually the HyperKähler quotient is a generalization of a similar Kähler quotient procedure, where the momentum map $\mathcal{P} : \mathcal{S} \to \mathbb{R} \otimes \mathbb{G}^*$ consists just of one hamiltonian function, rather than three. The Kähler quotient is related with either $\mathcal{N} = 1$, $\mathrm{D} = 4$ or $\mathcal{N} = 2$, $\mathrm{D} = 2$ supersymmetry, the reason being that, in these cases the vector multiplet contains just one real auxiliary field $\mathcal{P}$.

11.3 ALE Manifolds and the ADE Classification

ALE means asymptotically locally euclidian. This means that ALE manifolds are smooth 4-manifolds with euclidian signature and a metric leading to a self-dual curvature two-form:

$$\mathfrak{R}^{ab}_{ALE} = \frac{1}{2}\,\varepsilon^{abcd}\,\mathfrak{R}^{cd}_{ALE} \tag{11.3.9}$$

which, for large distances from a core, approaches the flat euclidian metric.

Actually ALE manifolds are all Ricci flat and constitute vacuum solutions of Einstein equations after Wick rotation. In this sense ALE-manifolds are gravitational instantons in the same way as the connections with a self dual field strength are gauge instantons.

The first instance of an ALE manifold was found by Eguchi and Hanson [68] in 1979 (see Fig. 11.3).

The fascination of ALE manifolds is that they happen to be in one-to-one correspondence with the finite subgroups $\Gamma \subset \mathrm{SU}(2)$ and are similarly classified by the ADE classification of simply-laced Lie algebras.

In 1989 Peter Kronheimer (see Fig. 11.4) succeeded in constructing all of them as HyperKähler quotients of suitably chosen flat HyperKähler manifolds dictated by the structure of the finite group Γ to which each of them corresponds.

The association between ALE manifolds, ADE singularities and subgroups $\Gamma \subset \mathrm{SU}(2)$ is not a superficial matter rather it is a very deep and structural one. The topological properties of the ALE four-manifold are identified with intrinsic numbers of the corresponding Lie algebra; for instance the Hirzebruch signature τ of the ALE coincides with the rank r of the corresponding Lie Algebra $\mathbb{G}$ and with the dimension of the chiral ring $\mathcal{R}_\Gamma$ associated with the singular potential W_Γ. On the other hand

Fig. 11.3 On the left Tohru Eguchi (1948), on the right Andrew J. Hanson. Eguchi is currently emeritus professor of the University of Tokyo, Yukawa Institute. He held positions at SLAC and at the Enrico Fermi Institute of Chicago University. Andrew J. Hanson received the BA degree in chemistry and physics from Harvard College in 1966 and the Ph.D. degree in theoretical physics from MIT in 1971. He is an Emeritus Professor of Computer Science in the School of Informatics and Computing at Indiana University, Bloomington. He worked in theoretical physics from 1971 until 1980, when he began working in machine vision, graphics, and visualization, first with the perception research group at the SRI Artificial Intelligence Center, and then at Indiana University from 1989 until his retirement in 2012. The Eguchi Hanson metric was derived by the two authors in 1978 when both of them were in California, the first in Stanford, the second in Berkeley

the same number r is also that of the non trivial conjugacy classes of Γ, apart of the identity class.

The catch of all this is encoded in a surprising correspondence between extended Dynkin diagrams and irreducible representations of the finite groups Γ that had been discovered years before Kronheimer by McKay [143]. Without any doubt McKay correspondence provided Kronheimer with an essential guideline for his construction.

A very important basis for Kronheimer work was encoded in the previous work on gravitational instantons conducted by Gibbons and Hawking [102] (see Fig. 11.5) and by Hitchin [116] (see also [30, 170]).

11.3.1 ALE Manifolds and ADE Singularities

ALE spaces are non–compact manifolds that have originally emerged in the literature as gravitational instantons. Indeed they are Riemannian 4–manifolds with an (anti)selfdual curvature 2-form and a metric that approaches the Euclidean metric at infinity. In polar coordinates (r, Θ) on $\mathbb{R}^4$, we have $g_{\mu\nu}(r, \Theta) = \delta_{\mu\nu} + O(r^{-4})$. This corresponds to the intuitive concept of an instanton as a defect which is localized in a finite region of space-time. This picture, however, is verified only modulo an addi-

Fig. 11.4 Peter Benedict Kronheimer (born 1963) is a British mathematician, known for his work on gauge theory and its applications to 3- and 4-dimensional topology. He is currently William Caspar Graustein Professor of Mathematics at Harvard University. He completed his PhD at Oxford University under the direction of Sir Michael Atiyah. Kronheimer's early work was on gravitational instantons, in particular the classification of HyperKähler four manifolds with asymptotical locally euclidean geometry (ALE spaces) leading to the papers *The construction of ALE spaces as hyper-Kahler quotients* and *A Torelli-type theorem for gravitational instantons*. He also contributed extensively to the topology of 4-manifolds and to the theory of Donaldson invariants. He and Nakajima gave a construction of instantons on ALE spaces generalizing the Atiyah-Hitchin-Drinfeld-Manin construction

Fig. 11.5 Gary William Gibbons (born 1946) is a British theoretical physicist. Gibbons was born in Coulsdon, Surrey. He was educated at Purley County Grammar School and the University of Cambridge, where in 1969 he became a research student under the supervision of Dennis Sciama. When Sciama moved to the University of Oxford, he became a student of Stephen Hawking, obtaining his PhD from Cambridge in 1973. Gibbons became a full professor in 1997, a Fellow of the Royal Society in 1999, and a Fellow of Trinity College, Cambridge in 2002. He has given outstanding contributions to the theory of quantum black holes and to the theory of gravitational instantons. His special interests in geometry in all of its aspects led him to contribute to many issues in string and M-theory compactifications

tional subtlety that is of utmost relevance in the present geometrical construction. The base manifold of the gravitational instanton has a boundary at infinity which, rather than a pure 3–sphere is:

$$\mathbb{S}^3/\Gamma \tag{11.3.10}$$

$\Gamma \subset \mathrm{SU}(2)$ being a finite subgroup of $\mathrm{SU}(2) \sim \mathbb{S}^3$. Therefore, outside the core of the instanton, rather than $\mathbb{R}^4$, the manifold looks like the quotient singularity $\mathbb{R}^4/\Gamma$. This is the reason for the name given to these spaces: the asymptotic behaviour is euclidean only *locally*.

For the sake of our purposes the most important aspect of ALE spaces is that they are complex 2–folds endowed with a HyperKähler structure and a trivial canonical bundle $c_1\,(ALE_\Gamma) = 0$. This makes ALE spaces the *non–compact* analogues of the $K3$ surface which, apart from the T^4 torus is the only compact Calabi–Yau 2–fold. Indeed viewed as a complex manifold, outside the core of the instanton, the ALE space looks like the quotient singularity

$$ALE_\Gamma \sim \mathbb{C}^2/\Gamma \quad ; \quad \Gamma \subset \mathrm{SU}(2) \tag{11.3.11}$$

where Γ is the above mentioned finite subgroup of SU(2). In this way we have explained the rationale for the subindex Γ attached to the symbol denoting an ALE space. Indeed it can be shown that the choice of the identification group at infinity completely fixes the *topological type* of the ALE manifold. These types are in one–to–one correspondence with the finite groups Γ which admit an ADE classification, like simple Lie algebras and simple singularities. The correspondence between the ADE classification of ALE spaces and that of simple singularities is shortly discussed below. On the contrary the whole topic of ALE manifolds, Kronheimer construction and quotient singularities is fully mathematically developed in my other book [90]. For the moment we note that the remaining ambiguity, once the identification group Γ has been fixed is given by the moduli of the self dual metric (i.e. of the HyperKähler structure) at fixed topological type.

In the HyperKähler quotient construction of the ALE spaces the complete set of the HyperKähler structure moduli can be seen as the *levels* of the *quaternionic momentum map*.

11.3.2 The McKay Correspondence for $\mathbb{C}^2/\Gamma$

The character table of any finite group Γ allows to reconstruct the decomposition coefficients of any representation D along the irreducible representations D_μ that for any finite group are as many as the conjugacy classes, i.e. $r + 1$:

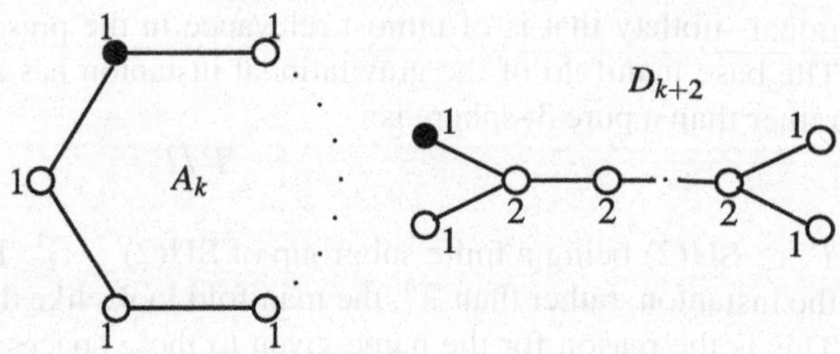

Fig. 11.6 Extended Dynkin diagrams of the infinite series

$$
\begin{aligned}
D &= \bigoplus_{\mu=0}^{r} a_\mu \, D_\mu \\
a_\mu &= \frac{1}{g} \sum_{i=0}^{r} g_i \, \chi_i^{(D)} \, \chi_i^{(\mu)\star}
\end{aligned}
\tag{11.3.12}
$$

For the finite subgroups $\Gamma \subset \mathrm{SU}(2)$ a particularly important case is the decomposition of the tensor product of an irreducible representation D_μ with the defining 2-dimensional representation $\mathcal{Q}$. It is indeed at the level of this decomposition that the relation between these groups and the simply laced Dynkin diagrams becomes explicit and it is named the McKay correspondence. As we already stressed, this decomposition plays a crucial role in the explicit construction of ALE manifolds according to Kronheimer. Setting:

$$
\mathcal{Q} \otimes D_\mu = \bigoplus_{\nu=0}^{r} A_{\mu\nu} \, D_\nu \tag{11.3.13}
$$

where D_0 denotes the identity representation, one finds that the matrix $\bar{c}_{\mu\nu} = 2\delta_{\mu\nu} - A_{\mu\nu}$ is the *extended Cartan matrix* relative to the *extended Dynkin diagram* corresponding to the given group. We remind the reader that the extended Dynkin diagram of any simply laced Lie algebra is obtained by adding to the *dots* representing the *simple roots* $\{\alpha_1 \ldots\ldots \alpha_r\}$ an *additional dot* (marked black in Figs. 11.6 and 11.7) representing the negative of the highest root $\alpha_0 = \sum_{i=1}^{r} n_i \, \alpha_i$ (the integers n_i are named the Coxeter numbers). Thus we see a correspondence between the non-trivial conjugacy classes $\mathscr{C}_i$ (or equivalently the non-trivial irrepses) of the group $\Gamma(\mathbb{G})$ and the simple roots of $\mathbb{G}$. In this correspondence the extended Cartan matrix provides the Clebsch-Gordon coefficients (11.3.13), while the Coxeter numbers n_i express the dimensions of the irreducible representations. All these informations are summarized in Figs. 11.6 and 11.7 where the numbers n_i are attached to each of the dots: the number 1 is attached to the extra dot since it stands for the identity representation.

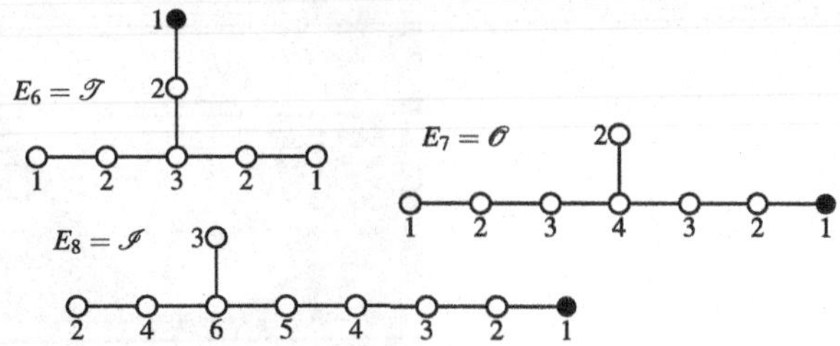

Fig. 11.7 Exceptional extended Dynkin diagrams

11.3.3 Philosophical Interlude

Let us pose for a moment and comment about the extraordinary philosophical message contained in the just explained correspondence.

We outlined in previous chapters the correspondence between the ADE classification of simply laced Lie Algebras and the classification of finite subgroups of the rotation group SO(3), ultimately the classification of platonic polyhedra. We emphasized that this correspondence is due to the identity of the two diophantine inequalities of which the two classifications are respectively solutions. Thanks to the McKay correspondence, discovered in the last decades of the XXth century, we understand that the identity of the two classifications is no pure coincidence. Indeed the Dynkin diagrams and the associated Coxeter numbers encode such a fundamental information about the finite Platonic Groups as the dimensions of their irreps. In this we observe the manifestation of a profound unity of apparently different algebraic structures which is a matter for further thinking and investigation. The fact that these structures are consistently utilized in an algorithm which is capable of constructing new manifolds of physical interest as the gravitational instantons and provides the clue to resolve quotient singularities, should attract our attention to the successful, vastly non galilean, path of discovery underlying these important developments in contemporary Geometrical Physics.

It is worth to consider briefly the personality and the way of thinking of John McKay (see Fig. 11.8) who is responsible for the opening up of such new horizons in geometric–physical thinking.

McKay's other most relevant contribution is related with the *Monstrous Moonshine* and provides another example of the discovery of an unexpected relation between seemingly different mathematical entities. The term Monstrous Moonshine was coined by the Princeton mathematician John Conway to describe the mysterious connection between sporadic simple groups and modular invariants initiated by an observation that McKay did in 1978. That year, reconsidering the classical modular function of weight zero, namely Klein's $J(\tau)$-function, McKay observed that in the Fourier expansion of this latter:

$$J(\tau) = \frac{1}{q} + \sum_{\ell=1}^{\infty} n_\ell \, q^\ell \quad ; \quad q = \exp\left[2\pi \, i\tau\right] \tag{11.3.14}$$

Fig. 11.8 John McKay (16 June 1939, Kent (England)) is currently Emeritus Distinguished Professor of Mathematics at the Concordia University in Canada. McKay went to Manchester University in 1958 from which he graduated in 1962. Next he went to Edinburgh University from which he obtained his Ph.D. in 1971. He works at the Concordia University since 1974. He was elected a fellow of the Royal Society of Canada in 2000, and won the 2003 CRM-Fields-PIMS prize. He is especially known for his discovery of the monstrous moonshine, his joint construction of some sporadic simple groups, for the McKay (McKay–Alperin) conjecture in representation theory, and for the McKay correspondence

the integer coefficients n_ℓ could be expressed as linear combinations with positive integer coefficients of the dimensions of the irreducible representations of the Monster Group, the largest sporadic simple group M of order $|M| \sim 10^{53}$. McKay guessed that there should be an infinite dimensional graded representation of the Monster Group whose lower grades decompose into irreps in the way shown by the coefficients of the J expansion. Such a representation was later constructed by Frenkel, Lepowsky and Meuman in 1988 in terms of primary conformal fields associated with a self-dual lattice in 24 dimensions and ultimately related with compactified bosonic string theory.

As his friend and colleague John Harnad says, McKay's peculiar genius lies in noticing connections that no one else has seen. *He doesn't solve new problems so much as make observations of surprising relations, and pose challenges that sometimes end up pointing to whole new domains of research,* Harnad said with admiration.

McKay attributes his ability to the fact that his knowledge is unusually broad in an era of specialization. *You have to have a lot of curiosity and do a lot of thinking,* he said simply. *There's a lot of pressure to find applications, but if you don't do the basics, you wont get the applications.*[2]

Personally I subscribe entirely to this way of thinking. The pressure for application is a *morbus* that can, in the long run, kill our modern science and also the economic

[2]These informations about John McKay are taken from an article by Barbara Black published on the web-site of Concordia University.

prosperity of our society which is built on the technological advances that came from it.

Yet, accepting the relevance of fundamental science, which many people do, is not yet sufficient to make it advance if we stick only to the galilean method and we ignore the methods of inquiry followed by scientists like John McKay. The following aphorism of his:

"Pure math is math that hasn't happened yet. We solve problems without understanding them".

encodes in a provocative quite vivid expression a great truth. Sooner or later all elegant and deep mathematical constructions found their place in theories that aim at the explanation of physical phenomena. The discovery of the possible physical meaning of mathematical constructions goes hand in hand with the discovery of their deeper mathematical and philosophical sense. As I said in earlier pages we can Interrogate Nature only in parallel with the Interrogation of Human Mathematical Thought, since the only language by means of which we can talk with Nature is indeed Mathematics, as Galileo properly stated, yet Mathematics is made by the human mind.

11.3.4 Sketch of Kronheimer's Construction

Given any finite subgroup of $\Gamma \subset \mathrm{SU}(2)$, we consider a space $\mathscr{V}$ whose elements are two-vectors of $|\Gamma| \times |\Gamma|$ complex matrices: $p \in \mathscr{V} = (A, B)$. The action of an element $\gamma \in \Gamma$ on the points of $\mathscr{V}$ is the following:

$$\begin{pmatrix} A \\ B \end{pmatrix} \xrightarrow{\gamma} \begin{pmatrix} u_\gamma & i\,\bar{v}_\gamma \\ i\,v_\gamma & \bar{u}_\gamma \end{pmatrix} \begin{pmatrix} R(\gamma)\,A\,R(\gamma^{-1}) \\ R(\gamma)\,B\,R(\gamma^{-1}) \end{pmatrix} \tag{11.3.15}$$

where the two-dimensional matrix on the right hand side is the realization of γ inside the defining two-dimensional representation $\mathscr{Q} \subset \mathrm{SU}(2)$, while $R(\gamma)$ is the regular, $|\Gamma|$-dimensional representation. The basis vectors in R named e_γ are in one-to-one correspondence with the group elements $\gamma \in \Gamma$ and transform as follows:

$$R(\gamma)\,e_\delta \;=\; e_{\gamma\cdot\delta} \quad \forall\gamma\,,\,\delta \in \Gamma \tag{11.3.16}$$

In mathematical notation the space $\mathscr{V}$ is named as:

$$\mathscr{V} \simeq \mathrm{Hom}\,(R, \mathscr{Q} \otimes R) \tag{11.3.17}$$

Next we introduce the space $\mathscr{S}$, which by definition is the subspace of Γ-invariant elements in $\mathscr{V}$:

$$\mathscr{S} \equiv \{p \in \mathscr{V} / \forall\gamma \in \Gamma, \gamma \cdot p = p\} \tag{11.3.18}$$

Explicitly the invariance condition reads as follows:

$$\begin{pmatrix} u_\gamma & i\bar{v}_\gamma \\ i v_\gamma & \bar{u}_\gamma \end{pmatrix} \begin{pmatrix} A \\ B \end{pmatrix} = \begin{pmatrix} R(\gamma^{-1}) A R(\gamma) \\ R(\gamma^{-1}) B R(\gamma) \end{pmatrix} \tag{11.3.19}$$

The decomposition (11.3.13) is very useful in order to determine the Γ-invariant flat space (11.3.18).

A two-vector of matrices can be thought of also as a matrix of two-vectors: that is, $\mathscr{P} = \mathscr{Q} \otimes \mathrm{Hom}(R, R) = \mathrm{Hom}(R, \mathscr{Q} \otimes R)$. Decomposing the regular representation, $R = \bigoplus_{\nu=0}^{r} n_\mu D_\mu$ into irrepses, using Eq. (11.3.13) and Schur's lemma, we obtain:

$$\mathscr{S} = \bigoplus_{\mu,\nu} A_{\mu,\nu} \mathrm{Hom}(\mathbb{C}^{n_\mu}, \mathbb{C}^{n_\nu}) \; . \tag{11.3.20}$$

The dimensions of the irrepses, n_μ are dispayed in Figs. 11.6 and 11.7. From Eq. (11.3.20) the real dimension of $\mathscr{S}$ follows immediately: $\dim \mathscr{S} = \sum_{\mu,\nu} 2A_{\mu\nu} n_\mu n_\nu$ implies, recalling that $A = 2 \times \mathbf{1} - \bar{c}$ [see Eq. (11.3.13)] and that for the extended Cartan matrix $\bar{c}n = 0$:

$$\dim_{\mathbb{C}} \mathscr{S} = 2 \sum_\mu n_\mu^2 = 2|\Gamma| \; . \tag{11.3.21}$$

In mathematical notation the space $\mathscr{S}$ is denoted as follows:

$$\mathscr{S} \simeq \mathrm{Hom}_\Gamma \, (R, \mathscr{Q} \otimes R) \tag{11.3.22}$$

So we can summarize the discussion by saying that:

$$\dim_{\mathbb{C}} \, [\mathrm{Hom}_\Gamma \, (R, \mathscr{Q} \otimes R)] \; = \; 2\,|\Gamma| \tag{11.3.23}$$

The manifold $\mathscr{S}$ defined in Eq. (11.3.20) is the flat HyperKähler manifold of which we are supposed to perform the HyperKähler quotient in order to obtain the ALE_Γ manifold. We need to know the isometry group $\mathscr{F}$ to quotient. This is mentioned below:

$$\mathscr{F} = \bigotimes_{\mu=0}^{r} \mathrm{U(n_\mu)} \bigcap \mathrm{SU}(|\Gamma|) \tag{11.3.24}$$

where the sum is extended to all the irreducible representations of the group Γ and n_μ are their dimensions. One should also take into account that the determinant of all the elements must be one, since $\mathscr{F} \subset \mathrm{SU}(|\Gamma|)$. Pictorially the group $\mathscr{F}$ has a $\mathrm{U(n_\mu)}$ factor for each dot of the diagram, n_μ being associated with the dots as in Figs. 11.6 and 11.7. $\mathscr{F}$ acts on the various *components* of $\mathrm{Hom}_\Gamma \, (R, \mathscr{Q} \otimes R)$ that are in correspondence with the edges of the diagram, see Eq. (11.3.20), as dictated by the diagram structure.

I do not add further details about Kronheimer's construction which is thoroughly presented and mathematically elaborated in [90]. What I collected here are just the formulae which I considered essential in order to explain the basic conception of such a construction and discuss its place in the current development of geometrical ideas.

In order to put such development into a historical perspective which looks at the future, I conclude by mentioning the generalization of the McKay correspondence and its use to resolve higher dimensional singularities by means of suitable Kähler quotients. I will say just few words, since these are topics of current research.

11.4 Generalization of the Correspondence: McKay Quivers for $\mathbb{C}^3/\Gamma$ Singularities

One can generalize the extended Dynkin diagrams obtained in the above way by constructing McKay quivers, according to the following definition:

Let us consider the quotient $\mathbb{C}^n/\Gamma$, where Γ is a finite group that acts on $\mathbb{C}^n$ by means of the complex representation $\mathcal{Q}$ of dimension n and let D_i, $(i = 1, \ldots, r+1)$ be the set of irreducible representations of Γ having denoted by $r+1$ the number of conjugacy classes of Γ. Let the matrix $\mathscr{A}_{ij}$ be defined by:

$$\mathcal{Q} \otimes \mathrm{D}_i = \bigoplus_{j=1}^{r+1} \mathscr{A}_{ij}\, \mathrm{D}_j \tag{11.4.25}$$

To such a matrix we associate a quiver diagram in the following way. Every irreducible representation is denoted by a circle labeled with a number equal to the dimension of the corresponding irrep. Next we write an oriented line going from circle i to circle j if D_j appears in the decomposition of $\mathcal{Q} \otimes \mathrm{D}_i$, namely if the matrix element $\mathscr{A}_{ij}$ does not vanish.

The analogue of the extended Cartan matrix discussed in the case of $\mathbb{C}^2/\Gamma$ is defined below:

$$\bar{c}_{ij} = n\, \delta_{ij} - \mathscr{A}_{ij} \tag{11.4.26}$$

and it has the same property, namely it admits the vector of irrep dimensions

$$\mathbf{n} \equiv \{1, n_1, \ldots, n_r\} \tag{11.4.27}$$

as a null vector:

$$\bar{c}.\mathbf{n} = \mathbf{0} \tag{11.4.28}$$

THE MCKAY QUIVER OF L_{168}

An example is for instance provided by the simple group of order 168, L_{168}, which has a complex three dimensional irreducible representation $\mathcal{Q}$ that can be used to construct the singularity $\mathbb{C}^3/\mathrm{L}_{168}$.

We calculate the McKay matrix defined by

$$\mathcal{Q} \otimes \mathrm{D}_i = \bigoplus_{j=1}^{6} \mathcal{A}_{ij}\, \mathrm{D}_j \tag{11.4.29}$$

where D_i denote the 6 irreducible representation ordered in the following standard way:

$$\mathrm{D}_i = \{\mathrm{D}_1, \mathrm{D}_6, \mathrm{D}_7, \mathrm{D}_8, \mathrm{D}_3, \mathrm{D}_{\bar{3}}\} \tag{11.4.30}$$

$$\mathcal{A} = \begin{pmatrix} 0\,0\,0\,0\,1\,0 \\ 0\,0\,1\,1\,0\,1 \\ 0\,1\,1\,1\,0\,0 \\ 0\,1\,1\,1\,1\,0 \\ 0\,1\,0\,0\,0\,1 \\ 1\,0\,0\,1\,0\,0 \end{pmatrix} \tag{11.4.31}$$

The matrix $\mathcal{A}$ admits the graphical representation displayed in Fig. 11.9, named the *McKay quiver* of the quotient $\mathbb{C}^3/\mathrm{L}_{168}$ The picture in Fig. 11.9 is a diagram with a loop and does not correspond to any root space. This is so because there are in the corresponding group more than three types of element-order. Indeed, in $\Gamma = \mathrm{L}_{168}$

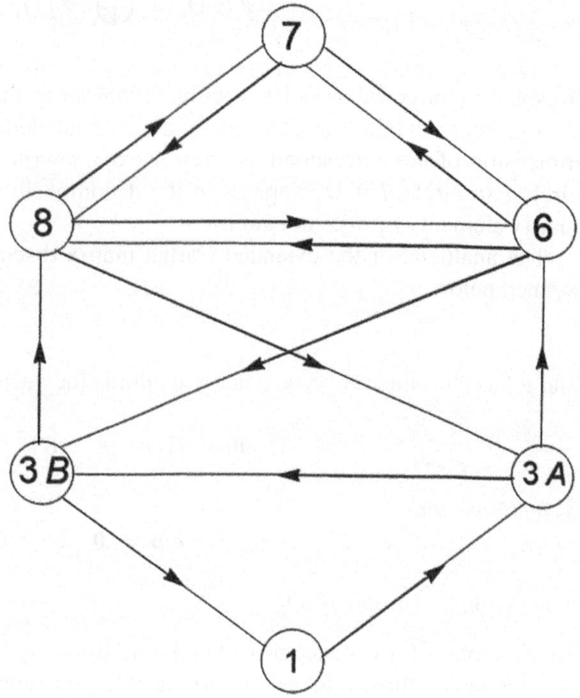

Fig. 11.9 The quiver diagram of the finite group $\mathrm{L}_{168} \subset \mathrm{SU}(3)$

we have elements of order 2, 3, 4, 7. Yet Kronheimer construction can be extended to cases like this.

Just as before we introduce the space:

$$\mathscr{S}_\Gamma = \bigoplus_{i,j} A_{i,j}\mathrm{Hom}(\mathbb{C}^{n_i}, \mathbb{C}^{n_j}) \tag{11.4.32}$$

and we can take its Kähler quotient with respect to the analogue of the group (11.3.24). In this case the quotient can be only Kählerian since no three-dimensional complex space can be HyperKählerian. The number of constraints in this case is not sufficient to produce a three-dimensional complex manifold, we need an extra holomorphic constraint. This construction was recently investigated and physically interpreted in [29]. A review is provided in the twin book [90].

11.5 Conclusion

In this chapter we have seen that the 2400 year old classification problem of platonic solids is still alive and able to produce very challenging modern fruits. We started in Chap. 4 with the diophantine equation that provides the ADE classification of finite rotation groups. In Sect. 5.5.2 we retrieved, via Dynkin diagrams, the same classification in terms of simply laced Lie algebras. In the present chapter we found a third incarnation of the same classification under the form of gravitational instantons associated with the resolution of singularities [7, 27, 159].

The relation between finite groups, Lie algebras and complex geometry have in the topics discussed in this chapter a most exciting illustration. Furthermore the profound role played by supersymmetry in bringing to the surface deep and unexpected connections is exemplified by the contents of the present chapter in a paradigmatic way.

Indeed one of the most fundamental question at stake in many problems of supergravity and superstring, in particular related with compactifications and with the AdS/CFT correspondence, is just the classical algebraic geometry problem of resolving quotient singularities. Under the inspired stimulus of supersymmetric theories a rich set of results were obtained by the mathematical community at the beginning of 1990s, those reviewed in this chapter being just the first ones in such a list.

Entering a more circumstantial analysis I have tried to emphasize where the catch of a such a stimulus is. The example of (Hyper)Kähler quotients is indeed paradigmatic. The whole story began from the physical interpretation of the mathematical notion of moment-map. Identifying the moment-maps with the auxiliary fields of supersymmetric gauge–theories new scenarios opened up. Extremization of the scalar potential, namely the physical problem of searching for classical vacua of a field–theory naturally produced the notion of (Hyper)Kähler quotient. It was once again a physical problem, that of instantons extended from gauge–theories to gravity, what motivated the consideration of ALE manifolds. Yet their construction as Hyper-

Kähler quotients would not have been possible without the further ingredient of the McKay correspondence. This latter came neither from physics nor from the solution of some mathematical problem posed in a standard way. It just came from that type of Interrogation of Mathematics rather than of Nature which we discussed above. It is looking for traces of unexpected correspondences that sometimes we uncover the deeper nature of certain mathematical structures we have known for long time. As a result of such discoveries we usually open new scenarios not only for Mathematics but also for Physics, where identifications, such as that of the moment-maps with the auxiliary fields, become possible with far reaching consequences of the type highlighted above.

From these considerations it is evident how wide and deep is the extent of the fertilizing influence exerted by *supersymmetry* on the development of modern Geometry. I believe that this latter has entered a new season of expansion and progress that can, in the long run, lead to new conceptions in Physics.

Epilogue

The theoretical understanding of the world, which is the aim of philosophy, is not a matter of great practical importance to animals, or to savages, or even to most civilised men.
Bertrand Russell

The patient reader who followed me through the previous pages went on a quite long journey, through different times and across different human ideas, a journey that embraces about twenty-five centuries. While talking about mathematical concepts and theories, my aim was that of telling a story, a long and complex one, not deprived, occasionally, of dramatic touches.

My main theme was the mathematical conception of symmetry and its development from Classical Antiquity to the most advanced constructions that have been vigorously springing up in the twilight of the XXth century and at the dawn of the new millenium.

Telling this story I tried to convince my reader that mathematics and physics have a status not so much different from that of other branches of culture, like literature, philosophy or the figurative arts. At the basis of the so named *exact sciences* there is a *cultural substratum* made of shared feelings, shared attitudes towards life and death, personal aspirations and ways of thinking that developed through history and have quite remote roots.

Modern Theoretical Science is an expression of Western Culture, whose firm basis consists of the Analytic Thought System developed by the ancient Greeks and rooted in their early development of individualism and democracy. Nothing similar to the mathematical physical science that we know today might have been developed by any of the Asian civilizations, notwithstanding their antiquity and their technical advances. Actually the system of thought pertaining to those civilizations might have produced a different kind of science, if the Western model did not prevail world-wide. Although the Arabs played an invaluable historical role in preserving, reorganizing and transmitting back to the Latin West the ancient hellenistic science, a process which took place in Sicily and Spain in the eleventh and twelfth century, yet a careful analysis reveals that the Arabic scholars, mostly Persians, Syrians and Jews were the late depositaries of Hellenistic Culture, who survived as they could, through the tempests of Islamic invasions and the imposition, by means of the sword, of an

P. G. Fré, *A Conceptual History of Space and Symmetry*,
https://doi.org/10.1007/978-3-319-98023-2

alien language and faith. The Byzantines, on the contrary, went the opposite way and, under the continuous pressure, wave after wave, of barbarian invaders, crystallized the vital Greek culture of Antiquity into an Asiatic hieratic system of values and paradigms, deprived of any innovative force. This, however, was useful to preserve some of the ancient heritage for better times.

Not surprisingly the resurrection of what we call mathematical and physical sciences took place during the thirteenth century in the environment of the Italian, commerce oriented, self governed cities, continuously fighting against each other and against the German Emperors. This was a historical setup which reproduced most of the conditions of ancient Greece, *mutatis mutandis*. May be the German Emperors played the same role as the Persian Empire of the Vth century B.C. and the inclination of the Italian Maritime Republics to navigate through the Mediterranean, disseminating strongholds and trade centers everywhere on its shores is somewhat similar to the Greek colonization movement of classical antiquity. In any case the development by Fibonacci of the *Abacus*, namely the first step in modern algebra, was mostly motivated by commercial accounting reasons, just as the cradle of ancient geometry was in the quite practical need to estimate the extension of land properties for fiscal imposition.

Also the second stage of development of modern mathematics and physics, which took place in the age of the Renaissance and in the following two centuries, displays many similarities with the impetuous growth of Ancient Science in the Hellenistic Age. The new national monarchies that were forming at the expenses of the declining feudal system, in particular France, England and the Northern States were in a status of continuous confrontation and warfare, more and more basing their power on the use of new science–derived technologies, just as it was the case of the Hellenistic States. Fortunately for Modern European Science no Roman Peace was imposed on the Continent by any prevailing superpower and the Analytic Critical Thinking of Greek origin could develop, without a stop, up to the stage reached in the Age of Enlightenment, eventually leading to the French Revolution.

The Napoleonic Age was really a most fertile one for mathematics and physics and France took the lead, propelled by the democratic ideals of the Revolution and by a new conception of the State whose newly perceived obligations to organize and finance the progress of science is clearly exemplified by the establishment of the École Polytechnique and of the École Normale Superiéure.

The most intense phase of the tale told in this book begins May 31st 1832 with the death of Évariste Galois. Through several chapters I tried to trace back the flow of ideas and of conceptions that finally issued the modern Theory of Groups and of their Linear Representations. It was a lengthy process whose principal actors, Camille Jordan, Felix Klein, Sophus Lie, Arthur Cayley, Joseph Sylvester, Wilhelm Killing, Élie Cartan, Ferdinand Frobenius, Issai Schur and finally Hermann Weyl were active in France, Germany, England and more lately in the United States of America. I tried to emphasize how many notions that we consider quite elementary and granted, like that of vector spaces, were instead characterized by a long gestation. From the ancient contemplative admiration of nice proportions, the notion of symmetry turned into an operative code of transformations active on intermediate mathematical objects that, according to Weyl's views, do not need to have any immediate correspondence with reality. Mathematical constructions are just sophisticated intellectual machines which, once correctly utilized, can produce some observable predictions. Comparison between theory and experiments can occur only at the level of such final predictions, not at any intermediate stage.

Along a parallel historical path which started in 1828 with Gauss' *Disquisitiones Generales circa superficies curvas* and lasted more than a century, the notion of Space developed from the apodictic view of Euclidian Geometry sustained by Kant into the modern notion of *differentiable manifolds* whose geometry is not known a priori rather it is dictated by a Riemannian or a pseudo Riemannian metric which, according to Einstein's General Relativity, might be determined by dynamical equations.

In our ideal historical journey we saw that the gestation of modern differential geometry, in particular of the notions of differentiable manifolds, fibre-bundles, connections and metrics was just as long as the gestation of linear algebra, passing through the truly genial work of Riemann, Klein, Ricci–Curbastro, Bianchi, Levi-Civita, Ehresman. The first part of this conceptual development was essential to Einstein and to his *geometrization of gravitation* that could not even be imagined without such a mathematical framework, slowly constructed in the course of about 50 years.

With the addition of topology, of characteristic classes and of harmonic integrals all the geometrical ingredients of the contemporary *episteme*, as I described it in points (A)–(E) of the first chapter, were essentially ready by the mid fifties of the XXth century but it took another forty years before they were consistently and consciously threaded into the fabrics of theoretical physics.

What happened in mathematics since the mid thirties of the XXth century to the early eighties of the same century is deeply characterized, in my opinion, by the following two highly momentous developments, one intrinsic to the mathematical community the other forced on it by the new visions of theoretical physics. These developments are the following ones:

(a) Starting with the monumental work of Cartan on symmetric spaces the *theory of symmetry*, meaning group theory, Lie Algebra theory and associated topics merged more tightly with the *theory of geometry*, meaning manifolds and fibre-bundles, their isometries, their holonomies and their topology.
(b) With the advent of supersymmetry and of its obligatory consequences, namely supergravity, superstrings and branes, what in geometry was so far generic, for instance the dimensions D of the space-time manifold or the possible scalar potentials ceased to be such and started being determined within finite ranges of choices that are dictated by a superior structure, at the same time very restrictive and surprisingly rich in its power to relate so far uncorrelated mathematical objects.

Before supersymmetry D might be any number, after supersymmetry it took the fixed values either $D = 11$, or $D = 10$, related to each other by a deep mechanism named duality. Before supersymmetry, all Riemannian spaces were equally interesting, after supersymmetry *special geometries* occupied the scene introducing new exciting mathematical structures that I have described at length in several chapters of another more technical book [90], also published by Springer. Before supersymmetry, exceptional Lie algebras were mathematical curiosities mostly disregarded by physicists, after supersymmetry all the exceptional Lie algebras fell into appropriate boxes specially prepared for them in a grandiose fresco which almost unexpectedly started revealing itself.

Looking at matters from a distance and with a mathematical attitude one gets the impression that supersymmetry played the role of that critical tile in a puzzle, putting which into its proper place, all the other tiles almost automatically find their way to their correct positions. Many examples can be made but one spectacular one might suffice to clarify this point.

The possible holonomy groups of Riemannian manifolds were classified before supersymmetry and fill a very short list. Generic manifolds have holonomy SO(n) in $d = n$ dimensions. In even dimensions $d = 2n$, manifolds with holonomy $\mathrm{U(n)} \subset \mathrm{SO(2n)}$ are the complex manifolds. Those among the complex manifolds that have holonomy $\mathrm{SU(n)} \subset \mathrm{U(n)} \subset \mathrm{SO(2n)}$ are the Kähler manifolds and here we meet with $\mathcal{N} = 1$ supersymmetry, as the attentive reader of my other book [90] knows. In $d = 4n$, manifolds with holonomy $\mathrm{USp(2n)} \subset \mathrm{U(2n)} \subset \mathrm{SO(4n)}$ are the Hyper-Kähler manifolds while those with holonomy $\mathrm{USp(2n)} \times \mathrm{SU(2)} \subset \mathrm{U(2n)} \subset \mathrm{SO(4n)}$ are the quaternionic Kähler manifolds. In both cases we meet here with $\mathcal{N} = 2$ supersymmetry, rigid in the first case, local in the second one. The list contained two more exceptional cases, the mysterious 7-dimensional manifolds with $\mathrm{G}_{2(-14)}$ holonomy and the 8-dimensional manifolds with $\mathrm{Spin}(7) \subset \mathrm{SO}(8)$ holonomy. Both cases were decoded by supergravity. The first was decoded by observing that $d = 7$ is the complement of $d = 4$ in compactifications of $D = 11$ supergravity and that $\mathrm{G}_{2(-14)}$ holonomy is the condition for a residual $\mathcal{N} = 1$ supersymmetry of the compactified vacuum. The second case was decoded considering M2-branes in $D = 11$ space-time, Spin(7)-holonomy of the 8-manifold transverse to the M2-brane being the condition for its $\mathcal{N} = 1$ supersymmetry.

Not only known mathematics found its interpretation within the framework of supersymmetry and supergravity but new entire chapters of geometry were constructed under the stimulus of supergravity. Most notable among them are some of the topics extensively discussed in my other book [90], namely:

1. Special Kähler Geometry.
2. The c and $c^\star$ maps from Special Kähler Geometry to quaternionic or pseudo quaternionic geometry.
3. The relations of the above constructions with the *Tits Satake projection*.
4. The systematics of Kähler and HyperKähler quotients leading, for instance, to the classifications and construction of all ALE manifolds
5. The σ-model approach to supergravity black-holes and the refinement of the theory of nilpotent orbits.

What are, at the end of this long historical and mathematical journey, the conclusions we might draw on the status of the *episteme* in the current year 2018?

After the spectacular detection of gravitational waves emitted from the coalescence of two massive black holes and the numerical verification of Einstein field equations, the points (A)–(E) introduced in the preface are firmly established, at least within our Western Analytical System of Thought. The choices of symmetries, bundles and potentials within such a framework have to be made in a way enlightened by the lesson of supersymmetry. It is not yet clear whether supersymmetry is realized in Nature in the way we think and it might take a quite long time before we are able to answer such a question in an experimental way, yet we cannot ignore the geometrical structures and the miraculous relations among them that supersymmetry has brought to the front stage. We have to continue the exploration of the new mathematics introduced by supergravity and superstrings to find new hidden clues, so far not yet observed.

Let me, at this point, summarize some of the general ideas I have put forward in Sect. 6.1.2 while making my own annotations to Weyl's *mathematical way of thinking*.

In our effort to understand Nature in purely rational terms we generalize the notion of *what exists* into a mathematically defined family of *what is possible*. Typically the *possible structures* can be thought of as points in a certain variety that we name *moduli space*. Our understanding of the *virtual*, i.e. of the aristotelian *potential*, is essentially encoded in our command over the *geometry of moduli spaces*. On the other hand, within the possible, we always would like to be able to select what indeed exists in Nature, i.e. to determine the aristotelian *actual*. Our ambition is to characterize a priori the *actual points* of moduli space as some special ones on the basis of some criterion. To this effect one resorts to *new functions* defined over moduli space, let us name them *hamiltonians*, whose *minima* can select *what exists in actuality*. The game starts at this point once again in the new rush to define the family of possible hamiltonians and their moduli spaces. In these games the fundamental issue is provided by *symmetries* and by their classification. The ultimate dream of many scientists is associated with *sporadic entities*, for instance sporadic groups. Because of their uniqueness they have no moduli and correspond to some end point in the conceptual chain. In some sense sporadic structures are the analogue, in mathematical thinking, of God or better of Plato's Demiurge.

Discovering hidden relations among quite different mathematical structures is probably the only possible way of reshuffling the formulation of physical laws into new terms that reveal new conceptions and open the way to new *moduli spaces* and new *hamiltonians* in our quest for the sporadic end-point.

In this vein, from generic choices we have been instructed, within the *superworld*, to look at special structures, restricted holonomy, for instance, exceptional Lie algebras, hyperbolic algebras, sporadic simple groups and the like, searching for new corners where other tiles of the mathematical puzzle might find their proper place. At the end of a long day it might happen that supersymmetry is only the tip of an iceberg and that in the deep waters under the cold sea surface there lies another mathematical logic able to lead us to a new physical vision and to new far reaching conclusions. Yet the tip is there, it was observed and one cannot avoid to explore further what lies underneath the surface of the sea.

References

1. Alekseevsky, D. Classification of quaternionic spaces with a transitive solvable group of motions. *Math. USSR Izvestija 9* (1975), 297–339.
2. Alekseevsky, D. V., Cortes, V., Devchand, C., and Van Proeyen, A. Polyvector superPoincare algebras. *Commun. Math. Phys. 253* (2004), 385–422.
3. Andrianopoli, L., Bertolini, M., Ceresole, A., D'Auria, R., Ferrara, S., Fre, P., and Magri, T. N = 2 supergravity and N = 2 super Yang-Mills theory on general scalar manifolds: Symplectic covariance, gaugings and the momentum map. *J. Geom. Phys. 23* (1997), 111–189.
4. Andrianopoli, L., D'Auria, R., Ferrara, S., Fré, P., Minasian, R., and Trigiante, M. Solvable Lie algebras in type IIA, type IIB and M-theories. *Nuclear physics B 493*, 1–2 (1997), 249–277. [arXiv:hep-th/9612202].
5. Andrianopoli, L., D'Auria, R., Ferrara, S., Fré, P., and Trigiante, M. R-R scalars, U-duality and solvable Lie algebras. *Nuclear Physics B 496*, 3 (1997), 617–629. [arXiv:hep-th/9611014].
6. Andrianopoli, L., D'Auria, R., Ferrara, S., and Trigiante, M. Extremal black holes in supergravity. *Lect. Notes Phys. 737* (2008), 661–727.
7. Arnold, V., Gusein-Zade, S., and Varchenko, A. *Singularities of differentiable maps*. Birkhäuser, 1975.
8. Atyah, M. Hermann Weyl 1885–1955: A biographical memoir. *National Academy of Sciences, Biographical Memoirs volume 82* (2002).
9. Bachet de Meziriac, C. G., and de Fermat, P. *Diophanti Alexandrini Arithmeticorum libri 6, et De numerismultangulis liber unus. Cum comm. C. G. Bacheti et observationibus P. de Fermat*. Acc. doctrinae analyticae inventum novum, coll. ex variis eiu Tolosae, 1670.
10. Bagger, J., and Witten, E. The Gauge Invariant Supersymmetric Nonlinear Sigma Model. *Phys. Lett. 118B* (1982), 103–106.
11. Belinsky, V., Khalatnikov, I., and Lifshitz, E. Oscillatory approach to a singular point in the relativistic cosmology. *Adv. Phys. 19* (1970), 525–573.
12. Belinsky, V., Khalatnikov, I., and Lifshitz, E. A general solution of the Einstein equations with a time singularity. *Adv. Phys. 31* (1982), 639–667.
13. Bellucci, S., Ferrara, S., Gunaydin, M., and Marrani, A. Charge orbits of symmetric special geometries and attractors. *Int. J. Mod. Phys. A21* (2006), 5043–5098.
14. Bellucci, S., Ferrara, S., and Marrani, A. On some properties of the attractor equations. *Phys. Lett. B 635* (2006), 172.
15. Bergshoeff, E., Chemissany, W., Ploegh, A., Trigiante, M., and Van Riet, T. Generating geodesic flows and supergravity solutions. *Nucl. Phys. B 812* (2009), 343.
16. Bertolini, M., and Trigiante, M. Microscopic entropy of the most general four-dimensional BPS black hole. *JHEP 10* (2000), 002.

P. G. Fré, *A Conceptual History of Space and Symmetry*,
https://doi.org/10.1007/978-3-319-98023-2

17. Bianchi, L. *Lezioni di Geometria Differenziale*. Enrico Spoerri, Libraio-Editore, Pisa, 1894.
18. Bianchi, L. Sugli spazii a tre dimensioni che ammettono un gruppo continuo di movimenti. *Soc. Ital. Sci. Mem. di Mat. 11* (1898), 267.
19. Bianchi, L. Sui simboli a quattro indici e sulla curvatura di riemann. *Rend. Accad. Naz. dei Lincei 11* (1902).
20. Borel, A., and Tits, J. Groupes réductifs. *Publications Mathémathiques de l'IHES 27* (1965), 55–151. Compléments à l'article: vol. 41, 1972, pp. 253–276.
21. Bossard, G. Extremal black holes and nilpotent orbits. In *Proceedings, 16th International Congress on Mathematical Physics (ICMP09): Prague, Czech Republic, August 3–8, 2009* (2009).
22. Bossard, G. The Extremal black holes of N = 4 supergravity from so(8, 2+n) nilpotent orbits. *Gen. Rel. Grav. 42* (2010), 539–565.
23. Bossard, G., and Nicolai, H. Multi-black holes from nilpotent lie algebra orbits. *Gen. Rel. Grav. 42* (2010), 509.
24. Bossard, G., Nicolai, H., and Stelle, K. Universal BPS structure of stationary supergravity solutions. *JHEP 0907* (2009), 003.
25. Bossard, G., and Ruef, C. Interacting non-BPS black holes. *Gen. Rel. Grav. 44* (2012), 21.
26. Breitenlohner, P., Maison, D., and Gibbons, G. W. Four-dimensional black holes from kaluza–klein theories. *Commun. Math. Phys. 120* (1988), 295.
27. Brieskorn, E. unknown. *Actes Congrés Intern. Math. Ann. t* (1970), 2.
28. Brown, J., Ganor, O. J., and Helfgott, C. M theory and E(10): Billiards, branes, and imaginary roots. *JHEP 08* (2004), 063.
29. Bruzzo, U., Fino, A., and Fré, P. The Kähler Quotient Resolution of $\mathbb{C}^3/\Gamma$ singularities, the McKay correspondence and D = 3 $\mathcal{N} = 2$ Chern–Simons gauge theories.
30. Calabi, E. Metriques kähleriennes et fibres holomorphes. *Ann. Scie. Ec. Norm. Sup. 12* (1979), 269.
31. Cartan, E. Über die einfachen transformationgruppen. *Leipz. Ber.* (1893), 395–420.
32. Cartan, E. *Sur la structure des groupes de transformations finis et continus*. Ph.D. thesis, Paris, Nony, 1894.
33. Cartan, E. Sur la structure des groupes infinis de transformations. *Annales Scientific de l'ENS 21* (1904), 153–206.
34. Cartan, E. Sur les equations de la gravitation d'Einstein. *J. Math. Pures Appl. (9) 1* (1922), 93–161.
35. Castellani, L., D'Auria, R., and Ferrara, S. Special geometry without special coordinates. *Class. Quant. Grav. 7* (1990), 1767–1790.
36. Castellani, L., D'Auria, R., and Ferrara, S. Special Kahler Geometry: An Intrinsic Formulation From $N = 2$ Space-time Supersymmetry. *Phys. Lett. B241* (1990), 57–62.
37. Cayley, A. On Jacobi's elliptic functions, in reply to the rev..; and on quaternions. *Philosophical Magazine 26* (1845), 208–211. Appendix reprinted in The Collected Mathematical Papers, Johnson Reprint Co., New York, 1963, p. 127.
38. Cayley, A. On the theory of groups, as depending on the symbolic equation $\theta^n = 1$. *Philosophical Magazine 7* (1854), 123–130.
39. Cayley, A. A memoir on the theory of matrices. *Philosophical transactions of the Royal society of London 148* (1858), 17–37.
40. Cecotti, S. Homogeneous Kahler Manifolds and T Algebras in $N = 2$ Supergravity and Superstrings. *Commun. Math. Phys. 124* (1989), 23–55.
41. Ceresole, A., Ferrara, S., and Marrani, A. Small N = 2 Extremal Black Holes in Special Geometry. *Phys. Lett. B693* (2010), 366–372.
42. Chemissany, W., Rosseel, J., Trigiante, M., and Van Riet, T. The full integration of black hole solutions to symmetric supergravity theories. *Nucl. Phys. B 830* (2010), 391.
43. Christoffel, E. Über die transformation der homogenen differentialausdrüucke zweiten grades. *J. für die Reine und Angewandte Mathematik 70* (1869), 46–70.
44. Cicero, M. T., and Baiter, J. G. Plato, Timaeus, 53c. In *M. Tullii Ciceronis de officiis ad Marcum filium libri tres; Paradoxa stoicorum; Timaeus*. ex officina B. Tauchnitz, 1865. Plato's dialogue Timaeus was translated from Greek into Latin by Cicero.

45. Cortés, V. Alekseevskian spaces. *Diff. Geom. Appl. 6* (1996), 129–168.
46. Cremmer, E., Kounnas, C., Van Proeyen, A., Derendinger, J. P., Ferrara, S., de Wit, B., and Girardello, L. Vector Multiplets Coupled to N = 2 Supergravity: SuperHiggs Effect, Flat Potentials and Geometric Structure. *Nucl. Phys. B250* (1985), 385–426.
47. Cremmer, E., and Van Proeyen, A. Classification of Kahler Manifolds in $N = 2$ Vector Multiplet Supergravity Couplings. *Class. Quant. Grav. 2* (1985), 445.
48. Crilly, A. J. *The mathematics of Arthur Cayley with particular reference to linear algebra.* Ph.D. thesis, Middlesex Polytechnic, 1981. (http://eprints.mdx.ac.uk).
49. Dabholkar, A. Microstates of nonsupersymmetric black holes. *Phys. Lett. B402* (1997), 53–58.
50. Damour, T. Cosmological singularities, Einstein billiards and Lorentzian Kac-Moody algebras. In *Miami Waves 2004: Conference on Geometric Analysis, Nonlinear Wave Equations and General Relativity Coral Gables, Florida, January 4-10, 2004* (2005). [J. Hyperbol. Diff. Equat. (2005)].
51. Damour, T. Poincare, relativity, billiards and symmetry. In *Proceedings, Symposium Henri Poincare, Brussels, Belgium, 8-9 Oct 2004* (2005), p. 149.
52. Damour, T., de Buyl, S., Henneaux, M., and Schomblond, C. Einstein billiards and overextensions of finite dimensional simple Lie algebras. *JHEP 08* (2002), 030.
53. Damour, T., Henneaux, M., Julia, B., and Nicolai, H. Hyperbolic Kac–Moody algebras and chaos in Kaluza–Klein models. *Phys. Lett. B509* (2001), 323–330.
54. Damour, T., Henneaux, M., and Nicolai, H. Cosmological billiards. *Class. Quant. Grav. 20* (2003), R145–R200.
55. D'Auria, R., Ferrara, S., and Fre, P. Special and quaternionic isometries: General couplings in n = 2 supergravity and the scalar potential. *Nucl. Phys. B359* (1991), 705.
56. D'Auria, R., Ferrara, S., and Trigiante, M. C - map, very special quaternionic geometry and dual Kahler spaces. *Phys. Lett. B587* (2004), 138–142.
57. D'Auria, R., and Fré, P. BPS black-holes in supergravity: duality groups, p-branes, central charges and entropy. In *Classical and Quantum Black Holes* (1999), P. Fré, V. Gorini, G. Magli, and U. Moschella, Eds., IOP Publishing Ltd., pp. 137–272.
58. de Buyl, S., Henneaux, M., Julia, B., and Paulot, L. Cosmological billiards and oxidation. *Fortsch. Phys. 52* (2004), 548–554. [PoSjhw 2003, 015 (2003)].
59. de Wit, B., Lauwers, P. G., Philippe, R., Su, S. Q., and Van Proeyen, A. Gauge and Matter Fields Coupled to N = 2 Supergravity. *Phys. Lett. 134B* (1984), 37–43.
60. de Wit, B., and Van Proeyen, A. Potentials and Symmetries of General Gauged N = 2 Supergravity: Yang-Mills Models. *Nucl. Phys. B245* (1984), 89–117.
61. de Wit, B., and Van Proeyen, A. Broken sigma model isometries in very special geometry. *Phys. Lett. B293* (1992), 94–99.
62. de Wit, B., and Van Proeyen, A. Special geometry, cubic polynomials and homogeneous quaternionic spaces. *Commun. Math. Phys. 149* (1992), 307–334.
63. de Wit, B., and Van Proeyen, A. Isometries of special manifolds. 1995.
64. de Wit, B., Vanderseypen, F., and Van Proeyen, A. Symmetry structure of special geometries. *Nucl. Phys. B400* (1993), 463–524.
65. Demaret, J., De Rop, Y., and Henneaux, M. Chaos in Nondiagonal Spatially Homogeneous Cosmological Models in Space-time Dimensions ≤ 10. *Phys. Lett. B211* (1988), 37–41.
66. Demaret, J., Hanquin, J. L., Henneaux, M., Spindel, P., and Taormina, A. The Fate of the Mixmaster Behavior in Vacuum Inhomogeneous Kaluza–Klein Cosmological Models. *Phys. Lett. B175* (1986), 129–132.
67. Demaret, J., Henneaux, M., and Spindel, P. Nonoscillatory behavior in vacuum Kaluza–Klein cosmologies. *Phys. Lett. 164B* (1985), 27–30.
68. Eguchi, T., and Hanson, A. J. Selfdual Solutions to Euclidean Gravity. *Annals Phys. 120* (1979), 82.
69. Ehresmann, C. Sur la théorie des espaces fibrés. *Colloque de topologie algébrique du C.N.R.S., Paris* (1947), 315.
70. Ehresmann, C. Les connexions infinitésimales dans un espace fibré différentiable. *Colloque de Topologie, Bruxelles* (1950), 2955.

71. Englert, F., Henneaux, M., and Houart, L. From very-extended to overextended gravity and M-theories. *JHEP 02* (2005), 070.
72. Erickson, J. K., Wesley, D. H., Steinhardt, P. J., and Turok, N. Kasner and mixmaster behavior in universes with equation of state $w \geq 1$. *Phys. Rev. D69* (2004).
73. Ferrara, S., Freedman, D. Z., van Nieuwenhuizen, P., Breitenlohner, P., Gliozzi, F., and Scherk, J. Scalar Multiplet Coupled to Supergravity. *Phys. Rev. D15* (1977), 1013.
74. Ferrara, S., Girardello, L., Kounnas, C., and Porrati, M. The Effective Interactions of Chiral Families in Four-dimensional Superstrings. *Phys. Lett. B194* (1987), 358–365.
75. Ferrara, S., and Kallosh, R. Supersymmetry and attractors. *Phys. Rev. D54* (1996), 1514–1524.
76. Ferrara, S., Kallosh, R., and Strominger, A. $N = 2$ extremal black holes. *Phys. Rev. D52* (1995), 5412–5416.
77. Ferrara, S., and Sabharwal, S. Quaternionic Manifolds for Type II Superstring Vacua of Calabi–Yau Spaces. *Nucl. Phys. B332* (1990), 317–332.
78. Fré, P. Gaugings and other supergravity tools of p-brane physics. *arXiv preprint* [arXiv:hep-th/0102114] (2001). Proceedings of the Workshop on Latest Development in M-Theory, Paris, France, 1–9 Feb 2001.
79. Fre, P., Gargiulo, F., Rosseel, J., Rulik, K., Trigiante, M., and Van Proeyen, A. Tits-Satake projections of homogeneous special geometries. *Class. Quant. Grav. 24* (2007), 27–78.
80. Fré, P., Gargiulo, F., and Rulik, K. Cosmic billiards with painted walls in non-maximal supergravities: a worked out example. *Nuclear Physics B 737*, 1 (2006), 1–48. https://doi.org/10.1016/j.nuclphysb.2005.10.023, [arXiv:hep-th/0507256].
81. Fré, P., Gili, V., Gargiulo, F., Sorin, A. S., Rulik, K., and Trigiante, M. Cosmological backgrounds of superstring theory and solvable algebras: Oxidation and branes. *Nucl. Phys. B685* (2004), 3–64.
82. Fré, P., and Sorin, A. The integration algorithm for nilpotent orbits of $g/h^\star$ lax systems: for extremal black holes. 2009.
83. Fré, P., and Sorin, A. Supergravity black holes and billiards and Liouville integrable structure of dual Borel algebras. *JHEP 03* (2010), 066.
84. Frè, P., Sorin, A., and Trigiante, M. Black hole nilpotent orbits and tits satake universality classes. 2011.
85. Frè, P., Sorin, A., and Trigiante, M. Integrability of supergravity black holes and new tensor classifiers of regular and nilpotent orbits. *JHEP 1204* (2012), 015.
86. Fré, P., and Sorin, A. S. Integrability of supergravity billiards and the generalized Toda lattice equation. *Nucl. Phys. B733* (2006), 334–355.
87. Frè, P., and Sorin, A. S. The Weyl group and asymptotics: All supergravity billiards have a closed form general integral. *Nucl. Phys. B815* (2009), 430–494.
88. Fré, P., Sorin, A. S., and Trigiante, M. The c-map, Tits Satake subalgebras and the search for $\mathcal{N} = 2$ inflaton potentials. *Fortsch. Phys. 63* (2015), 198–258.
89. Fré, P. G. *Gravity, a Geometrical Course*, vol. 1, 2. Springer Science & Business Media, 2012.
90. Fré, P. G. *Advances in Geometry and Lie Algebras from Supergravity*. Springer, 2018.
91. Frenet, J. F. Sur quelques proprétées des courbes à double courbure. *Journal des mathematiques pures et appliqueés* (1852).
92. Frobenius, F. G. Über lineare substitutionen und bilineare formen. *Journal fur die reine und angewandte Mathematik 84* (1878), 1–63. Reprinted in Gesammelte Abhandlungen Band I, pp. 343–405.
93. Fyodorov, E. S. Simmetriia pravil'nykh sistem figur (the symmetry of regular systems of figures). *Zapiski Imperatorskogo S. Petersburgskogo Mineralogichesgo Obshchestva (Proceedings of the Imperial St. Petersburg Mineralogical Society) series 2, vol. 28* (1891), 1146. English translation: David and Katherine Harker (trans.), Symmetry of Crystals, American Crystallographic Association Monograph No. 7 (Bufalo, N.Y. American Crystallographic Association, 1971) pages 50131.
94. Fyodorov, E. S. Simmetrija na ploskosti (symmetry in the plane). *Zapiski Imperatorskogo S. Petersburgskogo Mineralogichesgo Obshchestva (Proceedings of the Imperial St. Petersburg Mineralogical Society) series 2, vol. 28* (1891), 245–291.

95. Gaillard, M., and Zumino, B. Duality rotations for interacting fields. *Nucl. Phys. B 193* (1981), 221.
96. Gaiotto, D., Li, W., and Padi, M. Non-supersymmetric attractor flow in symmetric spaces. *JHEP 12* (2007), 093.
97. Galicki, K. A generalization of the momentum mapping construction for quaternionic Kahler manifolds. *Commun. Math. Phys. 108* (1987), 117.
98. Galois, E. Oeuvres mathématiques d'Évariste galois. *Journal des mathématiques pures et appliquees XI* (1846), 381–444.
99. Gauss, K. F. *Disquisitiones generales circa superficies curvas*. Dieterich Göttingen, 1828.
100. Ghosh, A., and Mitra, P. Counting of black hole microstates. *Indian J. Phys. 80* (2006), 867.
101. Gibbons, G., Kallosh, R., and Kol, B. Moduli, scalar charges, and the first law of black hole thermodynamics. *Phys. Rev. Lett. 77* (1996), 4992–4995.
102. Gibbons, G. W., and Hawking, S. W. Classification of Gravitational Instanton Symmetries. *Commun. Math. Phys. 66* (1979), 291–310.
103. Giryavets, A. New attractors and area codes. *JHEP 0603* (2006), 020.
104. Goldstein, K., Iizuka, N., Jena, R. P., and Trivedi, S. P. Non-supersymmetric attractors. *Phys. Rev. D72* (2005), 124021.
105. Gunaydin, M., Neitzke, A., Pioline, B., and Waldron, A. Bps black holes, quantum attractor flows and automorphic forms. *Phys. Rev. D73* (2006), 084019.
106. Gunaydin, M., Neitzke, A., Pioline, B., and Waldron, A. Quantum attractor flows. *JHEP 0709* (2007), 056.
107. Hamilton, W. R. Note, by Sir W. R. Hamilton, respecting the researches of John T. Graves, Esq. *Transactions of the Royal Irish Academy 21* (1848), 338–341.
108. Hamilton, W. R. Lectures on quaternions.
109. Hamilton, W. R. Elements of quaternions. Edited by William Edwin Hamilton, son of the deceased author.
110. Hassler, W. Differentiable manifolds. *Annals of Mathematics 37 (3)* (1936), 645–680. JSTOR 1968482.
111. Hawkin, T. Wilhelm killing and the structure of lie algebras. *Archive for Hist. Exact Sci. 26* (1982), 126–192.
112. Heath, S. T. *A History of Greek Mathematics*. Dover Publ. New York, 1921.
113. Helgason, S. A centennial: Wilhelm Killing and the exceptional groups. *The Mathematical Intelligencer, Volume 12, Number 3* (1990).
114. Henneaux, M., and Julia, B. Hyperbolic billiards of pure D = 4 supergravities. *JHEP 05* (2003), 047.
115. Higham, N. J. Cayley, Sylvester, and early matrix theory. MIMS EPrint 2007.119 (http://www.manchester.ac.uk/mims/eprints), 2007.
116. Hitchin, N. J. Polygons and gravitons. *Math. Proc. Cambridge Phil. Soc. 85* (1979), 465–476.
117. Hitchin, N. J., Karlhede, A., Lindström, U., and Roček, M. Hyperkahler Metrics and Supersymmetry. *Commun. Math. Phys. 108* (1987), 535.
118. Jordan, C. Traité des substitutions et des équations algébriques, Gauthiers–Villars, Parıs, 1870. *Reedition: Gabay, Paris* (1989).
119. Kähhler, E. Über eine bemerkenswerte hermitesche metrik. *Abhandlungen aus dem Mathematischen Seminar der Universität Hamburg 9 (1)* (1933), 173186.
120. Kallosh, R. New attractors. *JHEP 0512* (2005), 022.
121. Kallosh, R., Sivanandam, N., and Soroush, M. The non-BPS black hole attractor equation. *JHEP 0603* (2006), 060.
122. Kerr, R. P. Gravitational field of a spinning mass as an example of algebraically special metrics. *Physical Review Letters 11* (1963), 237–238.
123. Keurentjes, A. The Group theory of oxidation. *Nucl. Phys. B658* (2003), 303–347.
124. Keurentjes, A. The Group theory of oxidation 2: Cosets of nonsplit groups. *Nucl. Phys. B658* (2003), 348–372.
125. Killing, W. Die zusammensetzung der stetigen/endlichen transformationsgruppen. *Mathematische Annalen, Volume 31, Number 2 June* (1887), 252–290.

126. Killing, W. Die zusammensetzung der stetigen/endlichen transformationsgruppen. *Mathematische Annalen, Volume 33, Number 1 March* (1888), 1–48.
127. Killing, W. Die zusammensetzung der stetigen/endlichen transformationsgruppen. *Mathematische Annalen, Volume 36, Number 2 June,* (1889), 57–122.
128. Killing, W. Die zusammensetzung der stetigen/endlichen transformationsgruppen. *Mathematische Annalen, Volume 34, Number 1 March* (1890), 161–189.
129. Klein, F. *Gesammelte Mathematische Abhandlungen*, vol. Vol I-III. Verlag Von Julius Springer, Berlin, 1923.
130. Kronheimer, P. B. The construction of ALE spaces as hyper-Kähler quotients. *J. Differential Geom. 29*, 3 (1989), 665–683.
131. Kronheimer, P. B. A Torelli-type theorem for gravitational instantons. *J. Differential Geom. 29*, 3 (1989), 685–697.
132. Kruskal, M. D. Maximal extension of Schwarzschild metric. *Phys. Rev. 119* (1960), 1743.
133. Larsen, F. A String model of black hole microstates. *Phys. Rev. D56* (1997), 1005–1008.
134. Levi, E. E. Sulla struttura dei gruppi finiti e continui. *Atti della Reale Accademia delle Scienze di Torino XL* (1905), 551–565. JFM 36.0217.02, archived from the original on March 5, 2009 Reprinted in: Opere Vol. 1, Edizione Cremonese, Rome (1959), p. 101.
135. Levi Civita, T., and Ricci, G. Méthodes de calcul differential absolu et leurs applications. *Math. Ann B 54* (1900), 125–201.
136. Lie, S. Zur theorie des integrabilitetsfaktors. *Christiania Fohr* (1874), 242–254.
137. Lie, S. *Theorie der Transformationsgruppen I*. B. G. Teubner , Leipzig, 1888. Written with the help of Friedrich Engel.
138. Lie, S. *Theorie der Transformationsgruppen II*. B. G. Teubner , Leipzig, 1890. Written with the help of Friedrich Engel.
139. Lie, S. *Theorie der Transformationsgruppen III*. B. G. Teubner , Leipzig, 1893. Written with the help of Friedrich Engel.
140. Lie, S. *Vorlesungen Über continuierliche Gruppen*. B. G. Teubner , Leipzig, 1893. Written with the help of Georg Scheffers.
141. Maldacena, J. M., Strominger, A., and Witten, E. Black hole entropy in M theory. *JHEP 12* (1997), 002.
142. Maurer, L. W. Über continuirliche transformationsgruppen. *Mathematische Annalen 39* (1891), 409–440.
143. McKay, J. Graphs, singularities and finite groups. *Proc. Symp. Pure Math. 37* (1980), 183.
144. Pengelley, D. J. Arthur Cayley and the first paper on group theory. *Mathematical Association of America notes 68* (2005), 3.
145. Pioline, B. Lectures on black holes, topological strings and quantum attractors. *Class. Quant. Grav. 23* (2006), S981.
146. Plücker, J. System der geometrie des raumes in neuer analytischer behandlungsweise. *Crelle's Journal 1846* (1846).
147. Poncelet, J. Traité des propriétés projective des figures. *Crelle's Journal* (1822).
148. Poncelet, J. Théorie des polaires réciproques. *Crelle's Journal* (1829).
149. Qvintiliani, M. F. *Institvtio Oratoria*, vol. liber primvs. Trifon, 90 AD.
150. Ricci, G. Résumé de quelques travaux sur le systèmes variables de fonctions associées a une forme différentielle quadratique. *Bulletin des Sciences Mathematiques* (1892).
151. Ricci, G. *Atti R. Inst. Veneto 53 : 2* (1903–1904), 1233–1239.
152. Riemann, G. Über die hypothesen, welche der geometrie zu grunde liegen. In *Gesammelte Mathematische Werke*. 1866.
153. Russo, L. *La rivoluzione dimenticata*. Feltrinelli, 1996.
154. Satake, I. On representations and compactifications of symmetric Riemannian spaces. *Annals of Mathematics Second Series* (1960), 71.
155. Schappacher, N. Diophantus of Alexandria: a text and its history. Available from internet, April 2005.
156. Schwarzschild, K. Über das gravitationsfeld eines massenpunktes nach der einsteinschen theorie. *Sitzungsberichte der Koniglich Preussischen Akademie der Wissenschaften 7* (1916).

157. Serret, J. Sur quelques formules relatives à la théeorie des courbes à double courbure. *J. De Math. 16* (1851).
158. Sesiano, J. Books iv to vii of Diophantus' arithmetica in the arabic translation attributed to Qusta ibn Luqa. In *Sources in the History of Mathematics and Physical Sciences*, vol. 3. Springer-Verlag, New York-Heidelberg-Berlin, 1982.
159. Slodowy, P. Simple singularities and simple algebraic groups. *Lect. Notes in Math. 815* (1980).
160. Strominger, A. SPECIAL GEOMETRY. *Commun. Math. Phys. 133* (1990), 163–180.
161. Strominger, A. Black hole entropy from near horizon microstates. *JHEP 02* (1998), 009.
162. Strominger, A., and Vafa, C. Microscopic origin of the Bekenstein–Hawking entropy. *Phys. Lett. B 379* (1996), 99.
163. Tits, J. Classification of algebraic semisimple groups. In *Algebraic Groups and Discontinuous Subgroups Proc. Sympos. Pure Math., Boulder, Colo. 1965* (1966), American Mathematical Society, pp. 33–62.
164. Torretti, R. *Philosophy of Geometry from Riemann to Poincaré*. D. Reidel Publishing Company, Dordrecht, Holland, 1978.
165. Trigiante, M. *Dualities in supergravity and solvable Lie algebras*. Ph.D. thesis, Swansea U., 1998.
166. Tripathy, P., and Trivedi, S. Non-supersymmetric attractors in string theory. *JHEP 0603* (2006), 022.
167. van der Waerden, B. A history of algebra.
168. Weyl, H. *The Classical Groups, Their Invariants and Representations*. Princeton University Press, 1939.
169. Weyl, H. The Mathematical Way of Thinking. *Science, Volume 92, Issue 2394* (1940), 437–446.
170. Witten, E. Phases of N = 2 theories in two-dimensions. *Nucl. Phys. B403* (1993), 159–222. [AMS/IP Stud. Adv. Math. 1, 143 (1996)].
171. Zumino, B. Supersymmetry and Kahler Manifolds. *Phys. Lett. 87B* (1979), 203.

The manufacturer's authorised representative in the EU is Springer Nature Customer Service Centre GmbH, Europaplatz 3, 69115 Heidelberg, Germany. If you have any concerns regarding our products, please contact ProductSafety@springernature.com

Printed and bound by CPI Group (UK) Ltd, Croydon, CR0 4YY
30/09/2026
02225229-0011